CONTAMINATION OF GROUND WATER

CONTAMINATION OF GROUND WATER

Prevention, Assessment, Restoration

by

Michael Barcelona, Allen Wehrmann

Illinois State Water Survey
Champaign, Illinois

Joseph F. Keely

U.S. Environmental Protection Agency
Robert S. Kerr Environmental Research Laboratory
Ada, Oklahoma

Wayne A. Pettyjohn

Oklahoma State University
Stillwater, Oklahoma

NOYES DATA CORPORATION

Park Ridge, New Jersey, U.S.A.

Library of Congress Catalog Card Number: 90-31404
ISBN: 0-8155-1243-0
ISSN: 0090-516X
Printed in the United States

Published in the United States of America by
Noyes Data Corporation
Mill Road, Park Ridge, New Jersey 07656

10 9 8 7 6 5 4 3

Library of Congress Cataloging-in-Publication Data

Contamination of ground water : prevention, assessment, restoration /
by Michael Barcelona . . . [et al.] .
p. cm. -- (Pollution technology review, ISSN 0090-516X ; no. 184)
Includes bibliographical references.
ISBN 0-8155-1243-0 :
1. Water, Underground--Pollution. 2. Water, Underground--Management. I. Barcelona, Michael J. II. Series.
TD426.C67 1990
628.1'68--dc20 90-31404
CIP

Foreword

The need exists for a resource document that brings together available technical information on ground-water management in a form convenient for ground-water personnel at all levels of government, as well as in the private sector. The information contained in this handbook of ground-water contamination is intended to meet that need. The book covers measures for (1) prevention of contamination, (2) assessment of extent of contamination, and (3) restoration of ground-water quality.

The subsurface environment of ground water is characterized by a complex interplay of physical, geochemical and biological forces that govern the release, transport and fate of a variety of chemical substances. There are literally as many varied hydrogeologic settings as there are types and numbers of contaminant sources. In situations where ground-water investigations are most necessary, there are frequently many variables of land and ground-water use and contaminant source characteristics which cannot be fully characterized.

The impact of natural ground-water recharge and discharge processes on distributions of chemical constituents is understood for only a few types of chemical species. Also, these processes may be modified by both natural phenomena and man's activities so as to further complicate apparent spatial or temporal trends in water quality. Since so many climatic, demographic and hydrogeologic factors may vary from place to place, or even small areas within specific sites, there can be no single "standard" approach for assessing and protecting the quality of ground water that will be applicable in all cases.

Because contamination of ground water has occurred in every state and is being detected with increasing frequency, regulatory agencies and courts have been developing guidelines, laws and rules to protect this resource. Ground-water quality laws deal with both the prevention of ground-water contamination and assigning responsibility for ground-water protection or cleanup and legal liability for damages where contamination has occurred. Provisions aimed at prevention of contamination regulate the conduct of activities which could have the effect of polluting ground-water or posing risks to human health. Other statutory provisions call for government or private party response to incidents of contamination; they also may assign penalties or other legal liability to polluters. The operation of these provisions is generally triggered by the release of certain harmful substances, identified by statute or government regulation, into the environment.

The purpose of this document is to discuss measures that can be taken to ensure that uncertainties do not undermine our ability to make reliable predictions about the response of contaminants to various corrective or preventive measures. The book will provide research information to decision makers, field managers, and the scientific community; it will help satisfy the immediate need for technology transfer applicable to ground-water contamination control and prevention.

The information in the book is from *Handbook—Ground Water,* prepared by Michael Barcelona and Allen Wehrmann of the Illinois State Water Survey, Joseph F. Keely of the U.S. Environmental Protection Agency, and Wayne A. Pettyjohn of Oklahoma State University, for the U.S. Environmental Protection Agency, March 1987.

The table of contents is organized in such a way as to serve as a subject index and provides easy access to the information contained in the book.

Advanced composition and production methods developed by Noyes Data Corporation are employed to bring this durably bound book to you in a minimum of time. Special techniques are used to close the gap between "manuscript" and "completed book." In order to keep the price of the book to a reasonable level, it has been partially reproduced by photo-offset directly from the original report and the cost saving passed on to the reader. Due to this method of publishing, certain portions of the book may be less legible than desired.

Preface

Background and Regulatory Objectives

Because contamination of ground water has occurred in every state and is being detected with increasing frequency, regulatory agencies and courts have been developing guidelines, laws and rules to protect this resource.

Ground-water quality laws deal with both the prevention of ground-water contamination and assigning responsibility for ground-water protection or cleanup and legal liability for damages where contamination has occurred. Provisions aimed at prevention of contamination regulate the conduct of activities which could have the effect of polluting ground-water or posing risks to human health.

Other statutory provisions call for government or private party response to incidents of contamination: they also may assign penalties or other legal liability to polluters. The operation of these provisions is generally triggered by the release of certain harmful substances, identified by statute or government regulation, into the environment.

There is no federal law or program that directly and exclusively addresses control of ground-water pollution. However, EPA administers a number of federal environmental laws with varying requirements that do not exclusively address ground water, but do affect ground-water quality. Among these are the Clean Water Act (CWA), the Resource, Conservation and Recovery Act (RCRA), the Safe Drinking Water Act (SDWA) and the Comprehensive Environmental Response, Compensation, and Liability Act (CERCLA), commonly referred to as Superfund. Laws such as these and regulatory programs developed for their implementation have multiple purposes and objectives, including protection of land and surface water quality.

Two other laws which indirectly relate to ground-water quality are the Federal Insecticide, Fungicide and Rodenticide Act (FIFRA) and the Toxic Substances Control Act (TSCA). This legislation regulates the production, use and disposal of specific chemicals possessing an unacceptably high potential for contaminating ground water when released to the subsurface.

In addition, EPA has issued a policy document entitled "A Ground-Water Protection Strategy." This strategy embraces goals to: 1) foster stronger state programs for ground-water protection through existing Federal grant programs and provision of technical assistance; 2) study inadequately addressed problems of ground-water contamination; and 3) strengthen the "internal ground-water organization" within EPA by establishing an Office of Ground-Water Protection.

Many states, also, have acted in the last several years to address the problem. Prevention of ground-water contamination is the major thrust of most state programs. Elements of state prevention programs include developing background data on ground-water resources, establishing monitoring programs, and in some instances establishing permit and other regulatory requirements to control pollution discharges into aquifers. States have also enacted preventive legislation paralleling RCRA and SDWA, in order to qualify for federal delegation of authority under those acts.

States have also committed funds to the cleanup of hazardous waste pollution, including ground-water contamination. Most often, this state funding is provided as a condition of federal Superfund financing of the cleanup of priority hazardous waste sites in the state.

Summary of Federal Laws and Programs

CWA is one of the most far-reaching federal pollution control laws ever enacted. The Act has application to ground-water quality control in several ways. To the extent that surface and ground-water systems are hydrologically connected, protection of surface water quality beneficially affects ground water. Also, funding has been provided to states for water quality management planning and implementation, which includes ground water. In addition, where CWA funds are used to construct municipal sewage treatment plants using land application techniques, the municipalities are required to design the plants to ensure protection of ground water.

RCRA was enacted in 1976 after threats to human health and the environment posed by toxic and hazardous wastes had become matters of real public concern. The specific impetus for RCRA's passage was Congressional concern for the special dangers caused by unsound waste disposal practices, mainly in landfills and open generation, transportation, treatment, storage and disposal, as well as underground storage tanks.

SDWA was passed by Congress in 1974 to respond to accumulating evidence during the 1970's that called attention to the health threat posed by unsafe levels of contaminants in public drinking water supplies. Since about one-half of the nation's drinking water is drawn from underground sources, SDWA has obvious application to ground-water quality, although it applies to surface waters as well. The SDWA, as amended, provides protection to ground water through drinking water standards, sole source aquifer designation, protection of wellheads and the underground injection control program. FIFRA, first passed in 1947 and substantially amended in 1972, requires that before marketing a pesticide, the manufacturer must secure a registration of the product from EPA. In determining whether to issue a registration, EPA must find that the pesticide will not cause "unreasonable adverse effects on the environment" if used normally. FIFRA also imposes labeling and data reporting requirements on pesticide manufacturers. The Act authorizes EPA to suspend or cancel the registration of a pesticide where adverse environmental effects are shown to result from its use.

TSCA was enacted in 1976 in an effort to minimize risks to public health and the environment posed by the introduction into commercial use of a rapidly increasing number of chemical substances. To enable EPA to monitor the marketing of new chemicals, TSCA requires manufacturers to submit pre-manufacture notices on new chemical substances. EPA is authorized to take a variety of steps to protect against harmful effects caused by the introduction or unrestricted use of new chemicals. Such steps taken by EPA under TSCA include publication of the chemical inventory, which is a currently maintained list of all chemical substances manufactured or processed in the U.S., as well as information gathering authority, permitting access to manufacturing data which could assist in the development of source inventories for ground-water protection planning or investigation.

CERCLA was passed in 1980 to respond to the notorious Love Canal incident, which focused Congressional attention on the serious and widespread health threats posed by abandoned hazardous waste disposal sites. Congress established Superfund to enable the federal government to undertake prompt cleanup of especially dangerous abandoned sites, and later to seek reimbursement from the responsible parties. CERCLA applies cleanup, funding and liability provisions as triggered by a release or threat of release of a hazardous substance from a facility.

EPA's Ground-Water Protection Strategy sets out a "policy framework" to guide its programs affecting ground water. This framework involves classification of ground waters. Class I are those "special ground waters" in need of special protection because they are irreplaceable sources of drinking water or are otherwise ecologically vital, and are highly vulnerable to contamination because of hydrogeologic factors. With RCRA authority, EPA will ban siting of disposal facilities above these ground waters. The Agency will also continue to use the immediacy of a threat to ground water as a factor in selecting sites for Superfund cleanup. Further, EPA is considering developing special permit conditions for the underground injection control program to protect these waters. Class II ground waters are those used or potentially available for drinking water, though less vulnerable than Class I aquifers. Class II aquifers presently account for most of the country's ground water. As to these waters, EPA may impose facility siting restrictions under RCRA. Class III ground waters are classified as waters that are not potential sources of drinking water and are of limited beneficial use.

Purpose

The subsurface environment of ground water is characterized by a complex interplay of physical, geochemical and biological forces that govern the release, transport and fate of a variety of chemical substances. There are literally as many varied hydrogeologic settings as there are types and numbers of contaminant sources. In situations where ground-water investigations are most necessary, there are frequently many variables of land and ground-water use and contaminant source characteristics which cannot be fully characterized.

The impact of natural ground-water recharge and discharge processes on distributions of chemical constituents is understood for only a few types of chemical species. Also, these processes may be modified by both natural phenomena and man's activities so as to further complicate apparent spatial or temporal trends in water quality. Since so many climatic, demographic and hydrogeologic factors may vary from place to place, or even small areas within specific sites, there can be no single "standard" approach for assessing and protecting the quality of ground water that will be applicable in all cases.

Despite these uncertainties, investigations are under way and they are used as a basis for making decisions about the need for, and usefulness of, alternative corrective and preventive actions. Decision makers, therefore, need some assurance that elements of uncertainty are minimized and that hydrogeologic investigations provide reliable results.

A purpose of this document is to discuss measures that can be taken to ensure that uncertainties do not undermine our ability to make reliable predictions about the response of contamination to various corrective or preventive measures.

EPA conducts considerable research in ground water to support its regulatory needs. In recent years, scientific knowledge about ground-water systems has been increasing rapidly. Researchers in the Office of Research and Development have made improvements in technology for assessing the subsurface, in adapting techniques from other disciplines to successfully identify specific contaminants in ground water, in assessing the behavior of certain chemicals in some geologic materials and in advancing the state-of-the-art of remedial technologies.

An important part of EPA's ground-water research program is to transmit research information to decision makers, field managers and the scientific community. This publication has been developed to assist that effort and, additionally, to help satisfy an immediate Agency need to promote the transfer of technology that is applicable to ground-water contamination control and prevention.

The need exists for a resource document that brings together available technical information in a form convenient for ground-water personnel within EPA and state and local governments on whom EPA ultimately depends for proper ground-water management. The information contained in this handbook is intended to meet that need. It is applicable to many programs that deal with the ground-water resource. However, it is not intended as a guidance or support document for a specific regulatory program.

GUIDANCE DOCUMENTS ARE AVAILABLE FROM EPA AND MUST BE CONSULTED TO ADDRESS SPECIFIC REGULATORY ISSUES.

ACKNOWLEDGMENTS

Many individuals contributed to the preparation and review of this handbook. The document was prepared by JACA Corporation for EPA's Robert S. Kerr Environmental Research Laboratory, Ada, OK, and the Center for Environmental Research Information, Cincinnati, OH. Contract administration was provided by the Center for Environmental Research Information, Cincinnati, OH.

Authors:
Michael Barcelona - Illinois State Water Survey, Champaign, IL
Joseph F. Keely - EPA-RSKERL, Ada, OK
Wayne A. Pettyjohn - Oklahoma State University, Stillwater, OK
Allen Wehrmann - Illinois State Water Survey, Champaign, IL

Reviewers and Other Contributors:
Edwin F. Barth - EPA-OERR, Washington, DC
Stuart Z. Cohen - EPA-OTS, Washington, DC
Stephen Cordle - EPA-OEPER, Washington, DC
Mary Doyle - University of Arizona, Tucson, AZ
Gerald Grisak - Intera Technologies, Austin, TX
Kenneth Jennings - EPA-OWPE, Washington, DC
Jerry N. Jones - EPA-RSKERL, Ada, OK
Lowell E. Leach - EPA-RSKERL, Ada, OK
Joan Middleton - EPA-OSWER, Washington, DC
Marion R. Scalf - EPA-RSKERL, Ada, OK
Jerry T. Thornhill - EPA-RSKERL, Ada, OK
Calvin H. Ward - Rice University, Houston, TX

Contract Project Officer:
Carol Grove - EPA-CERI, Cincinnati, OH

Contents and Subject Index

PART I
FRAMEWORK FOR PROTECTING GROUND-WATER RESOURCES

PART II
SCIENTIFIC AND TECHNICAL BACKGROUND FOR ASSESSING AND PROTECTING THE QUALITY OF GROUND-WATER RESOURCES

Part I

Framework for Protecting Ground-Water Resources

1. Ground-Water Contamination

1.1 Definitions

Contaminant is defined by the Safe Drinking Water Act as "any physical, chemical, biological, or radiological substance or matter in water." Freeze and Cherry (1979) define as contaminants "all solutes introduced into the hydrologic environment as a result of man's activities regardless of whether or not the concentrations reach levels that cause significant degradation of water quality." For them, "pollution is reserved for situations where contaminant concentrations attain levels that are considered to be objectionable." Miller (1980) used a very similar definition: "Ground-water contamination is the degradation of the natural quality of ground water as a result of man's activities." According to Matthess (1982), "boundaries of polluted ground-water zones can be defined as the lines at which the concentration of all pollutants have fallen below the maximum permissible concentration for potable water, or where all water properties have taken on the normal values of the environment concerned."

Much current research is being devoted to defining just what "normal" ground-water quality is, or how it can best be defined. Ground water which naturally contains objectionable amounts of dissolved substances can properly be considered contaminated, as well as polluted; however, most regulatory functions focus on human activities which artificially introduce contaminants into the ground.

1.2 The Extent of Ground-Water Contamination

Contrary to what many people believe, ground-water contamination is not a new problem. Early investigations of ground-water contamination are abundant in scientific literature. The classic work of Dr. John Snow in 1854 (Prescott and Horwood, 1935; Mallman and Mack, 1961) first linked the contamination of wells by cholera to seepage from earth privy vaults even before the discovery of the microorganisms responsible for the disease. By 1959, a European publication (Michels *et al.*) cited 60 cases in which ground waters had become contaminated with petroleum products.

LeGrand, in his 1965 paper entitled "Patterns of Contaminated Zones of Water in the Ground" recognized the difficulty in predicting the spatial extent of a contaminated zone because of a number of interrelated factors including:

> "...the great variety of waste materials, their range in toxicity and adverse effects; man's variable pattern of waste disposal and of accidental release of contaminants in the ground; man's variable pattern of water development from wells; behavior of each contaminant in the soil, water, and rock environment; ranges in geologic and hydrologic conditions in space; and ranges in hydrologic conditions in time."

Additional problems include the fact that many potentially hazardous contaminants are colorless, odorless, and tasteless, and therefore difficult to detect by passive means. Many of the synthetic organic chemicals require sophisticated, expensive, sampling and analytical techniques burdening detection efforts.

It has been estimated that it will take 4 to 5 years to complete just one round of organic compound testing of the 3,400 public water supply wells in Illinois alone, given the present availability of personnel and laboratory facilities (Illinois EPA, 1986). Such an effort does not include the estimated 500,000 private wells in the State.

An assessment of the extent and severity of contamination is further complicated by the almost exponential growth of the synthetic organic chemistry industry in the U.S. since the early '40s (Figure 1-1). At least 63,000 synthetic organic chemicals are in common industrial and commercial use in the U.S. and this number continues to grow by approximately 500 to 1,000 new compounds every year (Epstein, 1979; U.S. EPA, 1979). Also, the human health effects of many of these chemicals, particularly over long periods of time at low exposure levels, is not known. It will take years to conduct the research necessary to properly test all these compounds and then be able to factor the results into a complete contamination assessment.

Figure 1-1 Growth of the synthetic organic chemical industry in the United States (from Senkan and Stauffer, 1981).

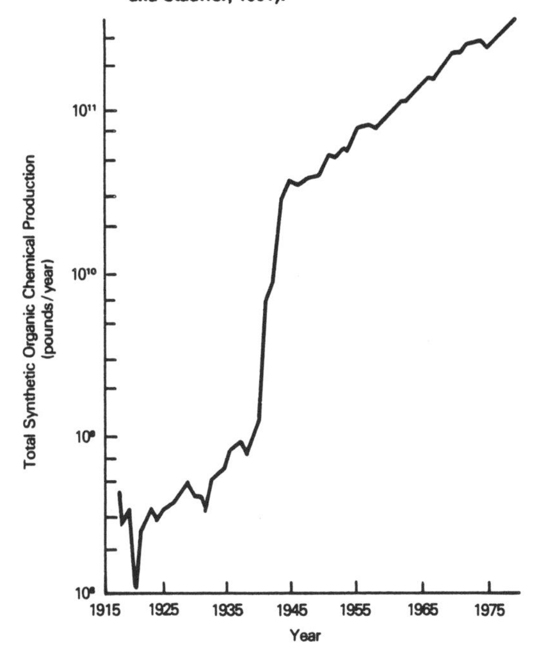

Though it has now been estimated that approximately 1 percent of the economically producible ground waters in the United States are contaminated (Lehr, 1982; Gass, 1980; Office of Technology Assessment, 1984), this estimate may not convey the problems associated with the coincidence of contamination and ground-water use. While on the whole, much of the ground water in the U.S. has not been affected by contamination, areas known to be contaminated are often densely populated areas where ground water is heavily used and depended upon as a drinking water source.

The presence of over 200 chemical substances in ground water has been documented (OTA, 1984). This number includes approximately 175 organic chemicals, over 50 inorganic chemicals (metals, nonmetals, and inorganic acids), and radionuclides. Many of these chemicals occur naturally in ground water, especially minerals dissolved from geologic earth materials in contact with the water. Many others have been introduced to the ground-water system by humans.

The detection of these substances has been biased by sampling and analytical limitations as well as the nature of the specific investigations which prompted the sampling and analysis to be conducted. The two most common circumstances under which substances (including naturally occurring minerals) have been detected in ground water are (a) regulatory compliance (e.g., Safe Drinking Water Act monitoring of public water supplies and Resource Conservation and Recovery Act (RCRA) monitoring at hazardous waste facilities); and (b) response to perceived quality problems, primarily citizen complaints. However, regulatory agencies have not historically sampled and analyzed for a wide range of potential contaminants, particularly synthetic organic chemicals, unless specific problems are suspected.

A study of the ground-water quality data base maintained in Illinois (O'Hearn and Schock, 1984) found that compliance monitoring for drinking water standards forms the basis for much of the 21,000 samples and 423,000 analytical determinations in this data base. However, *less than one-tenth of 1 percent* of all the samples in the data base had been analyzed for even a general indicator of organic contamination, total organic carbon (TOC).

Efforts have been made in recent years to assess the occurrence of organic chemicals in ground-water supplies. A survey conducted by the U.S. EPA, the Ground Water Supply Survey (GWSS), provided information on the frequency with which VOCs were detected in 466 randomly selected public ground-water supply systems (Westrick *et al.*, 1983). One or more volatile organic chemicals (VOCs) were detected in 16.8 percent of small systems and 28.0 percent of large systems sampled. The two VOCs found most often in this survey were trichloroethylene (TCE) and tetrachloroethylene (PCE). Two or more VOCs were found in 6.8 percent and 13.4 percent of the samples from small and large systems, respectively.

1.3 General Mechanisms of Ground-Water Contamination

Contaminant releases to ground water can occur by design, by accident, or by neglect. Most ground-water contamination incidents involve substances released at or only slightly below the land surface. Consequently, it is shallow ground water which is affected initially by contaminant releases. In general, shallow ground-water resources are considered more susceptible to surface sources of contamination than deeper ground-water resources. There are at least four ways by which ground-water contamination occurs: infiltration, direct migration, interaquifer exchange, and recharge from surface water. A general discussion of each of these mechanisms follows.

1.3.1 Infiltration

Contamination by infiltration is probably the most common ground-water contamination mechanism. A portion of the water which has fallen to the earth slowly infiltrates the soil through pore spaces in the soil matrix. As the water moves downward under the influence of gravity, it dissolves materials with which it comes into contact. Water percolating downward through a contaminated zone can dissolve contaminants, forming leachate. Depending on the composition of the contaminated zone, the leachate formed can contain a number of inorganic and organic constituents. Table 1-1 gives a general indication of the composition of leachate that has been found beneath sanitary landfills. The leachate will continue to migrate downward under gravity's influence until the saturated zone is reached. Once the saturated zone is contacted, horizontal and vertical spreading of the contaminants in the leachate will occur in the direction of ground-water flow (Figure 1-2). This process can occur beneath any surface or near-surface contaminant source exposed to the weather and the effects of infiltrating water.

Table 1-1 Representative Ranges for Inorganic Constituents in Leachate from Sanitary Landfills.

Parameter	Representative Range (mg/l)
K^+	200-1,000
Na^+	200-1,200
Ca^{2+}	100-3,000
Mg^+	100-1,500
Cl^-	300-3,000
SO_4^{2-}	10-1,000
Alkalinity	500-10,000
Fe (total)	1-1,000
Mn	0.01-100
Cu	<10
Ni	0.01-1
Zn	0.1-100
Pb	<5
Hg	<0.2
NO_3^-	0.1-10
NH_4	10-1,000
P as PO_4	1-100
Organic nitrogen	10-1,000
Total dissolved organic carbon	200-30,000
COD (chemical oxidation demand)	1,000-90,000
Total dissolved solids	5,000-40,000
pH	4-8

Source: Freeze and Cherry, 1979.

1.3.2 Direct Migration

Contaminants can migrate directly into ground water from below-ground sources (e.g., storage tanks, pipelines) which lie within the saturated zone. Leachate formation and downward movement through the unsaturated zone need not occur prior to contamination of nearby ground water. Much greater concentrations of contaminant may occur because of the continually saturated conditions. Storage sites and landfills excavated to a depth near the water table also may permit direct contact of contaminants with ground water. Another direct entry of contaminants from the surface to the ground-water system may be from the vertical leakage of contaminants through the seals around well casings or through improperly abandoned wells, or as a result of contaminant disposal through deteriorated or improperly constructed wells.

1.3.3 Interaquifer Exchange

Contaminated ground water can mix with uncontaminated ground water through a process known as interaquifer exchange in which one water-bearing unit "communicates" hydraulically with another. This is most common in bedrock aquifers where a well penetrates more than one water-bearing formation to provide increased yield. Each water-bearing unit will have its own head potential, some greater than others. When the well is not being pumped, water will move from the formation with the greatest potential to formations of lesser potential. If the formation with the greater potential contains contaminated or poorer quality water, the quality of water in another formation can be degraded.

Similar to the process of direct migration, old and improperly abandoned wells with deteriorated casings or seals are a potential contributor to interaquifer exchange. Vertical movement may be induced by pumping or may occur under natural gradients. For example, in Figure 1-3, an improperly abandoned well formerly tapping only a lower uncontaminated aquifer suffers from a corroded casing. This allows water from an overlying contaminated zone to communicate directly with the lower aquifer. The pumping of a nearby well tapping the lower aquifer creates a downward gradient between the two water-bearing zones. As pumping continues, contaminated water migrates through the lower aquifer to the pumping well. Downward migration of the contaminant may also occur through the aquitard (confining layer) separating the upper and lower aquifers. However, the rate of movement through the aquitard is often much slower than the rate at which contaminants move through the direct connection of an abandoned well.

Figure 1-2 Plume of leachate migrating from a sanitary landfill on a sandy aquifer using contours of chloride concentration (from Freeze and Cherry, 1979).

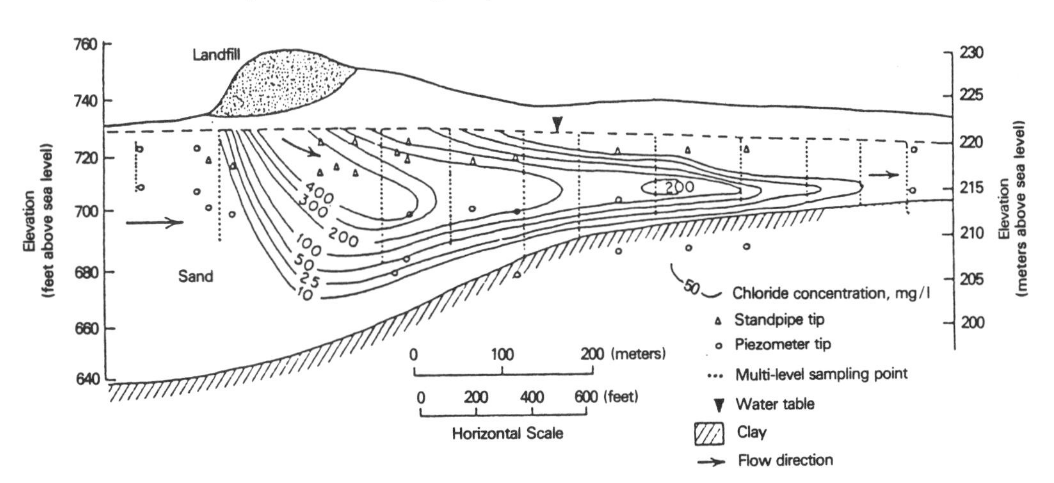

Figure 1-3 Vertical movement of contaminants along an old, abandoned, or improperly constructed well (from Deutsch, 1961).

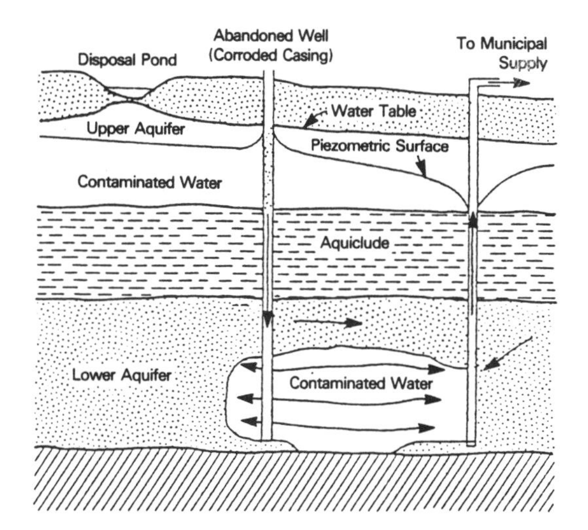

1.3.4 Recharge from Surface Water

Normally, ground water moves toward or "discharges" to surface water bodies (see discussion, Chapter 4). Occasionally, however, the hydraulic gradient is such that surface water has a higher potential than ground water (such as during flood stages), causing a reversal in flow. Contaminants in the surface water can then enter the ground-water system.

Reversal of flow can also be caused by pumping (Figure 1-4). Lowering the ground-water level to a level near a surface water body can induce leakage through the stream or lake bed. Contamination of a glacial sand and gravel aquifer by organic compounds present in an adjacent river in such a manner has been documented (Schwarzenbach *et al.*, 1983).

1.4 Sources of Ground-Water Contamination

A wide variety of ground-water contamination sources have been identified. As previously mentioned, contaminant releases to ground water can occur by design, by accident, or by neglect. The Office of Technology Assessment (OTA, 1984) grouped 33 types of ground-water contamination sources into six major categories (Table 1-2) based on the general nature of the contaminating activity. A number of these sources are depicted in Figure 1-5. *Category 1* includes sources that are intentionally designed to discharge substances. Subsurface percolation systems, such as septic tanks and cess pools, injection wells, and land application of wastewater or sludges fall within this category. Such

Figure 1-4 **Contaminated water induced to flow from surface water to ground water by pumping (from Miller, 1980).**

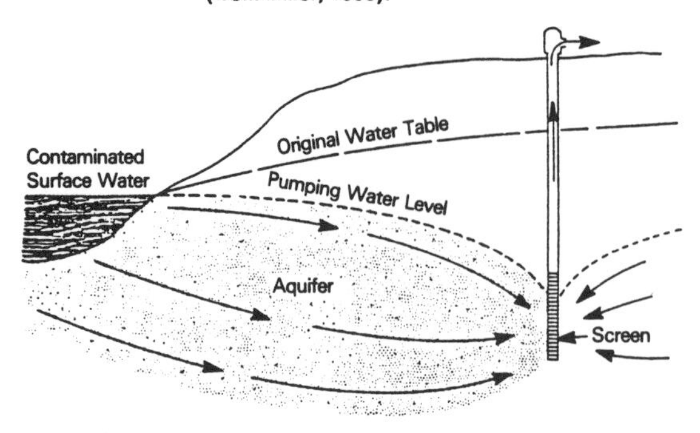

systems are primarily designed to use the natural capacity of the soil materials to degrade wastewaters. Injected wastewaters are often placed in unusable zones to be assimilated with poor quality ground water of natural origin. Septic tanks and cess pools have been estimated to discharge the largest volume of wastewater into the ground and are the most frequently reported source of ground-water contamination (Miller, 1980).

Injection wells are another major potential source of contamination. Although injection wells can be constructed and operated properly, contamination of ground water can occur in several ways (EPA, 1979):

- Faulty well construction (e.g., drilling and casing)
- The forcing upward of pressurized fluids into nearby wells and aquifers, and faults and fractures of confining beds
- The migration of fluids into hydrologically connected usable aquifers
- Faulty well closing.

The depths and operating procedures used in waste injection generally make monitoring and leak prevention very difficult to validate.

Land application is a popular, inexpensive alternative for wastewater and sludge treatment. The U.S. EPA (1983) estimated that 40 to 50 percent of the municipal sludge generated every year is applied to the land.

Category II includes sources that are designed to store, treat, or dispose of substances but are not designed to release contaminants to the subsurface. Landfills, open dumps, local residential disposal, surface impoundments, waste tailings and piles, materials stockpiles, graveyards, aboveground and underground storage tanks, containers, open burning sites, and radioactive disposal sites all fall into this broad category. It is important to note here that while a number of sources in this category are considered "waste" sources (e.g., landfills, dumps, impoundments, etc.), many others are "non-waste" related sources. Storage tanks, stockpiles, and a variety of containers with residues of commercial products have been found to contribute contaminants to ground water.

Category III consists of sources designed to retain substances during transport or transmission. Such sources primarily consist of pipelines and material transport or transfer operations. Contaminant releases generally occur by accident or neglect; for example, as a result of pipeline breakage or a traffic accident. Again, most substances which would be subject to release from sources within this category are not wastes but raw materials or products to be used for some beneficial purpose.

Category IV includes those sources discharging substances as a consequence of other planned activities. This category contains a number of agriculturally related sources such as irrigation return flows, feedlot operations, and pesticide and fertilizer applications. A number of sources related to urban activities such as highway desalting, urban runoff, and atmospheric deposition are included. Surface and underground mine-related drainage also fall within this category.

Category V comprises sources providing conduits or inducing discharge through altered flow patterns. For the most part, such sources are unintentional ground-water contamination sources and include water, oil, and gas production wells, monitoring wells, exploration holes, and construction excavations. The potential to contaminate ground water from production wells stems from poor installation and operation methods, and incorrect plugging or abandonment procedures. Such practices create opportunities for cross-contamination by vertical migration of contaminants.

Finally, *Category VI* includes naturally occurring sources whose discharge is created or made worse by human activity. Ground-water/surface water interactions, described in the previous section, and salt-water intrusion or upconing (ground-water movement upward as a result of pumpage) provide the basis for this category. Withdrawals significantly in excess of recharge can affect ground-water quality. Salt water intrusion in coastal areas and brine-water upconing from deeper formations in inland areas can occur when pumpage exceeds the aquifer's natural recharge rate.

Contaminant releases are also referred to as originating from point or nonpoint sources. Point sources are those which release contaminants from a discrete geographic location. Examples include leaking underground storage tanks, septic systems, and injection wells. Nonpoint contamination situations

Table 1-2 Sources of Ground-Water Contamination (from OTA, 1984).

Category I—Sources designed to discharge substances

- Subsurface percolation (e.g., septic tanks and cesspools)
- Injection Wells
  - Hazardous waste
  - Non-hazardous waste (e.g., brine disposal and drainage)
  - Non-waste (e.g., enhanced recovery, artificial recharge, solution mining, and in-situ mining)
- Land application
  - Wastewater (e.g., spray irrigation)
  - Wastewater byproducts (e.g., sludge)
  - Hazardous waste
  - Non-hazardous waste

Category II—Sources designed to store, treat, and/or dispose of substances; discharge through unplanned release

- Landfills
  - Industrial hazardous waste
  - Industrial non-hazardous waste
  - Municipal sanitary
- Open dumps, including illegal dumping (waste)
- Residential (or local) disposal (waste)
- Surface impoundments
  - Hazardous waste
  - Non-hazardous waste
- Waste tailings
- Waste piles
  - Hazardous waste
  - Non-hazardous waste
- Materials stockpiles (non-waste)
- Graveyards
- Animal burial
- Aboveground storage tanks
  - Hazardous waste
  - Non-hazardous waste
  - Non-waste
- Underground storage tanks
  - Hazardous waste
  - Non-hazardous waste
  - Non-waste
- Containers
  - Hazardous waste
  - Non-hazardous waste
  - Non-waste
- Open burning and detonation sites
- Radioactive disposal sites

Category III—Sources designed to retain substances during transport or transmission

- Pipelines
  - Hazardous waste
  - Non-hazardous waste
  - Non-waste
- Materials transport and transfer operations
  - Hazardous waste
  - Non-hazardous waste
  - Non-waste

Category IV—Sources discharging substances as consequence of other planned activities

- Irrigation practices (e.g., return flow)
- Pesticide applications
- Fertilizer applications
- Animal feeding operations
- De-icing salts applications
- Urban ruhnoff
- Percolation of atmospheric pollutants
- Mining and mine drainage
  - Surface mine-related
  - Underground mine-related

Category V—Sources providing conduit or inducing discharge through altered flow patterns

- Production wells
  - Oil (and gas) wells
  - Geothermal and heat recovery wells
  - Water supply wells
- Other wells (non-waste)
  - Monitoring wells
  - Exploration wells
- Construction excavation

Category VI—Naturally occurring sources whose discharge is created and/or exacerbated by human activity

- Groundwater—surface water interactions
- Natural leaching
- Salt-water intrusion/brackish water upconing (or intrusion and other poor-quality natural water)

Figure 1-5 Sources of ground-water contamination (from Geraghty and Miller, 1985).

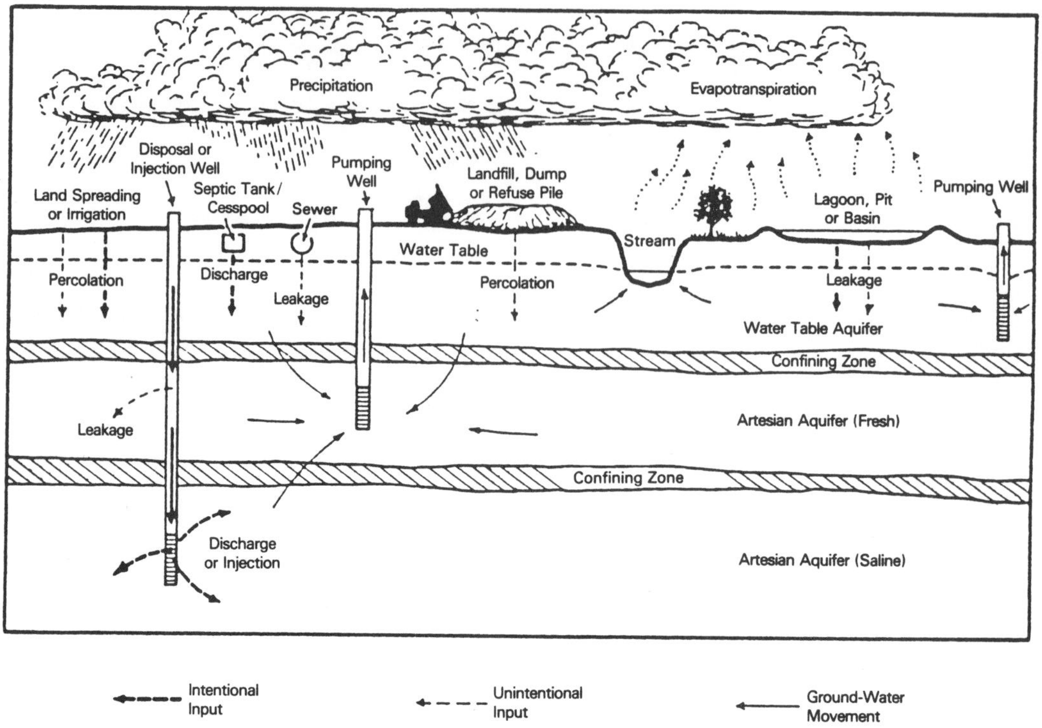

are more extensive in area and diffuse in nature. It is therefore difficult to trace contaminants from nonpoint sources back to their origin. Agricultural activities (i.e., application of pesticides and fertilizers), urban runoff, and atmospheric deposition are potential nonpoint contaminant sources.

1.5 Movement of Contaminants in Ground Water

1.5.1 Contaminant Migration

In broad terms, three processes govern the migration of chemical constituents in ground water: (1) advection, movement caused by the flow of ground water; (2) dispersion, movement caused by the irregular mixing of waters during advection; and (3) retardation, principally chemical mechanisms which occur during advection.

1.5.1.1 Advection

Ground water in its natural state is constantly in motion (advection), although in most cases it is moving very slowly (Todd, 1980). Ground-water movement is governed by the hydraulic principles discussed in Chapter 4.

For example, Darcy's Law states that the flow rate through any porous medium is proportional to the head loss and inversely proportional to the length of the flow path:

$$Q = -K \times A \times h_1/L \qquad (1\text{-}1)$$

where:

Q = ground-water flow rate, in gal/d

A = cross-sectional area of flow, in ft^2

h_1 = head loss, in feet, measured between two points L ft apart

K = hydraulic conductivity, a measure of the ability of the porous medium to transmit water, in $gal/d/ft^2$

Equation 1-1 can be rearranged in the following manner to produce the "bulk," or what is called Darcian, velocity:

$$v = 7.48\ (Q/A)$$
$$v = 7.48\ (-K)(h_1/L)$$
$$v = 7.48\ (-K)(dh/dl) \quad (1\text{-}2)$$

where:

v = Darcian velocity of ground-water flow, in ft/d

dh = the change in hydraulic head (head loss), in ft

dl = the distance or change in position (length) over which the head loss is measured, in ft.

The Darcian velocity assumes that flow occurs across the entire cross section of the porous material without regard to solid or pore spaces.

Actually, flow is limited to the pore space only, so the actual "interstitial" flow velocity is:

$$V_a = v/n = 7.48\ K/n \times dh/dl \quad (1\text{-}3)$$

where:

V_a = the actual ground-water flow velocity, in ft/d

n = the effective porosity, or the percent of the porous media which consists of interconnected pore spaces, the spaces which contribute to ground-water flow, unitless.

The hydraulic conductivity of a geologic formation depends on a variety of physical factors, including porosity, particle size and distribution, the shape of the particles, particle arrangement (packing), and secondary features such as fracturing and dissolution. In general, for unconsolidated porous materials, hydraulic conductivity values vary with particle size. Fine-grained, clayey materials exhibit lower values of hydraulic conductivity while coarse-grained sandy materials normally exhibit higher conductivities. Table 1-3 shows the range of values commonly exhibited by geologic materials.

The effective porosity is essentially an estimated parameter because the actual measurement of the volume of interconnected pore spaces in most porous media has not been conducted. Effective porosity is usually estimated as being somewhat less than the total porosity. Total porosity is calculated from ratios of the volumes of saturated and dry porous material. In coarse-grained materials which drain freely, the effective porosity is essentially equal to total porosity and is generally defined as the ratio of the volume of water which drains by gravity to the total volume of saturated porous material.

Equation 1-3 (interstitial velocity) has been used for determining the advective component of ground-water flow and as a conservative estimate of the rate of migration of dissolved constituents. The rate of movement of the front of a dissolved constituent "plume" by the process of advection can be calculated in a similar fashion.

Figure 1-6 shows the relative concentration of a dissolved constituent emanating from a constant source of contamination versus distance along the flow path. Figure 1-7 shows a similar plot for a discontinuous contaminant source which produced a single slug of dissolved contaminant. In both cases, advective movement causes the dissolved constituent to move with the ground water at the average rate described in Equation 1-3. Considering advective flow only, no diminution of concentration appears as a straight line moving at the rate of ground-water flow. Figure 1-8 shows the effect of advection on the movement of a contaminant in a regional ground-water flow field. Contaminants moving out of a leaking lagoon move horizontally and vertically following the pattern of flow established by ground water as it moves from an upgradient area of recharge to the zone of discharge at the river. Mechanisms influencing the spread of contaminant in the flow field are discussed in the following sections.

1.5.1.2 Dispersion

In natural porous materials, the pores possess different sizes, shapes, and orientations. Similar to stream flow, a velocity distribution exists within the pore spaces such that the rate of movement is greater in the center of the pore than at the edges. Therefore, in saturated flow through these materials, velocities vary widely across any single pore and between pores. As a result, a miscible fluid will spread gradually to occupy an ever increasing portion of the flow field when it is introduced into a flow system. This mixing phenomenon is known as dispersion. In this sense, dispersion is a mechanism for dilution.

Dispersion can occur both in the direction of flow and transverse (perpendicular) to it. Dispersion caused by microscopic changes in flow direction due to pore space orientation is depicted in Figure 1-9a. Macroscopic features, such as fingered lenses of higher conductivity, are shown in Figures 1-9b and 1-9c. Solution channeling and fracturing are other macroscopic features which may contribute to contaminant dispersion (Figure 1-10). Careful placement of wells is required when monitoring in complicated geologic systems such as those shown in Figures 1-9(b and c) and 1-10.

The effect of dispersion as a plot of relative constituent concentration versus distance along a flow path is shown in Figure 1-11. Notice that the front of the dissolved constituent distribution is no longer straight but rather appears "smeared." Some dissolved constituent actually moves ahead of what would have been predicted if only advection were considered.

Table 1-3 Range of Values of Hydraulic Conductivity (adapted rom Freeze and Cherry, 1979).

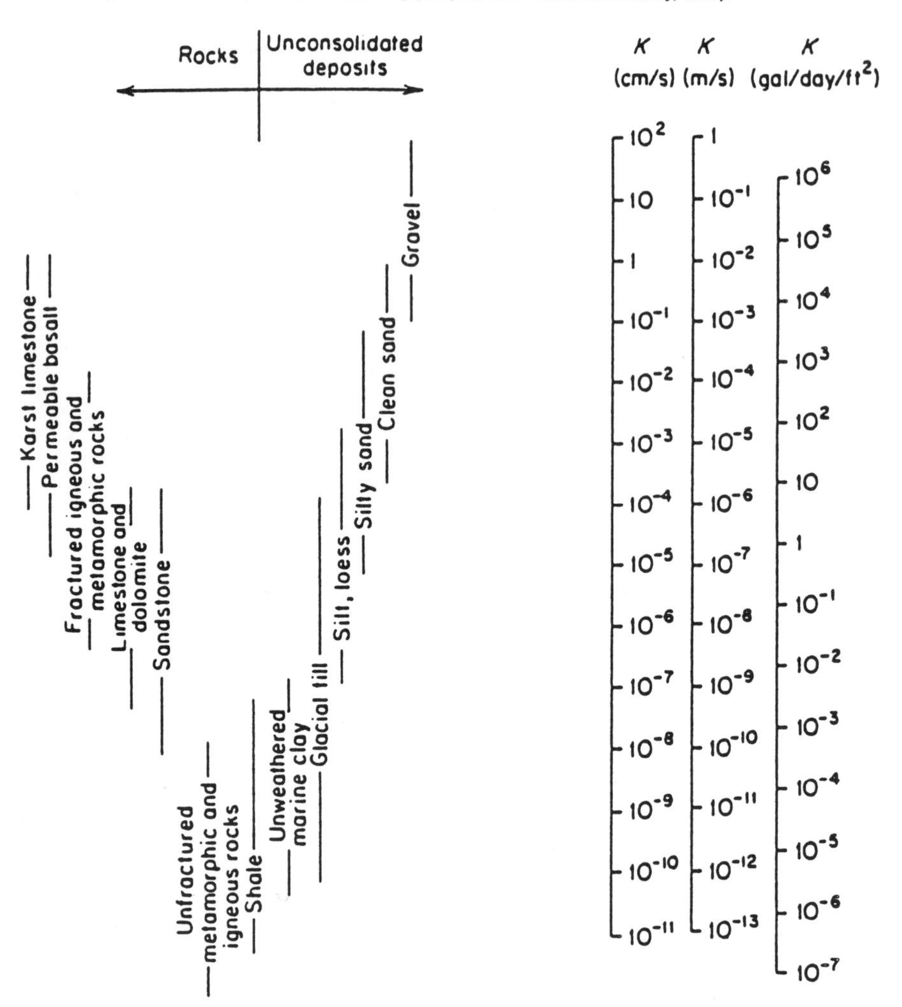

Figure 1-6 Movement of a concentration front by advection only.

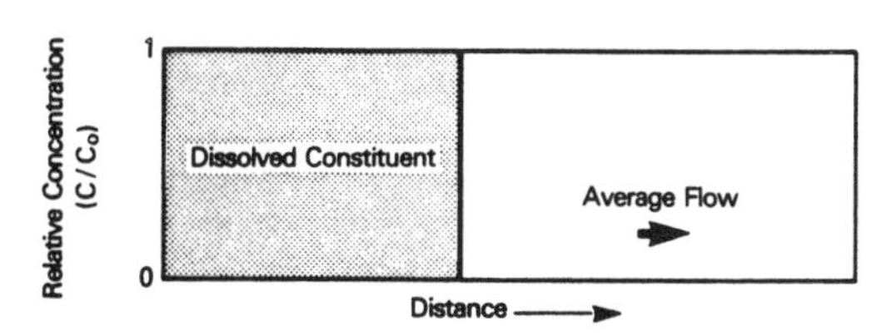

Figure 1-7 Movement of a dissolved constituent slug by advection only.

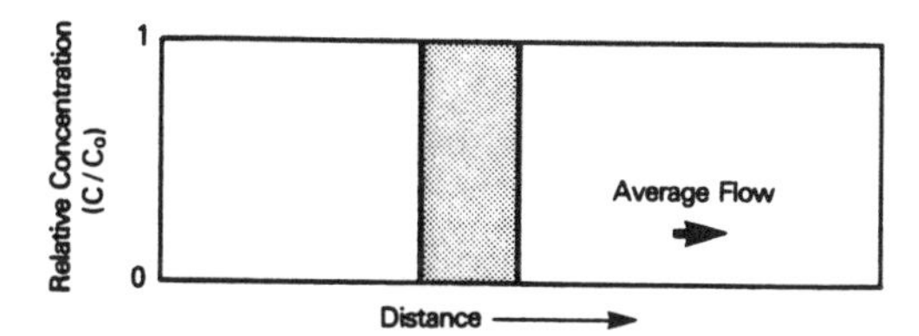

In a similar manner, the concentration of a slug of material introduced to a flow field will appear as shown in Figure 1-12 (a and b). The peak concentration is reduced over time and distance. In such a situation, the total mass of dissolved constituent remains the same; however, a larger volume is occupied, effectively reducing the concentration found at any distance along the flow path. In plane view, continuous and intermittent sources affected by dispersion will appear as shown in Figure 1-13.

1.5.1.3 Retardation

In ground-water contaminant transport, there are a number of chemical and physical mechanisms which retard, that is, delay or slow the movement of constituents in ground water. Four general mechanisms can retard the movement of chemical constituents in ground water: dilution, filtration, chemical reaction, and transformation.

Figure 1-14 illustrates the movement of a concentration front by advection only, and with dispersion, sorption, and biotransformation. The combined effects of advection, dispersion, sorption, and biotransformation on a slug of contaminant introduced into a flow system is also shown in Figure 1-14.

Dilution does not retard the movement of ground-water constituents. However, dilution may lessen the severity of contamination by reducing peak concentrations encountered in the ground-water system. For this reason, dilution, particularly by

Figure 1-8 Effect of leakage from a lagoon on a regional flow pattern (from Geraghty and Miller, 1985).

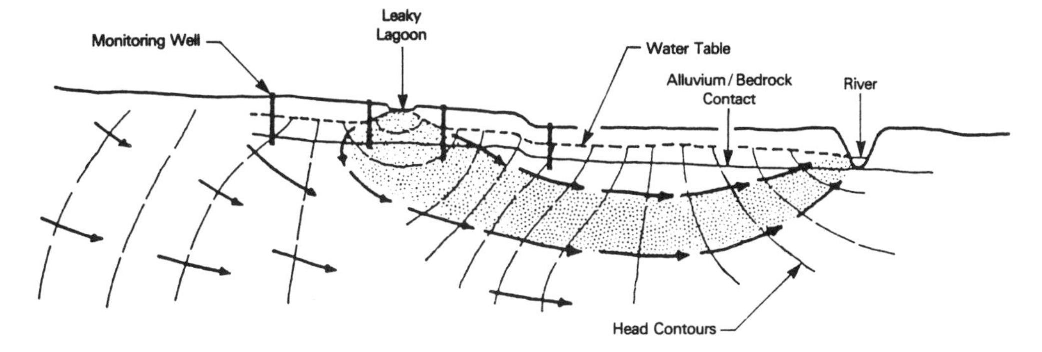

Figure 1-9 Comparison of advance of contaminant influenced by hydrodynamic dispersion (adapted from Freeze and Cherry, 1979).

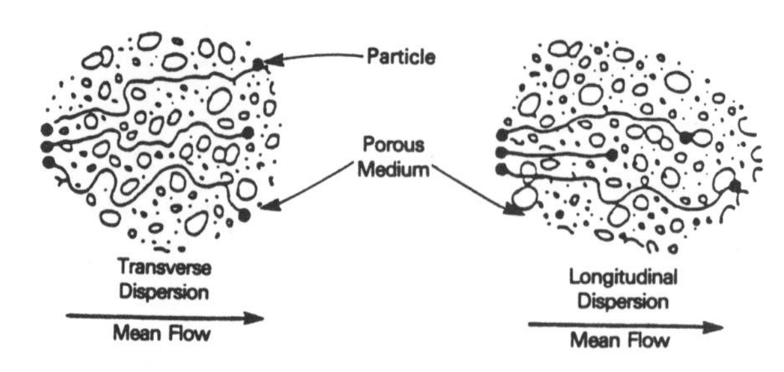

A. Microscopic scale of a granular medium

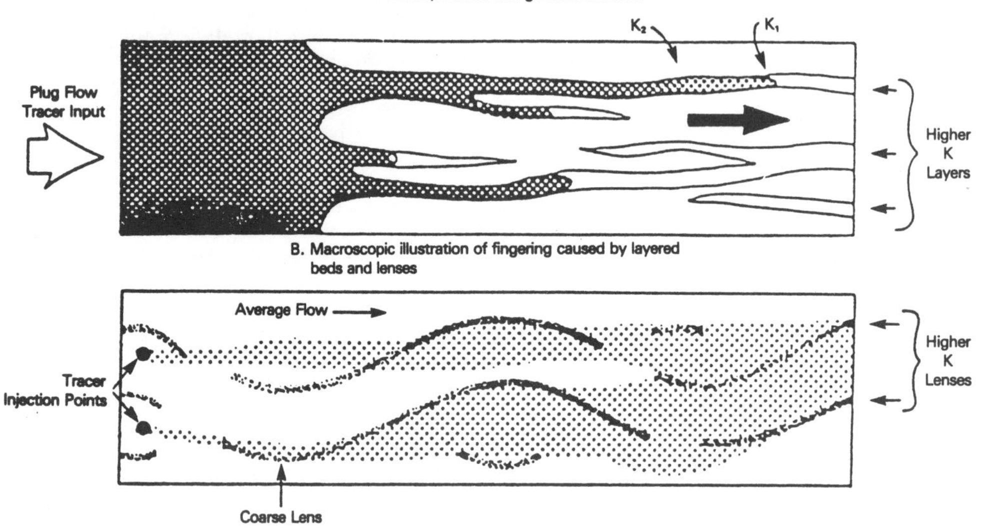

B. Macroscopic illustration of fingering caused by layered beds and lenses

C. Spreading caused by irregular lenses.

Figure 1-10 Flow of contaminated ground water in aquifer with solution porosity (from Geraghty and Miller, 1985).

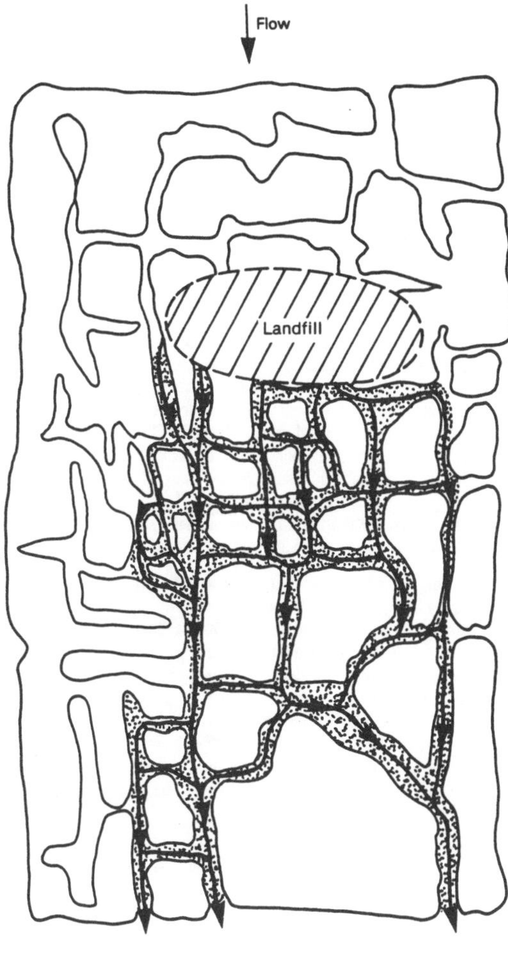

Flow direction of leachate

Leachate enriched ground water

Figure 1-11 Movement of a concentration front by advection and dispersion.

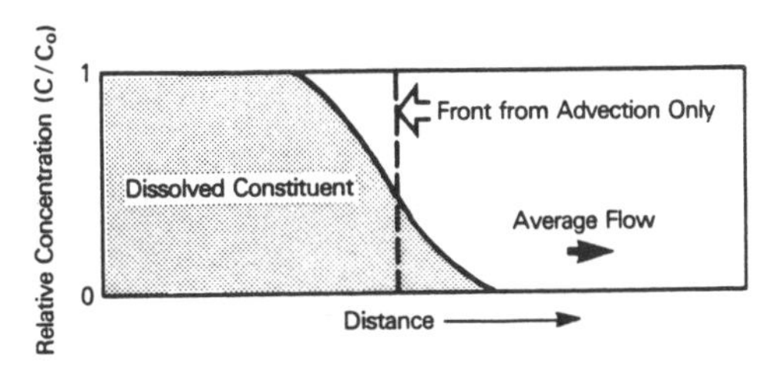

Figure 1-12 Movement of a dissolved constituent slug by advection and dispersion as it moves from time period (a) to (b).

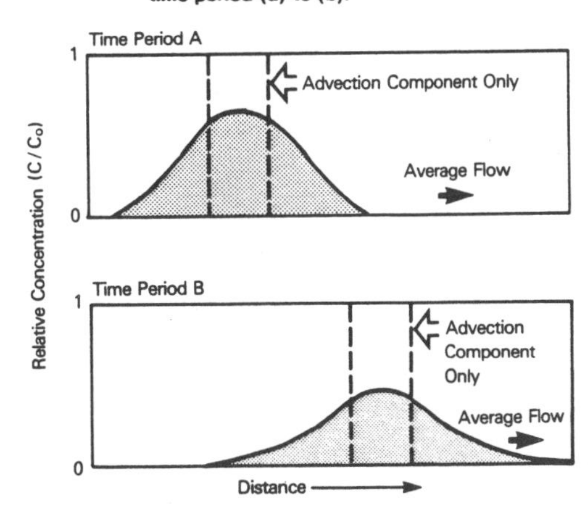

Figure 1-13 Continuous and intermittent sources affected by dispersion.

A. The development of a contamination plume from a continuous point source.

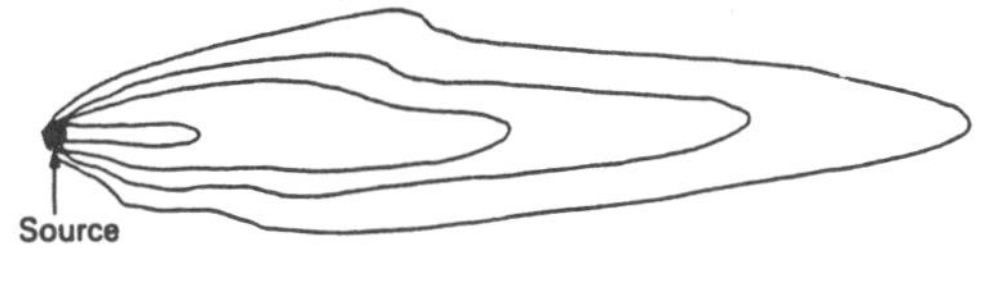

Flow

B. The travel of a contaminant slug(s) from a one-time point source or an intermittent source.

Figure 1-14 The influence of natural processes on levels of contaminants downgradient from continuous and slug-release sources.

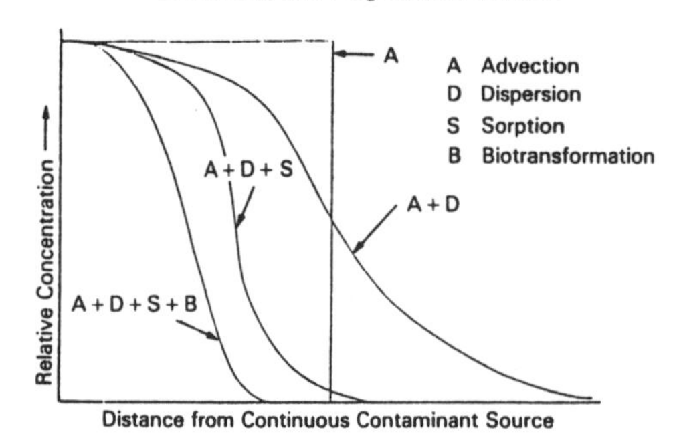

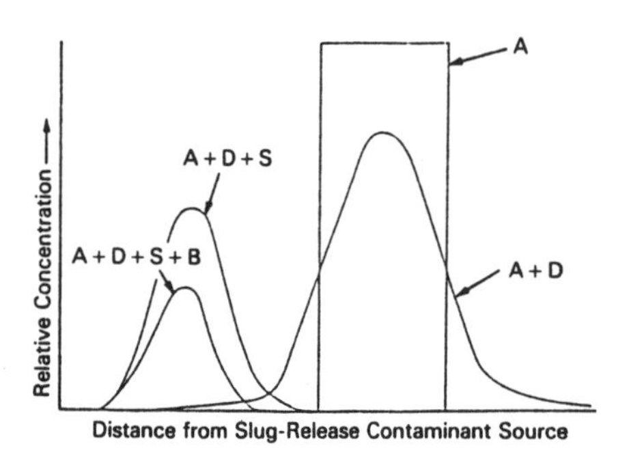

dispersive mechanisms, is included by many scientists in discussions of retardation.

Filtration occurs as dissolved and solid matter are trapped in the pore spaces of the soil and aquifer media, clogging the pore spaces and limiting flow. As the clogging process continues, a decrease in the hydraulic conductivity of the material is manifested. Chemical reactions can also take place which cause a dissolved molecule, for example, to combine with another such that the size of the new molecule is too large for the pore space and mechanical filtration occurs. Flocculation of colloidal material or gas bubble formation may cause eventual clogging of pore spaces resulting in a filtering effect. Microbial activity, especially when particulate matter and dissolved organic materials are present together, can enhance biological growth such that pore spaces become blocked, hindering the movement of dissolved constituents.

Ion exchange processes exert an important influence on retarding the movement of chemical constituents in ground water. In ground water systems, ion exchange occurs when ions (electrically charged particles) in solution displace ions associated with geologic materials. In Figure 1-15 the originally dissolved calcium (Ca^{2+}) ion becomes bound to the geologic media by displacing two sodium (Na^{+}) ions which have less affinity for the exchange sites on the geologic "matrix." This ion exchange process removes constituents from the ground water and releases others to the flow system. The calcium ion will stay bound to the site until another ion with a greater affinity for that site comes along or a shift in environmental conditions (such as a change in pH) causes the ion to release from its site. Ion exchange capacity is very dependent on pH; metal ions, in particular, may exchange onto geologic materials quite readily at neutral pH ($\sim$7) but will be displaced readily by hydrogen ions when the pH is lowered.

One major consideration in ion exchange is that the exchange capacity of a geologic material is limited. A measure of this capacity is quantified in a term called "ion exchange capacity" and is defined as the amount of exchangeable ions, in milliequivalents per 100 grams solids at pH 7. The exchange capacities of several different subsurface materials are given in Table 1-4 (from Matthess, 1982). Typically, clay minerals (e.g., montmorillonite) exhibit greater cation exchange capacities than other minerals such as quartz (the primary component of sand). This is because the available surface area of the clays is often much greater than other minerals.

It is important to recognize that the exchange capacity of a geologic material may retard contaminant movement from a waste source for years or even decades. However, if the source continues to supply a highly ionized leachate, it is possible to exceed the exchange capacity of the geologic material, eventually allowing unretarded transport. Changes in environmental conditions or ground-water solution composition can also cause the release of constituents formerly bound to the geologic materials.

Anionic exchange in aquifer systems is not as well understood as cationic exchange. Anions such as sulfate, chloride, and nitrate would not be expected to be retarded significantly by anion exchange because most mineral surfaces in natural water systems are negatively charged. Chloride ions may be regarded as conservative or non-interacting ions which move largely unretarded with the advective velocity of the ground water. An example of the copper metal ion (Cu^{2+}) being retarded along the flow path while the chloride ion (Cl^{-}) moves unretarded is shown in Figure 1-16.

The release of ions by exchange processes may aggravate a pollution problem. Increases in water hardness as a result of the displacement of calcium and magnesium ions from geologic materials by sodium or potassium in landfill leachate has been documented (Hughes *et al.*, 1971). The release of aluminum to solution, in addition to calcium and magnesium, from soils reacted with an industrial

Figure 1-15 **Ion exchange (from Geraghty and Miller, 1985).**

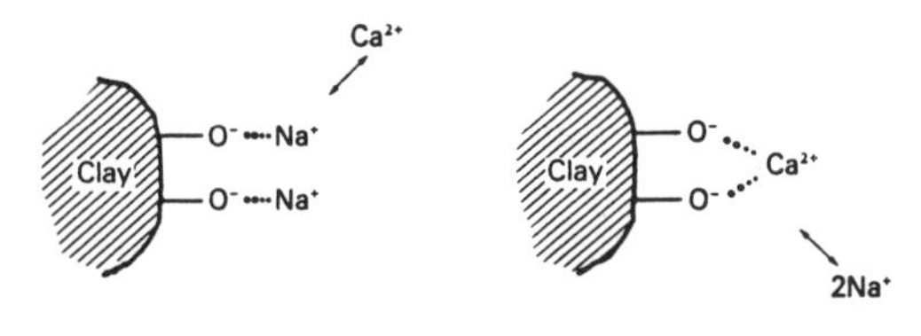

Figure 1-16 **Metal-ion movement slowed by ion exchange (from Geraghty and Miller, 1985).**

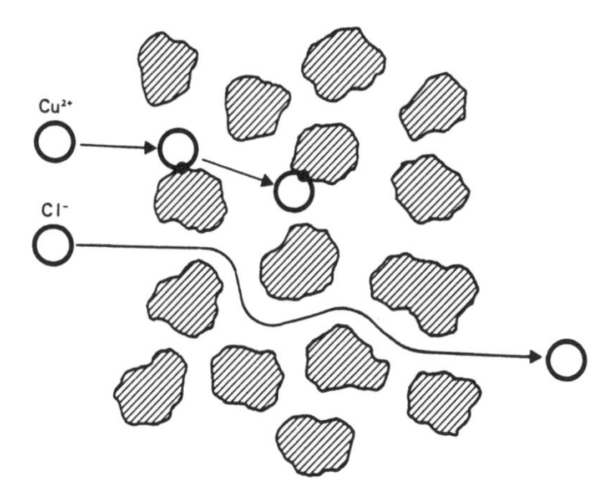

waste has also been documented (Rovers *et al.*, 1976).

A number of other chemical reactions can influence the movement of contaminants in ground water. These include precipitation and complexation. Chemical precipitation in waste leachates is controlled primarily by pH and ionic concentration products. Precipitation of metals as hydroxides, sulfides, and carbonates is very common. Complexation involves the formation of soluble charged or electrically neutral complexes called ligands, which form between metal ions and either organic or inorganic species. For example, the complexation of Cobalt-60 ions by both neutral and synthetic organic compounds enhanced subsurface mobility of the radionuclide (Killey *et al.*, 1984). Other metal species and organic pesticides have been observed to travel significant distances in ground water after the formation of soluble organic complexes with humic substances or organic solvents (Broadbent and Ott, 1957; Griffin and Chou, 1980; Duguid, 1975).

Other processes which may affect contaminant distribution and transport include volatilization as well as a number of transformation mechanisms. The process which occurs when a substance changes from the liquid phase to the gaseous phase is called volatilization. A number of organic compounds (benzene, TCE, and many other low molecular weight compounds) partition into and diffuse through soil gas as a result of their low aqueous solubility and high vapor pressure (low boiling point). Volatilization is enhanced by low soil moisture and high air porosity which generally occurs in coarse-textured materials such as sand and gravel. Remote detection techniques capable of locating subsurface volatile organic chemical plumes by analyzing the overlying soil gases have been devised to take advantage of the result of volatilization (Marrin, 1985). Hydrolysis or chemical reactions may also transform or partially degrade some components of a waste or contaminant mixture.

In addition, the transformation of carbonaceous and inorganic chemicals by microorganisms with evolution of CO_2, CH_4, H_2, H_2S, N_2, NH_3, and NO gases readily occurs in many landfill and other subsurface environments. Microbial processes may be a major factor in the transformation of organic materials present in ground water.

Under the appropriate circumstances pollutants can be completely degraded to harmless products. Under other circumstances, however, they can be transformed to new substances that are more mobile or more toxic than the original contaminant. Quantitative predictions of the fate of biologically reactive substances are at present very primitive, particularly compared to other processes that affect pollutant transport and fate. This situation resulted from the ground-water community's choice of an inappropriate conceptualization of the active processes: subsurface biotransformations were presumed to be similar to biotransformations known to occur in surface water bodies. Only very recently has detailed field work revealed the inadequacy of the prevailing view.

As little as 5 years ago ground-water scientists considered aquifers and soils below the zone of plant roots to be essentially devoid of organisms capable of transforming contaminants. However, recent studies have shown that water-table aquifers harbor appreciable numbers of metabolically active microorganisms, and that these microorganisms frequently can degrade organic contaminants. Thus, it became necessary to consider biotransformation as a process that affects pollutant transport and fate. Unfortunately, many ground-water scientists adopted the conceptual ideas most frequently used to describe biotransformations in surface waters. In ground water, contaminant residence time is usually long, at least weeks or months, and frequently years

Table 1-4 Exchange Capacities of Minerals and Rocks

Mineral	For Cations (meq/100 g) Grim (1968)	For Cations (meq/100 g) Carroll (1959)	For Anions (meq/100 g) Grim (1968)
Talc	–	0.2	–
Basalt	–	0.5-2.8	–
Pumice	–	1.2	–
Tuff	–	32.0-49.0	–
Quartz	–	0.6-5.3	–
Feldspar	–	1.0-2.0	–
Kaolinite	3-15	–	6.6-13.0
Kaolinite (colloidal)	–	–	20.2
Nontronite	–	–	12.0-20.0
Saponite	–	–	21.0
Beidellite	–	–	21.0
Pyrophyllite	–	4.0	–
Halloysite • $2H_2O$	5-10	–	–
Illite	10-40	10-40	–
Chlorite	10-40	10-40?	–
Shales	–	10-41.0	–
Glauconite	–	11-20	–
Sepiolite-attapulgite-palygorskite	3-15	20-30	–
Diatomite	–	25-54	–
Halloysite • $4H_2O$	40-50	–	–
Allophane	25-50	~70	–
Montmorillonite	80-150	70-100	23-31
Silica gel	–	80	–
Vermiculite	100-150	100-150	4
Zeolites	100-300	230-620	–
Organic substances in soil and recent sediments	150-500	–	–
Feldspathoids			
Leucite	–	460	–
Nosean	–	880	–
Sodalite	–	920	–
Cancrinite	–	1,090	–

Source: Matthess, 1982.

or decades. Further, contaminant concentrations that are high enough to be of environmental concern are often high enough to elicit adaptation of the microbial community. For example, the U.S. EPA maximum contaminant level (MCL) for benzene is 5 µg/l. This is very close to the concentration of alkylbenzenes required to elicit adaption to this class of organic compounds in soils. As a result, the biotransformation rate of a contaminant in the subsurface environment is not a constant, but increases after exposure to the contaminant in an unpredictable way. Careful field work has shown that the transformation rate in aquifers of typical organic contaminants, such as alkylbenzenes, can vary as much as two orders of magnitude over a meter vertically and a few meters horizontally. This surprising variability in transformation rate is not related in any simple way to system geology or hydrology.

Biological activity may, however, promote or catalyze chemical reactions. Stimulation of the native microbial population and the addition of contaminant specific "seed" microorganisms for the restoration of contaminated aquifers by in situ biological treatment is the subject of much current research (Canter and Knox, 1985).

1.5.2 Contaminant Plume Behavior

The physical mechanisms of advection and dispersion, as well as a variety of chemical and microbial reactions, will interact to influence the movement of contaminants in ground water. The degree to which these mechanisms influence contaminant movement is dependent on a number of factors:

a) *Geologic material properties*
The rate of ground-water movement is largely dependent on the type of geologic material through which it is moving. More rapid movement can be expected through coarse-textured materials such as sand or gravel than through fine-textured materials like silt and clay. The physical and chemical composition of the geologic material is equally important. Fine-grained materials present the most favorable environment for potential retardation.

b) *Hydrogen ion activity (pH)*
The pH of the geologic materials and the waste stream may be a major influence on actual retardation. The pH affects the speciation of many dissolved chemical constituents which determine solubility and reactivity. Ion exchange and hydrolysis reactions are also particularly sensitive to pH.

c) *Leachate composition*
The influence all other factors will have on contaminant migration ultimately depends on the composition of the leachate or contaminants entering the ground-water system. Similar contaminants may behave differently in the same environment due to the influence of other constituents in a complex leachate. Solubility (which affects the mobile concentration), density, chemical structure, and many other properties can affect net contaminant migration. For example, Figure 1-17 illustrates the appearance of two chemicals, benzene and chloride, in a monitoring well. Even though both contaminants may have entered the ground-water system at the same time and concentration, their detection in the monitoring well reveals significantly different migration rates. Chloride has migrated essentially unaffected while benzene has been retarded significantly. This type of relationship can be reversed if there is a solvent phase in the aquifer.

Sources releasing a variety of contaminants will create complex plumes composed of different constituents at downgradient positions. An idealized plume configuration composed of five different contaminants (A-E) moving at different rates through the ground-water system is shown in Figure 1-18. Because of this, "the onset of contamination at a supply well may mark the front of a set of overlapping plumes of different compounds advancing at different rates, which may affect the well in sequence for decades even if the original contaminant source is removed" (Mackay *et al.*, 1985).

The effect of contaminant density on transport in ground-water systems is presented in Figure 1-19. Substances with densities less than water may "float" on the surface of the saturated zone. Similarly, substances with densities greater than water can sink through the saturated zone until an impermeable layer is encountered. In the situation shown in Figure 1-19, the surface of an underlying, impermeable formation slopes opposite to the direction of ground-water flow in the overlying formation. Dense contaminant movement will follow the slope of the impermeable boundary while some dissolved product will move with the ground water.

d) *Source characteristics*
Source characteristics include source mechanism (i.e., infiltration, direct migration, interaquifer exchange, ground-water/surface water interaction), type of source (particularly point or nonpoint origination), and temporal features. Aspects of source mechanism and type have been discussed earlier. The manner in which a contaminant is released over time and the time which has elapsed since the contaminant was released will greatly affect the extent and configuration of the contaminated zone.

Figure 1-20 presents the effects caused by changes in the rate of waste discharge on plume size and shape. In the first case, plume enlargement results from an increase in the rate of waste discharge to the ground-water system. Similar effects can be

Figure 1-17 Benzene and chloride appearance in a monitoring well (from Geraghty and Miller, 1985).

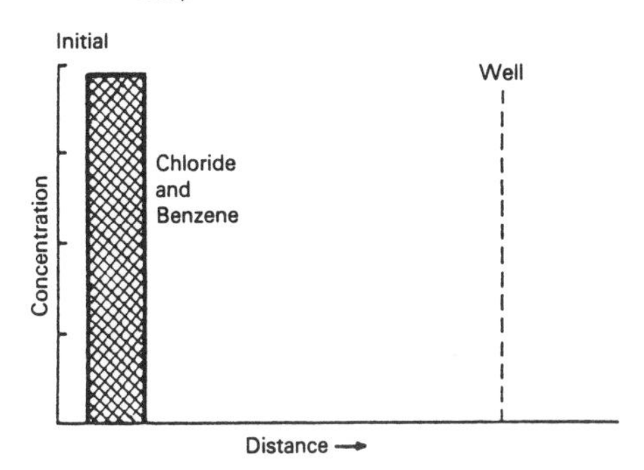

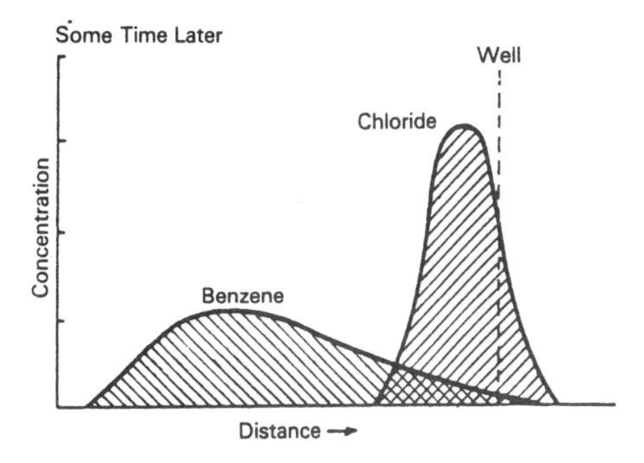

Figure 1-18 Constant release but variable constituent source (from LeGreud, 1965).

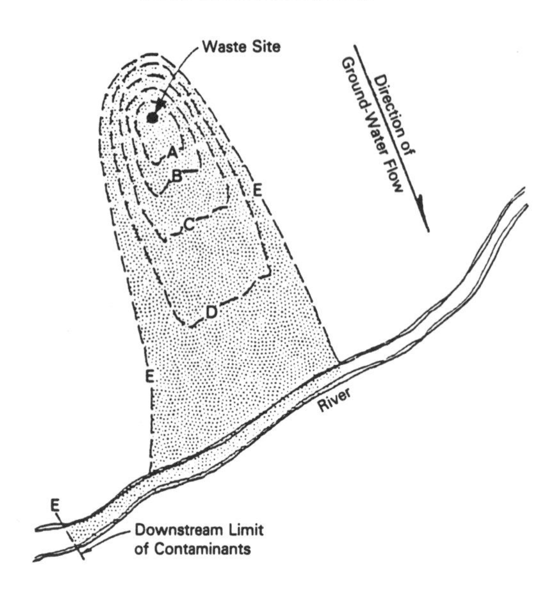

produced if the retardation capacity of the geologic materials is exceeded or if the water table rises closer to the source causing an increase in dissolved constituent concentration. Decreases in waste discharge, lowering of the water table, retardation through sorption, and reductions in ground-water flow rate can diminish the size of the plume. Stable plume configurations suggest that the rate of waste discharge is at steady state with respect to retardation and transformation processes. A plume will shrink in size when contaminants are no longer released to the ground-water system and a mechanism to reduce contaminant concentrations is present. Unfortunately, many contaminants, particularly complex chlorinated hydrocarbons and heavy metals, may persist in ground water for extremely long time periods without appreciable transformation. Lastly, an intermittent or seasonal source can produce a series of plumes which are separated by the advection of ground water during periods of no contaminant discharge.

1.6 Summary

To properly assess and predict the effect of ground-water contamination at a given site, detailed information about the nature of the suspected contaminants, the volume of contaminants disposed and released, the time period over which contaminants were released, and the areas in which contaminants were released is needed. For complex sites, such as industrial facilities and hazardous waste disposal landfills, this information may be limited. The transport and fate of contaminants in ground water must also be considered; these are often affected by a site-specific interrelationship of physical, chemical, biological, and temporal processes. Knowledge of site geology, hydrology, source characteristics and mechanisms must also precede an intelligent investigation of ground-water contamination.

1.7 References

Broadbent, F.E., and J.B. Ott. 1957. Soil Organic Matter - Metal Complexes: I. Factors Affecting Various Cations. Soil Science 83:419-427.

Brown, M. 1979. Laying Waste, The Poisoning of America by Toxic Chemicals. Pantheon Books, New York, NY.

Burmaster, D.E., and R.H. Harris. 1982. Groundwater Contamination: An Emerging Threat. Technology Review 85(5):50-62.

Canter, L.W., and R.C. Knox. 1985. Ground Water Pollution Control. Lewis Publishers, Inc, Chelsea, MI.

Coniglio, W. 1982. Criteria and Standards Division Briefing on Occurrence/Exposure to Volatile Organic Chemicals. U.S. Environmental Protection Agency, Office of Drinking Water, Cincinnati, OH.

Deutsch, M. 1961. Incidents of Chromium Contamination of Ground Water in Michigan. Proceedings of 1961 Symposium, Ground Water Contamination, U.S. Department of Health, Education, and Welfare, April 5-7, 1961, Cincinnati, OH.

Duguid, J.O. 1975. Status Report on Radioactivity Movement from Burial Grounds in Melton and Bethel Valleys. Environmental Science Publication No. 658, ORNL-5017, Oak Ridge National Laboratory, Oak Ridge, TN.

Epstein, S.S., L.O. Brown, and C. Pope. 1982. Hazardous Waste in America. Sierra Club Books, San Francisco, CA.

Epstein , S.S. 1979. The Politics of Cancer. Anchor Press/Doubleday. Garden City, NY.

Fetter, C.W., Jr. 1980. Applied Hydrogeology. Charles E. Merrill Publishing Company, Columbus, OH.

Freeze, R.A., and J.A. Cherry. 1979. Groundwater. Prentice-Hall, Inc., Englewood Cliffs, NJ.

Figure 1-19 Effects of density on migration of contaminants (from Geraghty and Miller, 1985).

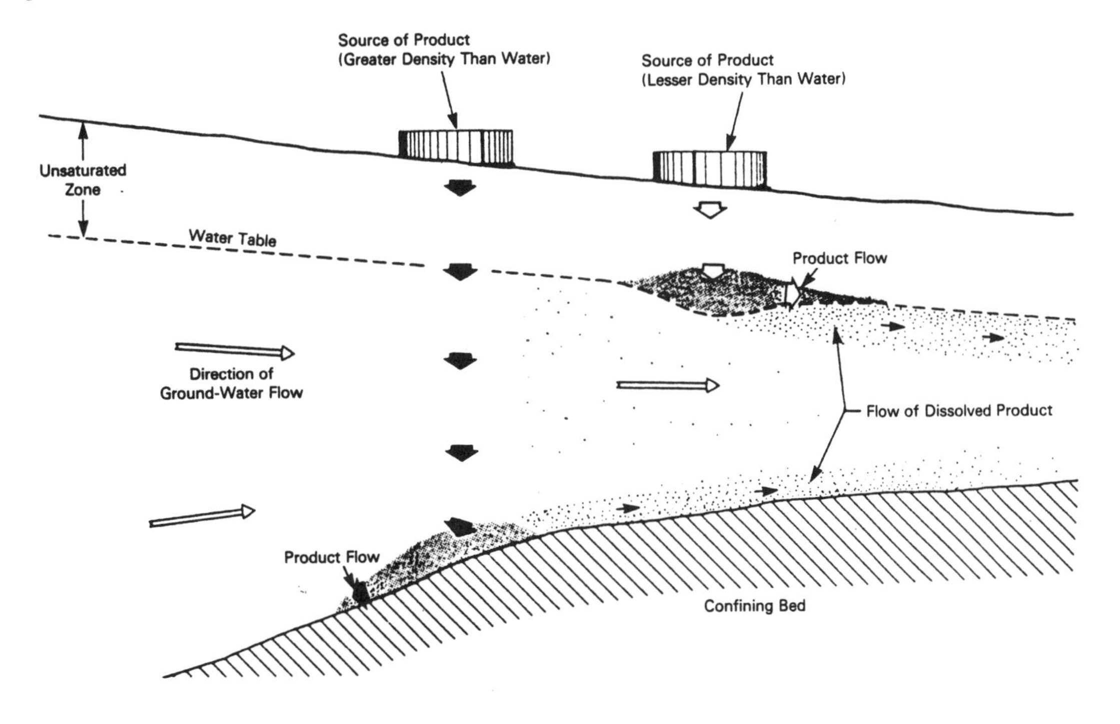

Figure 1-20 Changes in plumes and factors causing the changes (modified from U.S. EPA, 1977).

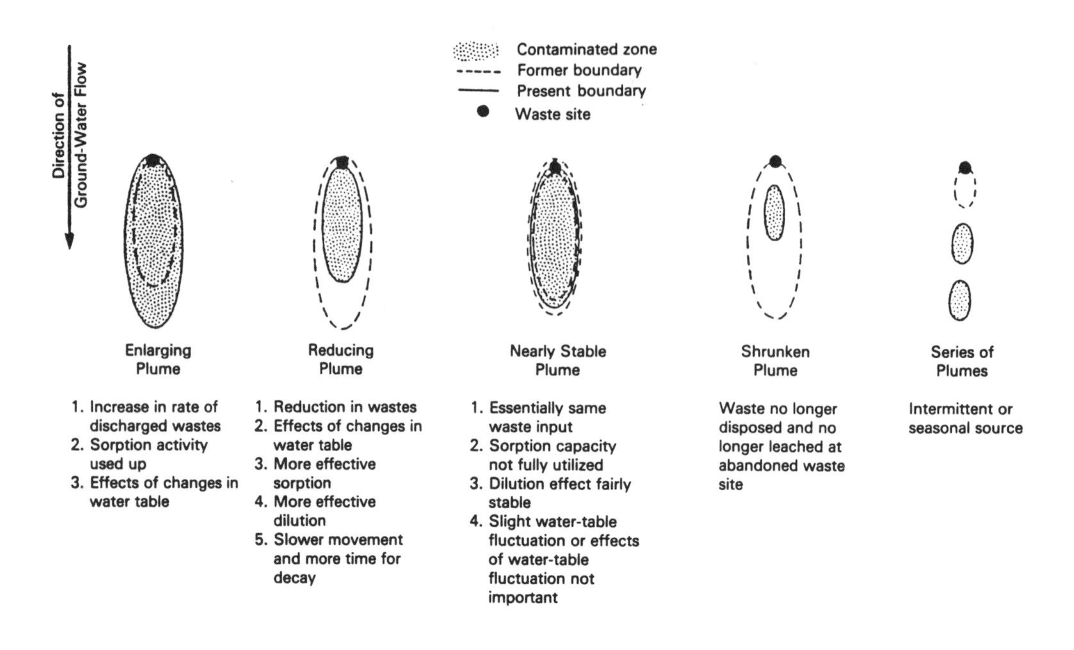

Gass, T.E. 1980. To What Extent Is Ground Water Contaminated? Water Well Journal 34(11):26-27.

Geraghty, J.J., and D.W. Miller. 1985. Fundamentals of Ground Water Contam ination, Short Course Notes. Geraghty and Miller, Inc., Syosset, NY.

Griffin, R.A., and S.F.J. Chou. 1980. Attenuation of Polybrominated Biphenyls and Hexachlorobenzene by Earth Materials. Environmental Geology Notes 87, Illinois State Geological Survey, Urbana, IL.

Griffin, R.A., K. Cartwright, N.F. Shimp, J.D. Steele, R.R. Buch, W.A. White, G.M. Hughes, and R.H. Gilkeson. 1976. Alteration of Pollutants in Municipal Landfill Leachate by Clay Minerals: Part I. Column Leaching and Field Verification. Illinois State Geological Survey Bulletin 78, Illinois State Geological Survey, Urbana, IL.

Hughes, G.M., R.A. Landon, and R.N. Farvolden. 1971. Hydrogeology of Solid Waste Disposal Sites in Northeastern Illinois. Solid Waste Management Series, Report SW-124, U.S. Environmental Protection Agency.

Illinois Environmental Protection Agency. 1986. A Plan for Protecting Illinois Groundwater. Illinois Environmental Protection Agency, Springfield, IL.

Killey, R.W., J.O. McHugh, D.R. Champ, E.L. Cooper, and J.L. Young. 1984. Subsurface Cobalt-60 Migration from a Low-Level Waste Disposal Site. Environmental Science and Technology 18(3):148-156.

Leckie, J.O., J.G. Pace, and C. Halvadakis. 1975. Accelerated Refuse Stabilization Through Controlled Moisture Application. Unpublished report, Department of Environmental Engineering, Stanford University, Stanford, CA.

LeGrand, H.E. 1965. Patterns of Contaminated Zones of Water in the Ground. Water Resources Research 1(1):83-95.

Lehr, J.H. 1982. How Much Ground Water Have We Really Polluted? Ground Water Monitoring Review 2(1):4.

Mackay, D.M., P.V. Roberts, and J.A. Cherry. 1985. Transport of Organic Contaminants in Groundwater Environmental Science and Technology. 19(5):384-392.

Magnuson, Ed. 1980. The Poisoning of America. Time (7):58.

Mallmann, W.L., and W.N. Mack. 1961. Biological Contamination of Ground Water. Proceedings of 1961 Symposium, Ground Water Contamination, U.S.

Department of Health, Education, and Welfare, April 5-7, 1961, Cincinnati, OH.

Marrin, D.L. 1985. Delineation of Gasoline Hydrocarbons in Groundwater by Soil Gas Analysis. Tracer Research Corporation, Tucson, AZ.

Matthess, G. 1982. Die Beschaffenheit des Grundwassers (The Properties of Groundwater). John Wiley and Sons, New York, NY.

Michels, Nabert, Udluft, and Zimmerman. 1959. Expert Opinion on Questions of Protection of Aquifers Against Contamination of Ground Water. Bundesministerium fur Atomkernenergie and Wasserwirtschaft, Bad Godesberg.

Middleton, M., and G. Walton. 1961. Organic Chemical Contamination of Ground Water. Proceedings of 1961 Symposium, Ground Water Contamination, U.S. Department of Health, Education, and Welfare, April 5-7, 1961, Cincinnati, OH.

Miller, D.W., ed. 1980. Waste Disposal Effects on Ground Water. Premier Press. Berkeley, California. 512 pp.

O'Hearn, M., and S.C. Schock. 1984. Design of a Statewide Ground-Water Monitoring Network for Illinois. Illinois State Water Survey Contract Report 354, Illinois State Water Survey, Champaign, IL.

Prescott, S.C., and M.P. Horwood. 1935. Sedgwick's Principles of Sanitary Science and Public Health. The MacMillan Company, New York, NY.

Rovers, F.A., H. Mooij, and G.J. Farquhar. 1976. Contaminant Attenuation - Dispersed Soil Studies. In: Residual Management by Land Disposal, edited by W.H. Fuller. EPA-600/9-76-015, U.S. Environmental Protection Agency, Cincinnati, OH.

Schiffman, A. 1985. The Trouble With RCRA. Ground Water 23(6):726-734.

Schwarzenbach, R., W. Giger, E. Hoehn, and J. Schneider. 1983. Behavior of Organic Compounds During Infiltration of River Water to Ground Water - Field Studies. Environmental Science and Technology 17(8):472-479.

Senkan, S.M., and N.W. Stauffer. 1981. What To Do With Hazardous Waste? Technology Review (11/12):34-37.

U.S. Congress. 1984. Protecting the Nation's Groundwater from Contamination, Vols. I and II. OTA-0-233 and OTA-0-276, Office of Technology Assessment, U.S. Government Printing Office, Washington, DC.

U.S. Department of Health, Education, and Welfare. 1961. Proceedings of 1961 Symposium, Ground Water Contamination, April 5-7, 1961, Cincinnati, OH.

U.S. Environmental Protection Agency. 1984. National Primary Drinking Water Regulations, Volatile Organic Chemicals. Federal Register, 49:24331-24355.

U.S. Environmental Protection Agency. 1983. Process Design Manual: Land Application of Municipal Sludge. EPA-625/1-83-016, U.S. Environmental Protection Agency, Municipal Environmental Research Lab, Cincinnati, OH.

U.S. Environmental Protection Agency. 1979. Environmental Assessment: Short-Term Tests for Carcinogens, Mutagens and Other Genotoxic Agents. Health Effects Research Laboratory, Research Triangle Park, NC.

U.S. Environmental Protection Agency. 1979. A Guide to the Underground Injection Program.

Westrick, J.J., J.W. Mello, and R.F. Thomas. 1983. The Ground Water Supply Survey: Summary of Volatile Organic Contaminant Occurrence Data. U.S. Environmental Protection Agency, Office of Drinking Water, Cincinnati, OH.

2. Ground-Water Quality Investigations

Within the last decade, a substantial number of ground-water quality investigations have been conducted. Most of these have centered on specific sites, sites that by one means or another were known or suspected to be contaminated. In general, the sites covered only several acres or a few square miles.

The cost of these investigations usually has been excesssive, largely because of analytical costs. The most disconcerting feature of many of them, however, is that to one degree or another they were found to be inadequate. This, in turn, necessitated additional work and expense in response to the ever present desire for additional information. It should be recognized that the data base will always be inadequate, and eventually there will be a finite sum that is dictated by time, common sense, and budgetary constraints. One simply has to do the best one can with what is available.

It is suspected that the major reason that many field investigations are both inadequate and expensive is that a comprehensive experimental work plan was not formulated before the project was initiated or if it was, then it probably was not followed. Any type of investigation must be carefully planned, keeping in mind the overall purpose, time limitations, and project funding. Moreover, the plan must be based on sound, fundamental principles and a practical approach. As far as ground-water quality investigations are concerned, the basic questions are (1) Is there a problem? (2) Where is it? and (3) How severe is it? A subsequent question may relate to what can be done to reduce the severity of the problem, that is, aquifer restoration.

2.1 Types of Ground-Water Quality Investigations

Ground-water quality investigations can be divided into three general types: regional, local, and site evaluations. The first, which may encompass several hundred or even thousands of square miles, is reconnaissance in nature, and is used to obtain an overall evaluation of the ground-water situation. A local investigation is conducted in the vicinity of a contaminated site, may cover a few tens or hundreds of square miles, and is used to determine local ground-water conditions. The purpose of the site evaluation is to ascertain, with a considerable degree of certainty, the extent of contamination, its source or sources, hydraulic properties, and velocity, as well as all of the other related controls on contaminant migration.

2.1.1 Regional Investigations

This broad brush type of investigation, which is reconnaissance in nature, can be the starting point for two general types of explorations. First, it can be carried out with the purpose of locating potential sources or sites of ground-water contamination. Second, it will provide an understanding of the occurrence and availability of ground water on a regional scale. The underlying objectives are first, to determine if a problem exists, and second, if necessary, to ascertain prevalent hydrologic properties of earth materials, generalized flow directions in both major and minor aquifers, the primary sources and rates of recharge and discharge, the chemical quality of the aquifers and surface water, and locations and yields of pumping centers. These data can be useful in more detailed studies because they provide information on the geology and flow direction, both of which impact the local situation.

2.1.2 Local Investigations

Investigations of this nature usually include a few square miles. The purpose is to define in greater detail the geology and hydrology in an area surrounding a specific site or sites of concern. Both the geology and hydrology are likely to exert some control on contaminant migration and nearby rock units may be impacted as well.

2.1.3 Site Investigations

The site investigation is the most detailed, complex, costly, and, from a legal and restoration viewpoint, the most critical of the three types of evaluations. This examination must address the local controls on contaminant migration, including the geology, soil, microbiology, geochemical interactions, and mass flow rate of contaminant to the water table, among others. At the same time, auxiliary investigations at the site might include tank inventories, toxicological evaluations, air pollution monitoring, manufacturing procedures, and manifest scrutiny, as well as many other studies, all of which will eventually interface in the development of a comprehensive report.

2.2 Conducting the Investigation

Regardless of the complexity or detail of the investigation, a logical series of steps should be followed. Of course, each inquiry is unique but, nonetheless, the general rules prevail. The steps are as follows:

1) Establish the objectives of the study.
2) Collect data.
3) Compile data.
4) Interpret data.
5) Develop conclusions.
6) Present results.

2.2.1 Establish the Objectives of the Study

Establishing the major objective of the study is paramount. The approach, time requirements, and funding can be vastly different between a regional reconnaissance evaluation and a site investigation. The former, which deals with gross features, may require only days while the latter, which necessitates minute detail, may demand years. The statement of the objective can be as simple as "Develop a general understanding of the regional ground-water situation" to "Evaluate the degradation and dispersion of selected organic compounds in the capillary zone at the A site."

In both of the above examples the objective is clearly stated and the complexity is evident.

Once the general objective is established, a number of secondary purposes must be considered. These involve the physical system and the chemical aspect. Secondary objectives include the following:

1) Determination of the thickness, soil characteristics, infiltration rate, and water-bearing properties of the unsaturated zone.

2) Determination of the geologic and hydrologic properties and dimensions of each unit in the geologic column that potentially could be impacted by ground-water contamination. This includes rock type, thickness of aquifers and confining units, areal distribution, structural configuration, transmissivity, hydraulic conductivity, storativity, water levels, infiltration or leakage rate, and rate of evapotranspiration, if appropriate.

3) Determination of recharge and discharge areas, if appropriate.

4) Determination of the direction and rate of ground-water movement in potentially impacted units.

5) Determination of the ground water and surface water relationships.

6) Determination of the background water quality characteristics of potentially impacted units.

7) Determination of potential sources of contamination and types of contaminants.

2.2.2 Data Collection

Data collection forms the basis for the entire investigation and, consequently, time must be expended and care exercised in carrying out this task. The amount and types of data to be collected are dictated by the objectives of the study. Before the field is ever visited, a thorough search should be made of files and the literature for pertinent information. Materials that should be collected, if available, include soil, geologic, topographic, county, and state maps, geologic cross sections, aerial photographs, satellite imagery, location of pumping centers and discharge rates, well logs, climatological and stream discharge records, chemical data, and the locations of potential sources of ground-water contamination.

Many of these data are readily available in files or report form and can be obtained from an assortment of state and Federal agencies. Personnel with these agencies also can be of great help owing to their knowledge of the state or county and their familiarity with the literature. Examples include the U.S. Geological Survey, which has at least one office in each state, the state geological survey and several state agencies that deal with water, such as the state water survey, water resources board, or the water commission. Some states have several commissions or boards that are involved with water. Other sources of information include the state or Federal department of agriculture, Soil Conservation Service, and weather service, among others.

Climatological data are important because they indicate precipitation rates and patterns, both of which influence surface runoff, runoff, and ground-water recharge. Additionally, climatological data include temperature measurements, which can be used in the evaluation of evapotranspiration. Evapotranspiration of shallow ground water can produce a significant effect on the water-table gradient, causing it to change in slope and direction, not only seasonally, but diurnally as well.

Stream discharge and chemical quality records can be used for several types of regional and local evaluations, as described in Chapter 4.

Soil types are related to the original rock from which they were derived. Consequently, soil maps can be used as an aid in geologic mapping. Soil information is necessary also to evaluate the potential for movement of organic and inorganic compounds through the unsaturated zone.

Exceedingly useful tools, both for office and field study, are aerial photographs and satellite imagery. The latter should be examined first in an attempt to detect trends of lineaments, which may indicate the presence of faults or joints. These may reflect zones of high permeability that exert a strong influence on fluid movement from the land surface or through the subsurface. Satellite imagery also can be used to detect the presence of shallow ground water owing to the subtle tonal changes and differences in vegetation brought about by the higher moisture content. Rock types may be evident also on imagery.

Aerial photographs, particularly stereoscopic pairs, should be an essential ingredient of any hydrologic investigation. They are necessary to further refine the trends of lineaments, map rock units, determine the location of cultural features and land use, locate potential drilling sites, and detect possible sources of contamination. Topographic and state and county road maps also are useful for many of these purposes.

Geologic reports, maps, and cross sections provide details of the surface and subsurface, including the areal extent, thickness, composition, and structure of rock units. It must be remembered that geology is the key to any ground-water investigation. These sources of information should be supplemented, if possible, by an examination of the logs of wells and test holes. Depending on the detail of the logs, they may provide a clear insight into the complexities of the subsurface.

Logs of wells and test holes are essential in ground-water investigations. They provide first-hand information on types and characteristics of rocks in the subsurface, their thickness, and areal extent. Logs also may describe drilling conditions that allow one to infer relative permeability values (see Chapter 9), describe well construction details, and report water-level measurements.

Chemical data may be available from reports, but the most recent information is probably stored in local, state, or Federal files. Concentrations of selected constituents, such as dissolved solids, specific conductance, chloride, and sulfate, should be plotted on base maps and used to estimate background quality and, perhaps, detect places of contamination. Both surface and ground-water quality data should be examined.

Chemical analyses that report concentrations of organic compounds are bound to be sparse and even those are likely to be questioned for one reason or another. Only within the last decade or so have organic compounds become of concern. The cost of analysis is high, and much remains to be learned about appropriate methods of collection, storage, analysis, and interpretation. Consequently, investigators, whenever possible, will need to rely on analyses of inorganic substances to detect sites of ground water contaminated by these complex substances. On the other hand, reliance on concentrations of inorganic constituents to evaluate contamination by organic compounds may not always be appropriate, possible, or desirable. In many situations, however, both organic and inorganic substances are present in a leachate.

2.2.3 Field Investigation

Once an exhaustive search of the literature, files, maps, aerial photography, and satellite imagery has been conducted and, at least to some extent, relevant information has been studied, it is appropriate to visit the field. During the office evaluation, several points should become reasonably clear. These include a general appreciation of the regional hydrogeology, geology, and water quality, as well as water use and areas of potential problems.

These characteristics should be verified during field examinations. Initially field work should be reconnaissance in nature. The complexity or detail of the field work will expand with time in response to increasing familiarity with the region, area, or site, as well as the objectives of the study.

During early field visits, particular attention should be paid to landforms, streams and stream patterns, locations of springs, seeps, and lakes, as well as vegetation. Landforms are controlled by the geology and many hills are capped by resistant strata, such as sandstone, while valleys are usually carved into soft, less resistant material, such as shale. Likewise, many changes in topographic slope are related to differences in rock type. These, in turn, provide a general impression of the types of rocks present, their areal extent, and composition. Rock exposures in stream channels and road cuts are very useful also when attempting to understand the local geology. Joint and fracture systems, their directional trends, density, and size can all be measured on rock outcrops. Fluid movement through joints and other fractures may control entirely, or nearly so, the migration of contaminants.

Stream patterns also are related to the geology, especially geologic structure and fracture or joint systems. A brief examination of streams in the region is useful since it provides an idea of the relative difference in discharge from one stream to another. As indicated in Chapter 4, streams can provide a wealth of information on basin permeability, shallow ground-water quality, and local sites where the ground water is contaminated.

Springs and seeps are zones of ground-water discharge. They should exist in the vicinity of strata of low permeability that are overlain by a unit of greater permeability, that is, an aquitard overlain by an aquifer. Rarely does water continually discharge from springs and seeps. Most commonly the discharge is greater during spring and early summer, during the fall rainy period, or during and after a period of precipitation. Following these intervals of ground-water recharge, the discharge of springs and seeps diminishes or ceases entirely as the water-bearing zone becomes depleted. Nonetheless, the area downslope from the discharge zone has a higher moisture content and commonly supports far more vegetation, both grasses and trees, than is present in adjacent areas. The presence of the vegetation may allow the mapping of certain rock types.

2.3 Regional Investigations

Regional investigations are conducted for many different purposes. One type is to detect potential sources and locations of ground-water contamination. An example was described in Chapter 4 in which surface water data were used to detect potential sources of contamination (abandoned and producing oil wells and salt-water disposal ponds) and locate relatively small areas in which the ground water was contaminated by these activities in Alum Creek basin in central Ohio.

Another type of exceedingly broad scope includes library searches. Examples include an early EPA effort to evaluate ground-water contamination throughout the United States (van der Leeden *et al.*, 1975; Miller and Hackenberry, 1977; Scalf *et al.*, 1973; Miller *et al.*, 1974; and Fuhriman and Barton, 1971). The reports are useful for obtaining a general appreciation of the major sources of contamination and the magnitude over a regionally extensive area.

An excellent description of the geology and hydrology of the Ohio River basin was prepared by Deutsch *et al.* (1969). The 10 volume manuscript depicted each subbasin in considerable though broad detail. It served as the basis for a subsequent report that related ground-water quality and streamflow throughout the Ohio River basin. The report was prepared for the Federal Water Pollution Control Agency and, although completed in 1968, the report, unfortunately, was never published; draft copies should be available in EPA files. Examination of any of the volumes of either of the reports would provide an investigator with many ideas on how to conduct a regional evaluation.

In 1980 individuals in EPA Region VII became aware of what appeared to be a large number of wells that contained excessive concentrations of nitrate. Suspecting a widespread problem, a regional reconnaissance investigation was initiated. The general approach consisted of a literature search, a meeting in each state with regulatory and health personnel, an evaluation of existing data, and an interpretation of all of the input values.

The fundamental principle guiding this study was the fact that abnormal concentrations of nitrate can arise in a variety of ways, both from natural and manmade sources or activities. The degradation may encompass a large area if it results from the overapplication of fertilizer and irrigation water on a coarse textured soil, from land treatment of wastewaters, or from a change in land use, such as converting grasslands to irrigated plots. On the other hand, it may be a local problem affecting only a single well if the contamination is the result of animal feedlots, municipal and industrial waste treatment facilities, or improper well construction/maintenance.

For the most part the data base for this study was obtained from STORET. First, nitrate concentrations in well waters were placed in a separate computer file. Two maps were generated from the file, the first showing the density of wells that had been sampled for nitrate, and the second showing the density of wells that exceeded 10 mg/l of nitrate (Figure 2-1). The maps were produced by the STORET routine, Multiple Station Plot. These maps indicated the areas of the most significant nitrate problems. In turn, the nitrate distribution maps were compared to geologic maps, which allowed some general identification of the physical system that was or appeared to be impacted (Figure 2-2).

Iowa, eastern Nebraska, northeastern Kansas, and the northern third of Missouri are characterized by glacial till interbedded with local deposits of outwash. Throughout the area are extensive deposits of alluvium. Many of the aquifers are shallow and wells are commonly dug, bored, or jetted. This area contained the greatest number of domestic wells with high nitrate concentrations. It also contained the

Figure 2-1 Location of wells with nitrate exceeding 10 mg/l in Region 7.

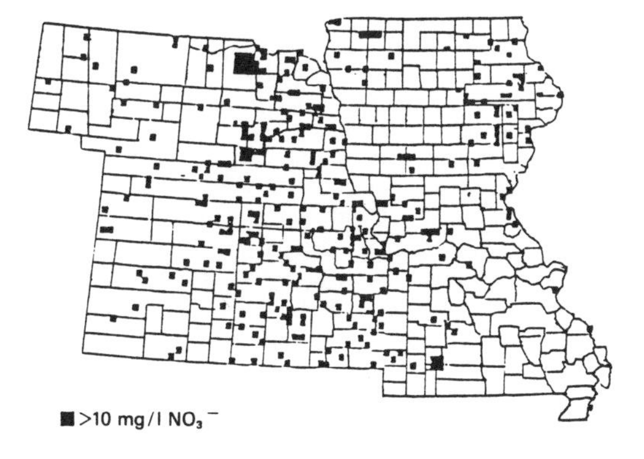

Figure 2-2 Generalized rock types with high nitrate concentrations in Region 7.

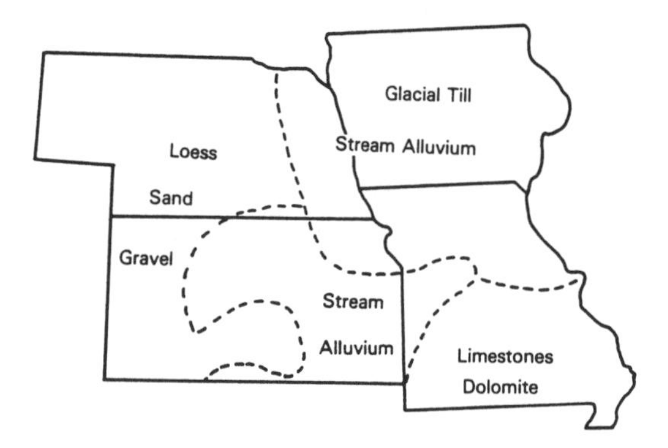

greatest number of municipal wells that exceeded the nitrate MCL (Maximum Contaminant Level). The cause of contamination in the shallow domestic wells was suspected to be poor well construction and maintenance, but this was possibly not the case for many of the generally deeper municipal wells, where the origin appeared to be from naturally occurring sources in the glacial till.

Most of Nebraska and western Kansas are mantled by sand, gravel, and silt, which allow rapid infiltration. The water table is relatively shallow. The irrigated part of this region, particularly adjacent to the Platte River and in areas of Holt County, Nebraska, contained the greatest regional nitrate concentrations in the four state area. This was brought about by the excessive application of fertilizers and irrigation waters in this very permeable area.

The remaining area in Kansas and an adjacent part of Missouri is underlain by sedimentary rocks across which flow many streams and rivers with extensive flood plains. Most of the contaminated wells tapped alluvial deposits. The primary cause of high nitrate in domestic wells was suspected to be poor well construction/maintenance or poor siting with respect to feedlots, barnyards, and septic tanks.

The southern part of Missouri is represented by carbonate rocks containing solution openings. Aquifers in these rocks are especially susceptible to contamination and the contaminants can be transmitted great distances with practically no change in chemistry other than dilution. The carbonate terrain is not easily manageable, nor is monitoring a simple technique because of the vast number of possible entry sites whereby contaminants can enter the subsurface.

The STORET file was also used to generate a number of graphs of nitrate concentration versus time for all of the wells that were represented by multiple samples. The graphs clearly showed that the nitrate concentration in the majority of wells ranged within wide limits from one sampling period to the next, suggesting leaching of nitrate during rainy periods from the unsaturated zone.

The state seminars were exceedingly useful because the personnel representing a number of both state and Federal agencies had a good working knowledge of the geology, water quality, and land-use activities of their respective states.

Although the study extended over several months, the actual time expended amounted to only a few days. The conclusions, for the most part, were straightforward and, in some cases, pointed out avenues for improvement in sample collection and data storage/access. The major conclusions are as follows:

1) High levels of nitrate in ground water appear to be randomly distributed through the region.

2) The most common cause of high nitrate concentration in well water appears to be related to inadquate well construction, maintenance, and siting. Adequate well construction codes could solve this problem. Dug wells, those improperly sealed, and wells that lie within an obvious source of contamination, such as a pig lot, should probably be abandoned and plugged.

3) In areas of extensive irrigation where excess water and fertilizer are applied to coarse textured soils, the nitrate concentration in ground water appears to be increasing.

4) In the western part of the region, changes in land use, particularly the cultivation or irrigation of grasslands, has resulted in leaching of substantial amounts of naturally occurring nitrate from the unsaturated zone.

5) The population that is consuming high nitrate water supplies is small, accounting for less than 2 percent of the population.

6) There have been no more than two reported cases of methemoglobinemia in the entire region within the past 15 years despite the apparent increase in nitrate concentration in ground-water supplies. This implies a limited health hazard.

7) State agency personnel are convinced that they do not have significant nitrate-related health problems.

8) Many of the wells that are used in the state and Federal monitoring networks are of questionable value because little or nothing is known about their construction.

9) The volume of chemical data presently in the files of most of the state agencies within the region is

not adequately represented in the STORET data system.

This cursory examination provided only a general impression of the occurrence, source, and cause of abnormal nitrate concentrations in ground water in the region. Nonetheless, it furnished a base for planning local or site investigations, was prepared quickly, and did not require field work or extensive data collection.

As mentioned previously, the source of excessive nitrate in many municipals could not be readily explained. There could be multiple sources related to naturally occurring high nitrate concentrations in the unsaturated zone or the glacial till, to contamination, or to poor well construction. Definitive answers would require more detailed local or site studies. The overall effect of changing from grazing land to irrigated agriculture, in view of the great mass of nitrate-bearing substances in the unsaturated zone that would be leached, clearly warrants additional local investigation. Although the concentration of nitrate in underlying ground water would increase following irrigation, it is likely that some control on the rate of leaching could be implemented by limiting the amount of water applied to the fields.

The obvious relationship between the application of excessive amounts of fertilizer and water on a coarse textured soil in Nebraska shows the need for experimental work on irrigation techniques in order to reduce the loading. Also implied is the necessity for developing educational materials and seminars, in order to offer means whereby irrigators can reduce water, pesticide, and fertilizer applications and yet maintain a high yield.

2.4 Local Investigations

Local investigations can be as varied in scope and areal extent as regional evaluations and the difference between the two is relative. For example, one might desire to obtain some knowledge of the hydrogeology of an area encompassing a few tens or several hundred square miles in order to evaluate the effect of oil-field brine production and disposal. Examples of this scope include Kaufmann (1978) and Oklahoma Water Resources Board (1975). The other extreme may center around a single contaminated well. In this case the local investigation would most likely focus on the area influenced by the cone of depression, the size of which depends on the geology, hydraulic properties, and well discharge. Consider an area in the Great Plains where a number of small municipalities have reported that some of their wells tend to increase in chloride content over a period of months to years. The increase in a few wells has been sufficient to cause abandonment of one or more wells in the field. Additionally, a number of wells when drilled yielded brackish or salty water necessitating additional drilling elsewhere. This is an expensive process that strains the operating budget of a small community.

In this case, the local investigation covered an area of 576 square miles, that is, 16 townships. A review of files and reports and discussions with municipal officials and state and Federal regulatory agencies indicated that the entire area had produced oil and gas for more than 30 years. Inadequate brine disposal was the most likely cause of the chloride problem.

During the initial stage of the investigation, all files dealing with the quality of municipal well water were examined and this task was followed by a review of the geology, which included a review of all existing maps, cross sections, and well logs, both lithologic and geophysical.

The chemical data clearly showed that the chloride content in some wells increased with time, although not linearly. The geologic phase of the study showed that the rocks consist largely of interbedded layers of shale and sandstone and that the sandstone deposits, which serve as the major aquifers, are lenticular and range from 12 to about 100 feet in thickness. The sandstones are fine-grained and cemented to some degree and, as a result, each unit will not yield a large supply. Resultingly, all sandstone bodies are screened.

Trending north-south through the east-central part of the area is an anticline (Figure 2-3) that causes the rocks to dip about 50 feet per mile either to the east or west of the strike of the structure (Figure 2-4). This means that a particular sandstone will lie at greater depths with increasing distances from the axis of the anticline. (Refer to Chapter 9).

In this example, the subsurface geology was based on an evaluation of geophysical and geologists' logs of wells and test holes, including oil and gas wells and tests. As shown in Figure 2-4, interpretation of the logs in the form of a geologic cross section brings to light an abundance of interesting facts. The municipal wells range in depth from 400 to 900 feet, but greater depth does not necessarily indicate a larger yield nor does depth imply a particular chemical quality. The difference in well depth and yield is related to the thickness and permeability of the sandstone units encountered within the well bore. Secondly, the volume of the sandstone components ranges widely, but the thinnest and most discontinuous units increase in abundance westward. More importantly, the mineral content of the ground water, which can be determined from geophysical logs, increases down the dip of the sandstone, from fresh in the outcrop area, to brackish, and finally to salt water (Figure 2-4). Notice also that brackish and saline water lie at increasingly shallower depths to the west of the outcrop area.

The position and depth of a few municipal wells and test holes are also shown on the cross section. Well

Figure 2-3 Generalized geologic map of a local investigation.

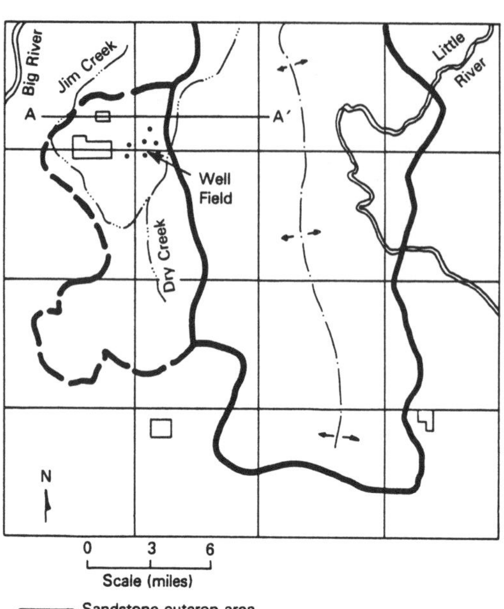

1 would be expected to have a small yield of brackish water. Well 2 is an abandoned test hole that penetrated a thick saline zone as well as a thick brackish water zone. In the case of Well 3, the fresh water derived from the thin, shallower sandstones is sufficient to dilute water derived from the more mineralized zones. On the other hand, as the artesian pressure in the shallow sandstones decreases with pumping and time, an increasing amount of the well yield might be derived from the deeper brackish layer, causing the quality to deteriorate.

The major conclusion derived from this study is that the most readily apparent source of high chloride content in municipal wells, that is, inadequate oil-field brine disposal, is not the culprit. Rather all of the problems are related to natural conditions in the subsurface, brought about by the downdip increase in dissolved solids content as fresh water grades into brackish and eventually into saline water. Deterioration of municipal well water quality is related to the different zones penetrated by the well and to a decrease in artesian pressure in fresh water zones brought about by pumping. The latter allows updip migration of brackish or saline water to the well bore or lateral or vertical leakage of mineralized water from one aquifer to another, which again is the result of a pressure decline in the fresh water zones. The problem could be diminished by constructing future wells eastward toward the axis of the anticline, limiting them to those areas either within the outcrop or where the thickness of the fresh water aquifers comprise a total thickness that exceeds 125 feet (Figure 2-3).

2.5 Site Investigations

Site investigations are ordinarily complex, detailed, and expensive. Furthermore, the results and interpretations are likely to be thoroughly questioned in meetings, interrogatories, and in court because the expenditure of large sums of money may be at stake. The investigator must exercise extreme care in data collection and interpretation. The early development of a flexible plan of investigation is essential and it must be based, at least in part, on guidelines established by the Environmental Protection Agency, such as the Technical Enforcement Guidance Document. State regulatory agencies may have even more stringent requirements.

The investigative plan needs to be flexible in a practical way. For example, the position of all test holes, borings, and monitoring wells should not be determined in the office at the start of the investigation. Rather, locations should be adjusted on the basis of the information obtained as each hole is completed. In this way, one can maximize the data acquired from each drill site and more appropriately locate future holes in order to develop a better understanding of the ground-water system.

In the case of Superfund and RCRA sites, the regulatory investigator probably will be required to work with or at least use data collected by consultants for the defendant. In some cases, the defendant conducts and pays for the entire investigation; regulatory personnel only modify the work plan so that it meets established guidelines. There are two points to consider in these situations. First, the consultant is hired by the defendant and should act in his best interest. This means that his interpretations may be slighted toward his client and concepts detrimental to the client are not likely to be freely given. Second, even though the regulatory investigator and the consultant, to some degree, are adversaries, this does not mean that the consultant is dishonest, ignorant, or that his ideas are incorrect. It must always be remembered that the entire purpose of the investigation is to determine, insofar as possible, what has or is occurring so that effective and efficient corrective action can be undertaken. In the long run cooperation leads to success.

Several generalized methods have been available for a number of years to evaluate a possible or existing site relative to the potential for ground-water contamination. These rating techniques are valuable, in a qualitative sense, for the formulation of a detailed

Figure 2-4 Geologic cross section showing downdip change in water quality.

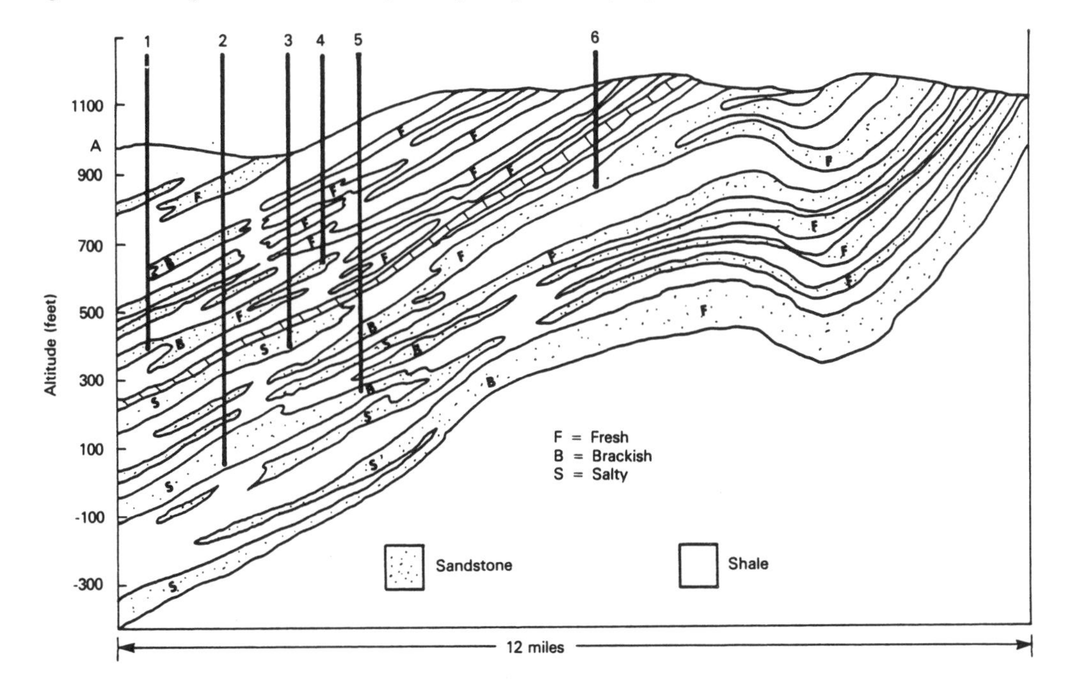

investigation. The most noted is probably the LeGrand (1983) system, which takes into account the hydraulic conductivity, sorption, thickness of the water table aquifer, position and gradient of the water table, stream density, topography, and distance between a source of contamination and a well or stream. The LeGrand system was modified by the U.S. Environmental Protection Agency (1983) for the Surface Impoundment Assessment.

Fenn and others (1975) formulated a water balance method to predict leachate generation at solid waste disposal sites. Gibb and others (1983) devised a technique to set up priorities for existing sites relative to their threat to health. An environmental contamination ranking system was contrived by the Michigan Department of Natural Resources (1983). On a larger scale is DRASTIC, which is a method, based on hydrogeologic settings, to evaluate the potential of ground-water contamination.

As an example of a ground-water quality site investigation, consider a rather small refinery that has been in existence for several decades. For some regulatory reason an examination of the site is required. The facility, which has not been in operation for several years, includes an area of about 245 acres. The geology consists of alternating deposits of sandstone and shale that dip slightly to the west; the upper 20 to 30 feet of the rocks are weathered.

Potential sources of ground-water contamination include wastewater treatment ponds, a land treatment unit, a surface runoff collection pond, and a considerable number of crude and product storage tanks. Line sources of potential contaminants include unimproved roads, railroad lines, and a small ephemeral stream that carries surface runoff from the plant property to a holding pond.

After considering the topography and potential sources of contamination, the locations of 11 test borings were established. The purpose of the holes was to determine the subsurface geologic conditions underlying the site. Following completion, the holes were geophysically logged and then plugged to the surface with a bentonite and cement slurry. The bore hole data were used to determine drilling sites for 20 observation wells, in order to ascertain the quality of the ground water, to establish the depth to water, and to determine the hydraulic gradient. Eight of the observation wells were constructed so that they could be used later as a part of the monitoring system. Two of the wells tapped the weathered shale, their purpose being to monitor the water table, evaluate the relation between precipitation and recharge, and ascertain the potential fluctuation of water quality in the weathered material in order to determine if it might serve as a pathway for contaminant migration from the surface to the shallowest aquifer. (From a technical perspective, the weathered shale and sandstone is not an aquifer, but from a regulatory point of view it could be considered a medium into which a release could occur and, therefore, might fall under RCRA guidelines.)

Regulations required that the shallowest aquifer be monitored, which in this case was a relatively thin, saturated sandstone. After the initial investigative information was available, all of the findings were used to design a ground-water monitoring system. This plan called for an additional 12 monitoring wells.

Graphics based on all of the drilling information (geologic and geophysical logs) included several geologic cross sections (Figure 2-5) and maps showing the thickness of shale overlying the aquifer (Figure 2-6), thickness of the aquifer, and the hydraulic gradient (Figure 2-7). The major purpose of the first map was to show the degree of natural protection that the shale provided to the aquifer relative to infiltration from the surface. The aquifer thickness map was needed for the design of monitoring wells. The water-level gradient map was necessary to estimate ground-water velocity and flow direction. During the drilling phases, cores of the aquifer and the overlying shale were obtained for laboratory analyses of hydraulic conductivity, porosity, specific yield, grain size, mineralogy, and general description. Aquifer tests were conducted on 20 of the wells.

The cross sections and maps indicate that the sandstone dips gently eastward and nearly crops out in a narrow band along the western margin of the facility. Elsewhere, owing to the change in topography and the dip of the aquifer, the sandstone is overlain by 25 feet or more of shale; throughout nearly all of the site the shale exceeds 50 feet in thickness. Consequently, only one small part of the aquifer, its outcrop and recharge area, is readily subject to contamination.

The water-level map indicates that the hydraulic gradient is not downdip but rather about 55 degrees from it. It is controlled by the topography off site. The average gradient is about 0.004 feet per foot, but from one place to another it differs to some extent, reflecting changes in aquifer thickness.

The topographic map shows that surface runoff from the entire facility is funneled down to a detention pond. The pond and the lower part of the drainage way lie in the vicinity of the aquifer's recharge or outcrop area.

Logs of the drill holes list specific depths in six of the holes in which hydrocarbons were present. All were reported in the unsaturated zone at depths of 2 to 9 feet with thicknesses ranging from a half inch to nearly a foot. At these locations the shale overlying the aquifer exceeded 55 feet in thickness.

Figure 2-5 Geologic cross section for the site investigation.

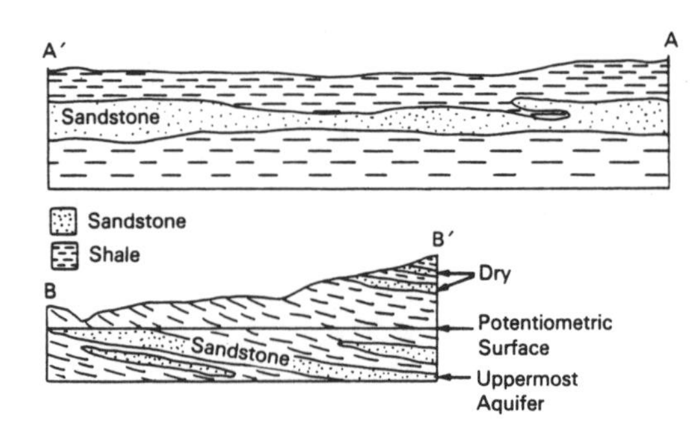

Chemical analyses of water from the observation wells indicated, with one exception, that the quality was within background concentrations and no organic compounds were present. The exception was an observation well near the surface runoff retention pond.

Precipitation in the area and the hydrograph of a well, 14 feet deep, in the weathered shale is shown in Figure 2-8. These data show that the weathered shale is shown in Figure 2-8. These data show that the weathered material responds quickly to rainfall events, despite the fact that laboratory values of hydraulic conductivity were exceedingly low. This strongly suggests that the weathered shale is quite permeable, the permeability being related to fractures. Therefore, from a hydraulic perspective, the weathered shale appears to form a medium that would allow the migration of contaminants from the surface.

The relation between precipitation and nitrate, which is a good tracer, in the weathered material is shown in Figure 2-9. Annually the nitrate concentration fluctuates between about 4 and 11 mg/l, which is typical of an area characterized largely by grasslands. The nitrate is of natural origin and the range in concentration only indicates the variation in

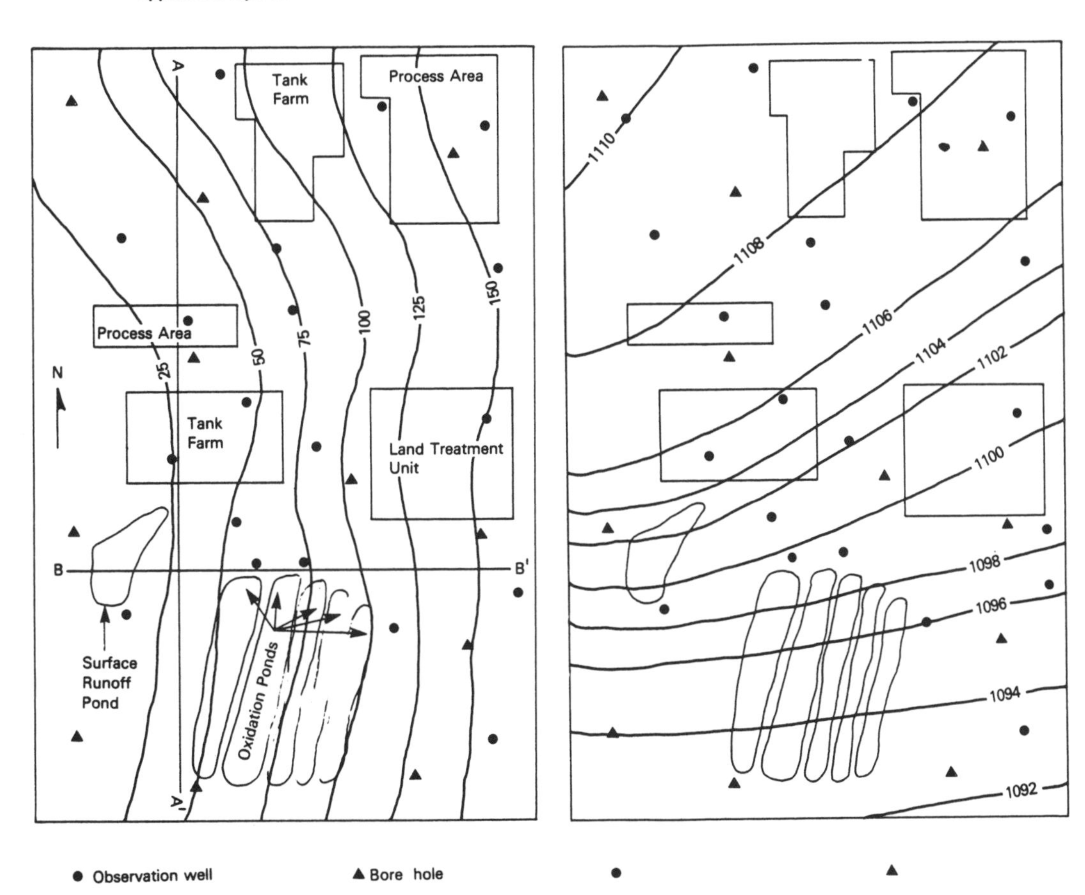

Figure 2-6 Map showing thickness of shale overlying the uppermost aquifer.

Figure 2-7 Potentiometric surface of the uppermost aquifer.

● Observation well ▲ Bore hole

Figure 2-8 Relation between precipitation and water level.

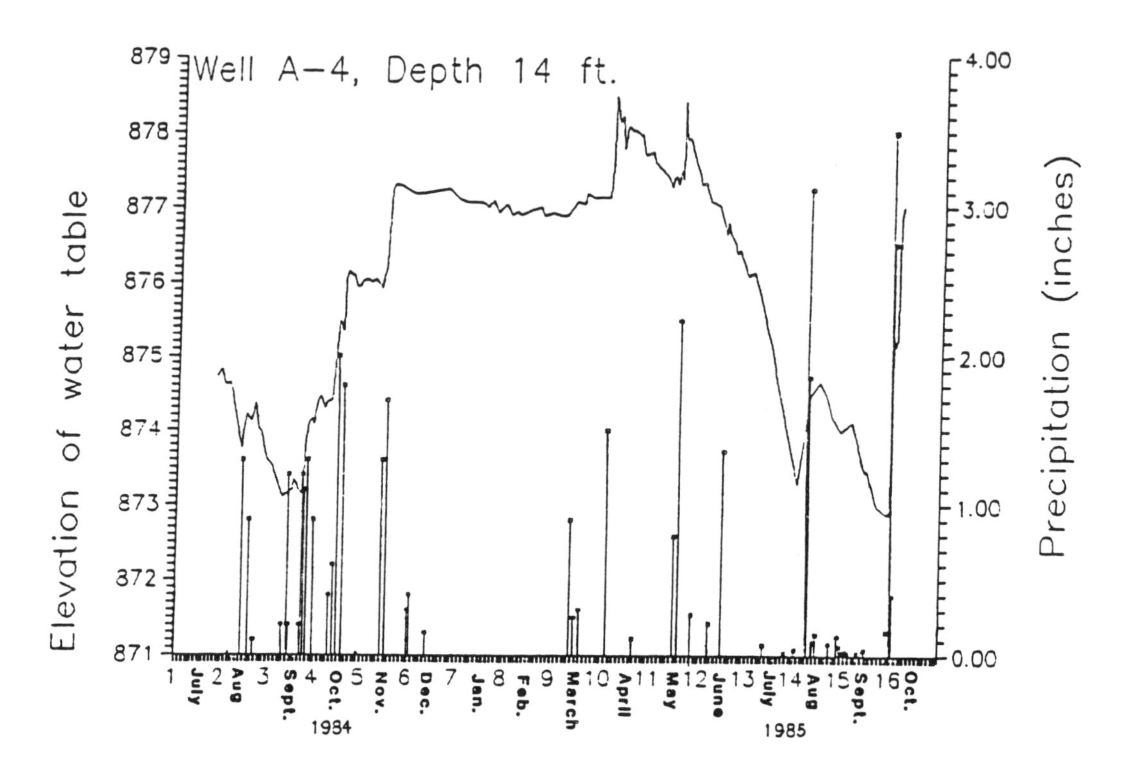

Figure 2-9 Relation between precipitation and nitrate concentration.

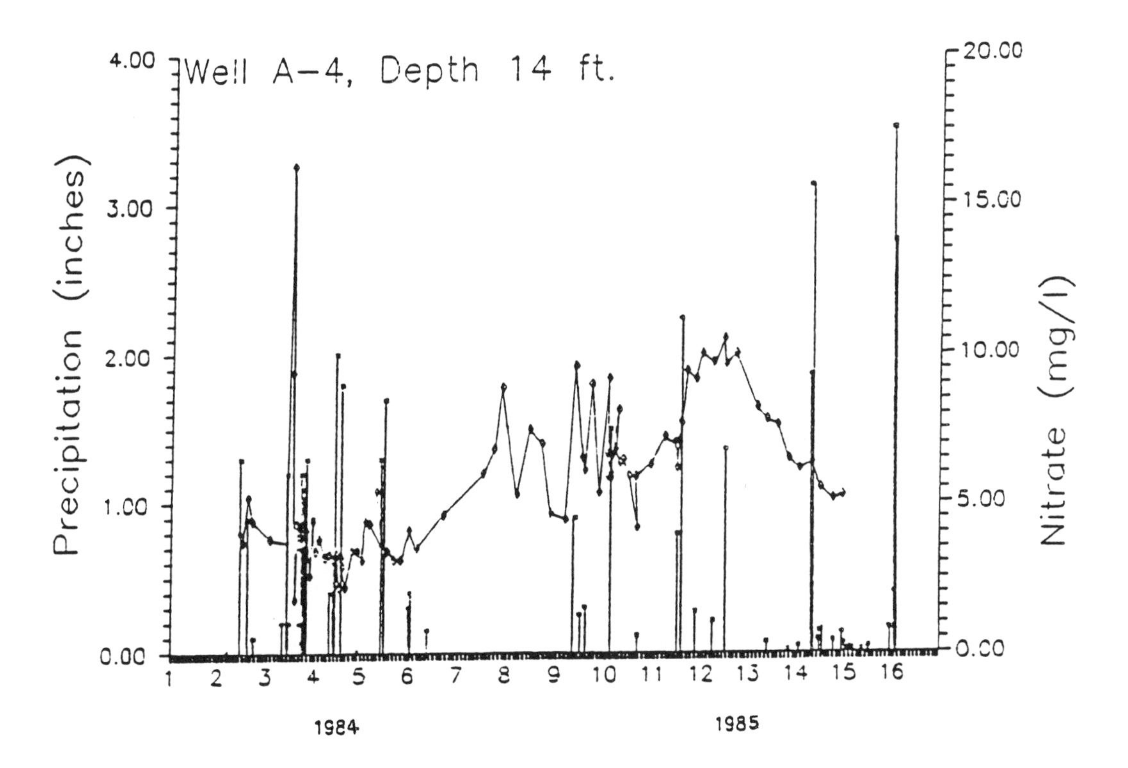

background. In this case as in every other, background concentration is not a finite number, but rather a range. The graph does not indicate a good correlation between nitrate and rainfall, but here are a few periods when the relationship is close.

Multiple analyses of nitrate in the sandstone aquifer showed that nitrate ranged only between 2 and 4 mg/l over a period of 14 months. This suggests that substances that originate from the surface or unsaturated zone do not impact the sandstone aquifer. More likely they migrate laterally to points of diffuse discharge along hillsides where the water is lost by evapotranspiration. The hydrograph also suggests that the water table declines rapidly in response to evapotranspiration.

Evaluation of all of the data indicated two potential problems -- hydrocarbons in the unsaturated zone and ground-water contamination in the vicinity of the surface runoff detention pond. Since the plant had been in operation more than 50 years, the hydrocarbons had migrated from the surface into the weathered shale no more than 9 feet, and there was a minimum of at least 45 feet of tight, unfractured shale between the hydrocarbons and the shallowest aquifer, it did not appear that the soil contamination would present a hazard to ground water.

The existence of contaminated ground water, however, was a problem that needed to be addressed even though the sandstone aquifer is untapped and is never likely to serve as a source of supply. Four additional monitoring wells were installed downgradient in order to determine the size of the plume and its concentration. Corrective action called for removal of sediment and sludge from the pond, backfilling with clean material, a cap, and pumping to capture the plume. The contaminated water was treated on site with existing facilities.

2.6 Summary

Each ground-water quality investigation is unique, although general guidelines need to be followed for all of them. The investigator must first clearly define the objectives of the study, for these will determine the complexity, time element, and cost of the project. Specific techniques that might be required are described in other chapters in this report.

2.7 References

Deutsch, M., P. Jordan, and J. Wallace. 1969. Ohio River Basin Comprehensive Report, Appendix E., Ground Water. Corps of Engineer Division, Ohio River, Cincinnati, OH.

Fenn, D.G., K.J. Hanley, and T.V. DeGeare. 1975. Use of the Water Balance Method for Predicting Leachate Generation from Solid Waste Disposal Sites. U.S. Environmental Protection Agency Solid Waste Report No. 168, Cincinnati, OH.

Fuhriman, D.K., and J.R. Barton. 1971. Ground Water Pollution in Arizona, California, Nevada and Utah. 16060 ERU 12/71, U.S. Environmental Protection Agency.

Gibb, J.P., M.J. Barcelona, S.C. Schock, and M.W. Hampton. 1983. Hazardous Waste in Ogle and Winnebago Counties, Potential Risk Via Ground Water Due to Past and Present Activities. Doc. No. 83/26, Illinois Department of Energy and Natural Resources.

Kaufmann, R.F. 1978. Land and Water Use Effects on Ground-Water Quality in Las Vegas Valley. EPA/600/2-78/179, U.S. Environmental Protection Agency.

LeGrand, H.E. 1983. A Standardized System for Evaluating Waste-Disposal Sites. National Water Well Association, Worthington, OH.

Michigan Department of Natural Resources. 1983. Site Assessment System (SAS) for the Michigan Priority Ranking System under the Michigan Environmental Response Act. Michigan Department of Natural Resources.

Miller, D.W., F.A. DeLuca, and T.L. Tessier. 1974. Ground Water Contamination in the Northeast States. EPA/660/2-74/056, U.S. Environmental Protection Agency.

Miller, J.C. and P.S. Hackenberry. 1977. Ground-Water Pollution Problems in the Southeastern United States. EPA/600/3-77/012, U.S. Environmental Protection Agency.

Oklahoma Water Resources Board. 1975. Salt Water Detection in the Cimarron Terrace, Oklahoma. EPA/660/3-74/033, U.S. Environmental Protection Agency.

Scalf, M.R., J.W. Keeley, and C.J. LaFevers. 1973. Ground Water Pollution in the South Central States. EPA/R2-73/268, U.S. Environmental Protection Agency.

U.S. Environmental Protection Agency. 1983. Surface Impoundment Assessment National Report. EPA/570/9-84/002, U.S. Environmental Protection Agency.

van der Leeden, F., L.A. Cerrillo, and D.W. Miller. 1975. Ground-water Pollution Problems in the Northwestern United States. EPA/660/3-75/018, U.S. Environmental Protection Agency.

3. Ground-Water Restoration

A number of techniques are available to either contain a pollutant and/or treat the ground water and at least partially clean up a contaminated aquifer. These techniques range from removal of the polluted material and physical, chemical, or biological treatment on the surface, to physical containment and in-situ treatment with chemicals or microbes. Most of the available technologies have been developed through remedial activities in the Superfund program.

The major emphasis of this chapter will be an overview of the remedial and restoration technology which can be considered for application in aquifer clean-up operations, with special emphasis on ground-water pumping systems and in-situ bioreclamation. Most of the discussions will concern hydrocarbons because there is more information on this particular contaminant. A number of the newer technologies, such as various in-situ biodegradation techniques, where applicable, are indicated as potentially very cost effective. The most significant benefit of in-situ treatment technologies is that physical removal of contaminated soils and pollutants is eliminated; this significantly reduces cost and public health risk.

3.1 Subsurface Effects on Contaminant Mobility

The movement of most ground-water contaminants is controlled by gravity, the permeability and wetness of the geological materials receiving them, and the miscible character of the contaminants in ground water. When material, particularly a hydrocarbon, is released to the soil, it is actively drawn into the soil by capillary attraction and by gravity. As the main body of materials moves down into the moister regions of the soil, the capillarity becomes less important and the materials move through the most favorable channels by displacing air, eventually reaching the water table where components less dense than water spread laterally along the air-water interface. In instances involving a heavier contaminant, the material continues to move downward in the saturated zone. In both cases the contaminants migrate downgradient with the natural ground-water flow (Wilson and Conrad, 1984).

The quantity of a contaminant such as hydrocarbons that reaches the water table is dependent both on the quantity involved and the nature of the earth materials. The coarser the earth materials, the larger the amount that will reach the ground water. The entire volume of hydrocarbon may be immobilized in the unsaturated zone, although it may continue to migrate downgradient where it becomes a threat to the quality of ground water. Material immobilized in the vadose zone may remain there unless it is physically, chemically, or biologically removed.

The hydrocarbon liquid phase is generally referred to as being immiscible with both water and air. However, it is important to realize that various components of the hydrocarbon volatilize into the air phase and dissolve into the water phase. A halo of dissolved hydrocarbon components precedes the immiscible phase, some of which is trapped in the pore space and left behind. The trapped hydrocarbon remains as pendular rings and/or isolated immobile blobs. Even when the so-called residual immiscible hydrocarbon is exhausted by immobility, ground water coming into contact with the trapped material leaches soluble hydrocarbon components and continues to contaminate ground water.

With two fluid phases, water and hydrocarbon, the residual hydrocarbon is trapped by one or both of two mechanisms known as by-passing and snap-off. Snap-off depends strongly on pore shape and wetability of the soil particles. In high aspect ratio pores, in which the pore throats are much smaller than pore bodies, snap-off is common. Snap-off occurs as water moving through the small throats begins to go around the outer surface of larger droplets, thereby isolating them in the larger voids. By-passing occurs as hydrocarbons are routed around soil grains where the branched pore canals are of unequal size. The velocity of water passing through the smaller diameter branched pore is faster and therefore travels around the grain faster in one branch than the other. This results in trapping of hydrocarbon in the larger slower moving pore canal (Wilson and Conrad, 1984).

Residual hydrocarbon can occupy from 15 to 40 percent or more of the pore space as a result of these trapping processes depending on several physical characteristics of the subsurface. The ability to design and conduct successful remediation strategies depends in large part on the ability to understand, predict and enhance the mobility of both liquid and dissolved hydrocarbons. Theoretically, there are a number of ways the trapped residuals can be mobilized and caused to concentrate where much lower content of residual product will have to be dealt with. Obviously the most used mobilization technique is to increase the hydraulic gradient, usually by pumping, thereby increasing the Darcy Velocity of the water phase in the saturated zone. When the velocity is increased sufficiently some of the blobs begin to move. A critical element in this mobilization process is the length of the blob in the direction of flow. The gradient must be high enough to squeeze the blobs through pore throats. After they are mobilized, blobs do not maintain their size and shape. The large mobilized blobs break up into smaller blobs with a significant fraction being only temporarily mobilized. Blobs may also coalesce and become trapped at greater distances from the source. Active research in physical mobilization technology is progressing rapidly but much remains to be learned (Wilson and Conrad, 1984).

3.2 Physical Containment Techniques

3.2.1 Removal

The purpose of removing contaminated soil and ground water, associated with a plume of contamination, would be to treat and/or relocate the wastes to a better engineered and controlled, or environmentally more favorable disposal site. Conceptually, removal and reburial of the contaminated material to a more controlled situation appears to solve the contamination problem. In practice, however, there are many considerations to deal with before excavation and reburial are used as a remedial action technique. Considerations include (1) excavation of bulky, partially decomposed or hazardous wastes; (2) distance to acceptable reburial site; (3) condition of roads between sites; (4) accessibility of both sites; (5) political, social, and economic factors associated with locating a new site; (6) disposition of contaminated ground water; (7) control of nuisances and vectors during excavation; (8) reclamation of the excavated site; and (9) costs (Tolman *et al.*, 1978). Due to these considerations and especially the cost of excavation, transportation and new site preparations, removal and reburial should be considered as a last resort or in cases of severe pollution where cost is not significant compared to the importance of the resource being protected. In some cases removal and reburial in an approved facility is simply transferring a problem from one location to another.

3.2.2 Barriers to Ground-Water Flow

Subsurface barriers are designed to either prevent or control ground water flow into or through desired locations. The types of barriers used include slurry trench walls, grout curtains, vibrating beam walls, sheet piling, bottom sealing, block displacement and passive interceptor systems (Knox *et al.*, 1984; Ehrenfeld and Bass, 1984).

A slurry trench wall is constructed by excavating a vertical trench to a desired depth while throughout the excavation process the trench is kept filled with a clay slurry composed of a 5 to 7 percent by weight suspension of bentonite in water. The bentonite slurry maintains the vertical stability of the trench walls by exerting a greater hydrostatic pressure against the walls than the surrounding ground water, and also by forcing bentonite into the pores of the soil in the trench walls thus forming a low permeability layer of soil and bentonite called a "filter cake."

As the slurry trench is being excavated it is simultaneously being backfilled with an engineered material that forms the final wall. The three major types of slurry trench backfill mixtures are (1) soil bentonite, (2) cement bentonite, and (3) concrete. The type and ratios of backfill chosen depend upon the specific site characteristics as well as the desired properties of the slurry trench wall including permeability, strength, compatibility with contaminants and cost. Although costly, slurry trench walls are generally the least expensive form of passive ground-water barrier. A properly designed slurry trench can lower the water table by providing a complete seal down to a low permeability layer or by increasing the length of the ground-water flow path and thereby creating an energy loss.

Grouting can be defined as the pressure injection of a stabilizing material into subsurface soils or rock in order to fill, and thereby seal the voids, cracks, fissures or other openings in the soil or rock strata. Grout curtains are fixed, underground physical barriers formed by injecting grout, either particulate (i.e., Portland cement) or chemical (i.e., sodium silicate), though tubes which are driven into the ground on two to three foot centers and withdrawn slowly during injection. Two or more rows of grout are normally needed to provide a good seal. Like a slurry trench the grout curtains are normally emplaced down to an impermeable layer. The rate of injection of the grout material is determined by site-specific characteristics. If the injection rate is too slow, premature grout/soil consolidation occurs, and if the rate is too fast, fracturing of the soil formation may result. There are many available grouts; however, the selection of grout material depends on site specific factors such as soil permeability, soil grain size, rate of ground water flow, chemical constituents of soil and ground-water, required grout strength, and cost. Because a grout curtain can be as much as three times as costly as a slurry wall, it is rarely used when ground water has to be controlled in soil or loose overburden. The major use of grout curtains is to seal voids in porous or fractured rock where other methods of ground water control are impractical.

A variation of a grout curtain is the vibrating beam technique for emplacing thin (approximately 4 in) curtains or walls. Although it is sometimes called a slurry wall technique, it is more closely related to a grout curtain since the slurry is injected through a pipe similar to grouting. A suspended I-beam, connected to a vibrating driver-extractor, is vibrated through the ground to the desired depth. As the beam is raised at a controlled rate, slurry is injected through a set of nozzles at the base of the beam, filling the void left by the withdrawal of the beam. The entire process is repeated with subsequent placements of the I-beam overlapping the previous placement to provide continuity. The vibrating beam technique is most efficient in loose, unconsolidated deposits such as sands and gravels. Where suitable conditions exist the vibrating beam technique has been used to depths of 80 feet. Costs using the vibrating beam technique are comparable to conventional slurry trenching methods.

Sheet piling cutoff walls can be made of wood, reinforced concrete or steel; however, steel sheet piles represent the most effective material for constructing a ground-water barrier. Construction of a steel sheet pile cutoff wall involves driving lengths of steel sheets through unconsolidated deposits with a pile driver. The individual steel sheet piles are connected along the edge of each pile through various types of interlocking joints. These joints provide permeable pathways for ground-water movement which may or may not become watertight naturally depending on the soil characteristics. It may be necessary to fill these joints with an impermeable material such as a grout; however, the ability to ascertain the success of the grouting operation is questionable. Steel is a readily corrodible material and therefore the lifetime of the steel sheet piles is dependent on the corrosive nature of the soil, ground water, and contaminants with which the steel piles come in contact. A common recommendation is that steel pilings be chemically coated or electrically protected so as to minimize corrosion. Although there are limitations, sheet piling cutoff walls may be used to contain contaminated ground water, divert a contaminant plume to a treatment facility, and divert ground-water flow around a contaminated area.

Block displacement is a new plume management method where a slurry is injected in such a manner that it forms a subsurface barrier around and below a specific mass or "block" of earth. Continued pressure injection of the slurry produces an uplift force on the bottom of the "block" which results in a vertical displacement proportional to the slurry volume pumped, thus the name block displacement. This technology is still in the developmental stages, especially verification of the bottom barrier, so cost data are not published.

Membrane and synthetic sheet curtains can be used in applications similar to grout curtains and sheet piling. The membrane is placed in a trench surrounding or upgradient from the plume of interest, thereby enclosing the contamination or diverting the ground-water flow. Placing a membrane liner in a slurry trench application has also been tried on a limited basis. Attaching the membrane to an impervious layer and having perfect seals between sheets is difficult but necessary in order for membranes and other synthetic sheet curtains to be effective. Impermeable synthetic membranes have also been used on the downgradient side of interceptor trenches to stop the migration of petroleum products for subsequent recovery.

Passive interceptor systems consist of trenches excavated to a depth below the water table with the possible placement of a collection pipe in the bottom of the trench. These interceptor systems can be used as preventive measures (i.e., leachate collection systems), abatement measures (i.e., interceptor drains), or in product recovery from a ground water (i.e., oil, gasoline). Interceptor drains are generally used to either lower the water table beneath a contamination source or to collect ground water from an upgradient source in order to prevent leachate from reaching uncontaminated wells or surface water. Interceptor systems are relatively inexpensive to install and operate and provide a means for leachate collection without impermeable liners. On the other hand, interceptor systems are not well suited to poorly permeable soils and the systems require continuous and careful monitoring to assure adequate leachate collection.

3.2.3 Surface Water Controls

Surface water control measures are used to minimize the infiltration of surface water or direct precipitation onto a waste site, thereby minimizing the amount of leachate produced. There are three basic technologies used to control surface water in a particular area or waste site. The first is changing the contour and runoff or runon characteristics of the site. The second is providing a cover barrier to infiltration by reducing the permeability of the land surface (surface sealing or capping). The third is revegetating the site so that the waste site cover material is stabilized, seasonal evapotranspiration is increased, and infiltration is decreased due to vegetation interception of direct precipitation.

Changing the contour and runoff or runon characteristics of a particular site can be accomplished by several standard engineering techniques. Some of the more common techniques include dikes and berms, ditches, diversions, waterways, terraces, benches, chutes, downpipes, levees, seepage basins, sedimentation basins, and surface grading.

Surface sealing is accomplished by covering or capping a waste site with a low permeable material to prevent water from entering the site, thus reducing leachate generation and also controlling vapor or gas produced. Covers or caps can be constructed from native soils, clays, synthetic membranes, soil cement, bituminous concrete, certain waste materials, or asphalt/tar materials. Capping is normally an economical technique, and because the surface is accessible, the cap can be monitored, maintained, and repaired.

Revegetation may be a cost effective method to stabilize the surface of a waste site, especially when preceded by capping and contouring. Vegetation reduces raindrop impact, reduces runoff velocity, and strengthens the soil mass, thereby reducing erosion by wind and water, and improves the site aesthetically.

3.2.4 Limitations of Physical Containment

Along with the positive attributes, each of the physical containment techniques have certain limitations. Removal is an extremely expensive and difficult procedure often plagued with political, social and economic constraints. Construction of barriers to ground-water movement can also present many problems both site and technique related. Slurry walls are limited by the availability of bentonite and the patents associated with several aspects of the construction procedures. Chemical grouts are expensive, some grouting techniques are proprietary, and grouting in general is limited to soils with permeabilities 10^{-5} cm/sec or greater. Sheet piling is not initially watertight, ineffective where large rocks are present, and is subject to corrosion depending upon site characteristics. Block displacement is an untried technique in its infancy and needs more verification studies especially concerning the bottom barrier. Passive interceptor systems are not well suited to slightly permeable soils and require continuous monitoring and maintenance. Limitations associated with surface water controls include availability of cover material to develop contours, availability of natural clay deposits for caps, expense of manmade cover materials (concrete) and synthetic membrane liners for caps, and initial time period and cover required for vegetation.

The various limitations illuminate the fact that each type of physical containment must be considered on a case by case basis taking into consideration all the many different site specific variables. It is possible that even though a specific technique may be expensive or the raw material may not be readily available, when all site variables are considered, that particular technique for physical containment may be the only viable alternative.

3.3 Hydrodynamic Controls

Hydrodynamic controls are employed to isolate a plume of contamination from the normal ground-water flow regime in order to prevent the plume from moving into a well field, another aquifer, or surface water. Isolation of the contaminated plume is accomplished when uncontaminated ground water is circulated around the plume in the opposite direction of the natural ground-water flow. The circulated zone creates a ground-water (hydrodynamic) barrier around the plume. Ground water upgradient of the plume will flow around the circulated zone while ground water downgradient will be essentially unaffected.

3.3.1 Well Systems

Well systems are used for hydrodynamic control of contaminated plumes by manipulating the hydraulic gradient of ground water through injection and/or withdrawal of water. The three general classes of well systems include (1) well point systems, (2) deep well systems, and (3) pressure ridge systems. All three types of well systems may require the installation of several wells at selected sites.

Well point systems consist of several closely-spaced, shallow wells connected to a main header pipe which is connected to a suction lift pump. Well point systems are used only for shallow aquifers because of the drawdown limitations as determined by the static water level and the limits of the pump. These systems should be designed so that the drawdown of the system completely intercepts the plume of contamination.

Deep wells are similar to well point systems except they are used for greater depths and are normally pumped individually. These wells are used in consolidated formations where the water table is too deep for economical use of suction life systems. Since the maximum depth for suction lift is around 25 feet, deep wells normally employ jet ejector or submersible pumps, or eductor well points.

Pressure ridge systems are produced by injecting uncontaminated water into the subsurface, through a line of injection wells, either up-gradient or down-gradient from a plume of contamination. Up-gradient ridges or mounds are used to force up-gradient uncontaminated ground water to flow around a contaminant plume while the contaminants are being collected by a line of down-gradient pumping wells. The procedure increases the velocity of ground water into the plume and to the recovery wells, and serves to wash the aquifer. Pressure ridge systems located down-gradient are normally used in combination with up-gradient pumping wells which supply uncontaminated injection water. In either case the injection of fresh water produces an uplift or mound in the original water table which acts as a barrier by forming a ridge which pushes the contaminated plume away from the mound.

Due to the economic incentives and the large number of cases of hydrocarbon leakage from storage tanks, considerable work has been directed toward applying well system technology to hydrocarbon recovery. Many variations of hydrocarbon recovery systems have been proposed centered around single or multiple pump systems and recovery wells. The type of recovery well used is dependent upon site specific characteristics and cost.

Hydrodynamic control systems offer a high degree of design flexibility and compared to passive containment can be easily constructed at minimal expense. These are also moderate to high operational flexibility which allows the system to meet increased or decreased pumping demands. When rapid response to a contamination problem is needed, pumping and injection wells can be installed relatively quickly as compared to certain passive barriers (i.e, slurry trenches, grout curtains, etc.). If the contamination threat is considered an emergency condition, then a hydrodynamic control system may be the temporary answer.

3.3.2 Limitations of Hydrodynamic Control

Even with the advantages discussed, well systems are not a permanent panacea to a ground-water contamination problem. Well systems simply are methods to stop the migration of a plume until a permanent solution can be decided upon. Some of the more specific limitations include (1) higher operation and maintenance costs than passive barriers including electrical power and manpower, (2) system failures, due to breakdown of equipment or power outages, can lead to contaminant movement, (3) flexibility is reduced in fine silty soils, and (4) incorrect pumping rates can draw a significant portion of the plume into the wells making treatment necessary before recharge into the aquifer.

3.4 Withdrawal and Treatment

Withdrawal and treatment of contaminated ground water is one of the most often used processes or current technology for cleaning up aquifers. The type of contamination and the cost associated with treatment will determine what specific treatment technology will be used. There are three broad areas of treatment possibilities, namely (1) physical, which includes adsorption, density separation, filtration, reverse osmosis, air and stream stripping, and incineration, (2) chemical, which includes precipitation, oxidation/reduction, ion exchange, neutralization, and wet air oxidation, and (3) biological, which includes activated sludge, aerated surface impoundments, land treatment, anaerobic digestion, trickling filters, and rotating biological discs.

3.4.1 Physical

The two major *adsorption methods* receiving the greatest attention as treatment methods are granular (to include powdered activated carbon) activated carbon (GAC) and synthetic resin adsorption. Both GAC and resins remove dissolved contaminants from water by adsorbing specific molecules. GAC is by far the most widely used adsorbent because synthetic resins are extremely costly and are still somewhat in the developmental stages of normal use. Synthetic resins trap contaminants within the chemical structure of the resin whereas GAC traps contaminants within the physical pore structure of the carbon. Typical adsorption sytems, whether GAC or resin, consist of a large vessel partially filled with adsorbent. There is an inlet for contaminated water and an outlet for treated water. Influent water enters and is in contact with the adsorbent for a specified period of time and then exits for collection, recharge or further treatment. Often systems are arranged with several tanks in parallel or in series to allow for the most efficient treatment possible. Once the micropore surfaces of the GAC are saturated with contaminants the GAC must either be replaced or thermally regenerated. GAC is an effective and reliable means of removing low solubility organics and some metals and inorganic species. It can be used for treating a wide range of contaminants over a broad concentration range.

A part of many treatment operations for contaminated ground water is the technique of *density separation* where suspended solids and water are separated depending upon their individual densities. If suspended solids are present, often common wastewater treatment operations such as clarifiers, settling chambers, and sedimentation basins are employed. Gravity separation is used when two-phased aqueous wastes are present. Gravity separation is a purely physical phenomenon in which one phase (i.e., oil, hydrocarbons) is allowed to separate from the other phase (i.e., water) in a conical tank and then discharged accordingly.

Filtration is a physical process whereby suspended solids are removed from solution by forcing the fluid through a porous medium. The filter media consists of a bed of granular particles (typically sand or sand with anthracite or coal). Filters are often preceded by sedimentation basins, and often precede biological or activated carbon units in order to decrease the suspended solids load. Filtration is a reliable and effective means of removing low levels of solids provided the solids content does not vary greatly and the filter is backwashed at appropriate intervals.

In the process of osmosis a solvent spontaneously flows from a dilute solution, through a semipermeable membrane, to a more concentrated solution by osmotic pressure. If enough pressure is placed on the concentrated solution to overcome osmotic pressure then water will flow toward the dilute phase thereby creating reverse osmosis. In *reverse osmosis* the contaminants are allowed to build up in a circulating bath on one side of the membrane while relatively pure water passes through the membrane.

Basically a reverse osmosis unit is composed of the membrane, a membrane support structure, a containing vessel, and a high pressure pump with the most critical elements being the membrane and the membrane support structure. The process is used to separate ions and small molecules in true solution from water, and to decrease the dissolved solids concentration, both organic and inorganic. Advances in membrane technology have made it possible to remove low molecular weight organics such as alcohols, ketones, amines, and aldehydes. If pretreatment measures are performed such as removal of suspended solids, pH adjustments, removal of oxidizers, oil, and grease, then reverse osmosis has been shown to be an effective treatment technology.

Stripping is a mass transfer process whereby volatile contaminants are removed from aqueous wastes by passing air or stream through the wastes. Air stripping has been directly applied to ground- water treatment in removing trichloroethylene (TCE), trihalomethane (THM), and hydrogen sulfide. Removal rates as high as 99+ percent for TCE from ground water, and 90+ percent for ammonia from wastewater has been observed.

Air Stripping is frequently accomplished in a stripping lagoon or more commonly in a packed tower equipped with an air blower. The packed tower works on the principle of countercurrent flow. The water stream flows down through the packing while the air flows upward, and is exhausted through the top to the atmosphere or to emission control devices (e.g., condensors, carbon adsorption filters). The volatile substances tend to leave the aqueous stream for the gas phase. In the cross-flow tower, water flows down through the packing as in the countercurrent packed column, however, the air is pulled across the water flow path by a fan. The coke tray aerator requires no blower. The water being treated trickles through several layers of trays. This produces a large surface area for gas transfer. Diffused aeration and induced draft stripping use aeration basins or lagoons similar to standard wastewater treatment technology. Water flows through the basin from top to bottom or from side to side with the air dispersed through diffusers at the bottom of the basin. The air-to-water ratio is significantly lower in the basins than in either the packed column or cross-flow towers.

Temperature has an effect on the mass transfer coefficient of substances. This is an important point when contaminated ground water contains compounds that are very soluble (i.e., compounds with low Henry's Law constants). High water solubility makes their removal by ambient temperature air stripping almost impossible. It has been shown that removal efficiency increases dramatically with temperature and less sharply with the air-to-water ratio. Therefore, high temperature air stripping, or steam stripping, offers increased flexibility and should be investigated for each case as necessary.

Incineration is a treatment method which employs high temperature oxidation under controlled conditions to decompose a substance into products that generally include CO_2, H_2O vapor, SO_2, NO_x, HCl, and products of incomplete combustion require air pollution control equipment to prevent release of undesirable species into the atmosphere. Incineration methods can be used to destroy organic contaminants in liquid, gaseous and solid waste streams.

The most common incineration technologies are liquid injection, rotary kiln, fluidized-bed, and multiple hearth. Rotary kiln and multiple hearth incinerators can be used with most organic wastes including solids, sludges, liquids and gases, while liquid injection incinerators are limited to pumpable slurries and liquids. Fluidized-bed incinerators work well for organic liquids, gases and granular or well processed solids. Incineration offers one of the most effective technological methods for complete destruction of organic compounds.

3.4.2 Chemical

Contaminated ground water can be withdrawn and treated chemically by various techniques. Among the more common chemical treatment technologies are neutralization, precipitation, oxidation and reduction, ion exchange, and chemical fixation.

Neutralization is merely a process whereby an acid or base is added to a waste in order to adjust the pH. Neutralization is a relatively simple unit process which can be performed using ordinary and commonly available treatment equipment. It is often used prior to other treatment processes where the pH of the waste is critical (e.g., biological treatment and carbon adsorption).

Precipitation is a physiochemical process whereby a substance in solution is transformed to the solid phase. Precipitation can be accomplished by (1) adding a chemical that will react with the contaminant in a solution forming a sparingly soluble compound, (2) adding a chemical which changes the solubility equilibrium of a waste thus reducing the solubility of the specific contaminants, and (3) changing the temperature to decrease the solubility of the contaminants. Removal of metals as carbonates, hydroxides, or sulfides is the most common application of precipitation in wastewater treatment. Many precipitation reactions (e.g., metal sulfides) do not readily form floc (large fluffy precipitates) particles, but rather precipitates very fine and relatively stable colloidal particles. In these cases flocculating agents (e.g., alum and/or polyelectrolytes) must be added to cause flocculation of the metal sulfide precipitates. The effectiveness of precipitation/flocculation reactions is dependent upon the nature and concentration of the contaminants, and upon the process design. The process design must consider the optimum chemicals and dosages, suitable chemical addition systems, optimum pH and mixing requirements, sludge production, and sludge flocculation, settling and dewatering characteristics.

Oxidation/reduction processes are employed to raise (oxidation) or lower (reduction) the oxidation state of a substance or substances in order to reduce toxicity or solubility, or to transform the substance to a form which can be more easily handled. Commonly used reducing agents include sulfite salts, sulfur dioxide and the base metals (i.e., iron, aluminum and zinc). Chemical reduction is used primarily for the reduction of hexavalent chromium, mercury and lead. There are currently no practical applications involving reduction of organic compounds. Oxidation, however, has found extensive use in treatment of organic wastes. Oxidizers which are most often used in wastewater treatment include oxygen or air, ozone, ozone with ultraviolet light, chlorine gas, hypochlorites, chlorine dioxide, and hydrogen peroxide.

Ion exchange is a process whereby toxic ions are removed from the aqueous phase by being exchanged with relatively harmless ions held by the ion exchange material. Ion exchange is used to remove a broad range of ionic species from water to include (1) all metallic elements when present as soluble species, either anionic or cationic, (2) ionorganic ions such as halides, sulfates, nitrates, cyanides, etc., (3) organic acids such as carboxylics, sulfonics, and some phenols, and (4) organic amines in sufficient acidity to form the acid salt (De Renzo, 1978). Ion exchange systems will function well in dilute waste streams of variable composition provided the effluent is monitored to determine when ion exchange resin bed exhaustion has occurred.

Solidification/stabilization technologies reduce leachate production potential by physically and/or chemically binding a waste in a solid matrix. Wastes are mixed with a binding agent to produce a solid form. Solidification/stabilization processes include (1) cementation, using Portland cement, (2) pozzolanic cementation, (3) thermoplastic binding, (4) organic polymer binding, (5) surface encapsulation, and (6) glassification. Cementation and pozzolanic cementation are generally the most widely applicable to a wide range of waste compositions. Most solidification/stabilization technologies are designed for inorganic wastes and can be seriously affected by high concentrations of organic wastes (i.e., increased cure, set inhibition, flashset, etc.). However, research is currently being conducted into specific interference effects caused by particular types of wastes, and into the solidification/stabilization of certain organic wastes. Important waste characteristics that impact solidification/stabilization processes include pH, buffer capacity, water content, organic concentrations, and specific inorganic constituents.

3.4.3 Biological

The function of biological treatment is to remove organic matter from the waste stream through microbial degradation. A number of biological treatment processes exist which may be applicable to the treatment of contaminated ground water, including various forms of activated sludge, surface impoundments, trickling filters, rotating biological discs, fluidized bed reactors, land treatment, and anaerobic digestion.

Activated sludge treatment consists of passing the contaminated waste stream through an aeration basin where it is aerated for several hours. During this time an active microbial population develops which degrades organic matter in the waste. In the process a portion of the activated sludge is recycled and along with new developing cells the microbial population is maintained for further degradation of the waste stream. Various versions of the activated sludge process (e.g., oxygen, oxygen-enriched, extended aeration, contact stabilization, etc.) are simply a result of the type of aeration, time of aeration, and the contact time with the activated sludge.

Surface impoundments or lagoons are similar to activated sludge units without sludge recycle. Surface impoundments are similar to a natural eutrophic lake in that natural processes of microbial oxidation, photosynthesis, and sometimes anaerobic digestion combine to degrade organic wastes. Aeration may be supplied passively by wind action or, in aerated surface impoundments, by mechanical aerators.

Trickling filters are a form of biological treatment in which a liquid waste ($<1\%$ suspended solids) is trickled over a bed of rocks or synthetic material upon which a slime layer of microbial organisms develops. The microbes in the slime layer metabolize the organics in the waste while oxygen to the microorganisms is provided as air moves countercurrent to the water flow.

A modification of the trickling filter is the *biological tower*. The medium (e.g., of polyvinyl chloride, polyethylene, polystyrene, or redwood) is stacked into towers which typically reach 16 to 20 feet. The contaminated water is sprayed across the top of the tower and, as it moves downward, air is pulled upward through the tower. A slime layer of microorganisms develops on the media and removes the organic contaminants as the water flows over the slime layer.

Another fixed film biological treatment process, similar in operating principle to trickling filters, is the *rotating biological disc system*. This system consists of a series of rotating discs, connected by a shaft, set in a basin or trough. Approximately 40 percent of each disc's surface area is submerged in the basin and as the contaminated water passes through the basin, the microorganisms growing on the disc metabolize the organics in the wastewater. As the discs rotate the microorganisms are brought in contact with the air where oxygen is obtained for growth.

Yet another fixed film process is the *fluidized bed reactor*. Particles of substances such as sand or coal are fluidized by the action of the aeration gas st ream and the wastewater stream. These particles support a dense growth of microorganisms which give rapid treatment to the wastewater. This process is still largely experimental.

Land treatment is the mixing or dispersion of wastes into the upper zone of the soil-plant system with the objective of microbial stabilization, adsorption, and immobilization leading to an environmentally acceptable degradation of the waste. The four major land treatment options are: (1) irrigation, (2) overland flow, (3) infiltration-percolation, and (4) leachate recycle. Before waste is applied, the assimilative capacity of the land treatment plant-soil system must be determined for each contaminant present, considering the nature of the pollutant (i.e., biodegradability, mobility, uptake, toxicity, etc.) and also the site characteristics (i.e., soil type, hydrogeology, meteorology).

Anaerobic digestion is another biological treatment process which can be used for organic contaminant degradation. Whether anaerobic lagoons or anaerobic digestors (totally enclosed) are used, the anaerobic process merits consideration due to ease of operation, minimal sludge production, and energy efficiencies.

3.4.4 Limitations of Withdrawal and Treatment Techniques

As with any treatment process, the importance of limitations associated with each process is determined by the urgency of treatment, the importance of the resource, and the availability of funding for the treatment. Some of the more important limitations characteristic of the previously mentioned physical, chemical, and biological treatment processes include:

Physical

1. Carbon adsorption is intolerant of high suspended solids; can be poisoned by high heavy metals concentrations; requires pretreatment for oil and grease >10 ppm; and has high operation and maintenance costs.
2. Resin adsorption is more expensive and usually has less capacity than carbon adsorption; resin is intolerant of strong oxidizing agents and suspended solids.
3. Density separation often yields incomplete removal of hazardous compounds and generates large quantities of contaminated sludges.
4. Filter clogging (in filtration process) due to high suspended solids causes reduced run lengths and requires frequent backwashing or replacement of the filter.
5. Reverse osmosis units are subject to chemical attack, fouling, and plugging, and can require extensive pretreatment.
6. Stripping is sensitive to pH, temperature, and fluxes in hydraulic load; may be cost prohibitive at temperatures below freezing; and may cause air pollution problems.
7. Incineration may require thickening and dewatering pretreatment; may pose air pollution problems; produces an inorganic ash (possibly hazardous); and may require costly fuel or power for operation.

Chemical

1. Precipitation can be limited by the presence of complexing agents in the waste and the precipitate itself may be a hazardous waste.
2. Reduction is used primarily for reducing hexavalent chromium, mercury and lead. There are no current applications for reducing organic compounds.
3. Chemical oxidation costs are generally higher than biological treatment. Some organics are resistant to most oxidants and in some cases partial oxidation generates toxic compounds.
4. The effectiveness of ion exchange is reduced by high suspended solids and/or high concentrations of certain organics.
5. Design considerations should be made to accommodate the corrosive nature of some neutralization reagents.
6. Solidification/stabilization techniques result in increased volume and weight for disposal and are still subject to leaching of contaminants. Certain wastes cause interferences with the solidification/stabilization processes.

Biological

1. Activated sludge costs are high with intensive operation and maintenance costs; is sensitive to suspended solids and metals; generates sludge high in metals and refractory organics; and is fairly energy intensive.

2. Surface impoundments are sensitive to shock loadings and temperature effects. Gas generation and chemical volatilization are problems with anaerobic lagoons.
3. Trickling filters, rotating biological discs, and fluidized bed reactors require an energy source and are vulnerable to below freezing temperatures; have potential for odor problems; require long recovery times if disrupted; and have limited flexibility (i.e., minimize variations in operating conditions such as flow and composition).
4. Land treatment requires large areas of land; has the potential for ground-water contamination; and the ground water must be monitored.

3.5 In-Situ Treatment Techniques

3.5.1 Chemical/Physical

In-situ physical and chemical techniques, for the most part, are not well developed and are highly dependent on a number of physical factors, including aquifer permeability and the nature of specific contaminants, as well as the natural geochemistry of the earth materials.

In-situ techniques are certainly more aesthetically desirable than most other alternatives since they require minimal surface facilities and minimize public exposure to pollutants. Costs are quite variable and directly relate to the contaminant constituents present their required control agents, hydrogeologic conditions of the aquifer, aerial extent of the pollution source and physical accessibility to the site.

In-situ *chemical detoxification* techniques include injecting neutralizing agents for acid or caustic leachates, adding oxidizing agents to decompose organics or precipitate inorganic compounds, adding agents that promote other natural degradation processes, bonding contaminants, and immobilization or reaction in treatment beds. These techniques should only be considered in cases where specific contaminants, their concentration levels and the extent of the contamination plume in the aquifer are well defined. The treatment agents are specific for the class of contaminant. For example, metals may be rendered insoluble and immobile with alkalines or sulfides, and cyanides can be oxidized using strong oxidizing agents such as sodium hypochlorite or by encapsulation in an insoluble matrix. Cations may be precipitated by injecting various anions or by in-situ aeration. Hexavalent chromium could be made insoluble by injecting specific reducing agents. Free fluorides can be insolubilized by the injection of solutions containing the calcium ion.

In-situ physical/chemical treatment processes generally entail installing a series of *wells for chemical injection* at the head of or within the plume of contaminated ground water. An alternate technique that has been used in shallow aquifers is *in-situ permeable treatment beds*. These are often used to detoxify migrating leachate plumes in ground water from landfills. Trenches are filled with a reactive permeable medium; contaminated ground water entering the trench reacts with the medium to produce a nonhazardous soluble product or a solid precipitate. Among the materials commonly used in permeable bed trenches are limestone to neutralize acidic ground water and remove heavy metals; activated carbon to remove nonpolar contaminants such as carbon tetrachloride (CCl_4), polychlorinated biphenyls (PCBs) and benzene by adsorption; and zeolites and synthetic ion exchange resins for removing solubilized heavy metals.

Permeable treatment beds are applicable only in relatively shallow aquifers because the trench must be constructed down to the level of the bedrock or an impermeable clay. They are also often effective for only a short time because they lose their reactive capacity or become plugged with solids. Over-design of the system or replacement of the permeable medium can lengthen the time period over which permeable treatment is effective.

There are a number of disadvantages associated with both of these techniques. In permeable treatment beds, plugging of the bed may divert contaminated ground water, or channeling through the bed may occur. Changing hydraulic loads and contaminant levels may mean that detection times in the beds are inadequate. In the chemical injection technique, pollutants may be displaced to adjacent areas when chemical solutions are injected under higher hydraulic heads. In addition, hazardous compounds may be produced by reaction of injected chemical solutions with waste constituents other than the treatment target.

Mobilization of contaminants by *injecting surfactants* during soil washing is possible. Surfactants and *alkaline flooding* for enhanced secondary oil recovery are being used experimentally with moderate success. Most oil field surfactants are expensive refined biodegradable organics, while alkaline floods produce lye. This approach does not appear promising for aquifer restoration because of the addition of potentially hazardous materials or the creation of hazardous degradation by-products which would then have to be dealt with.

To recover hydrocarbons, there are three possible physical-chemical methods. At shallow depths, *thermal or steam floods* may be helpful. On a larger scale, *alcohol flooding* may be feasible. Alcohol is easily produced and dissolves the hydrocarbon. Theoretically, if an entire polluted zone is flooded with alcohol, all of the residual hydrocarbon can be removed. Limitations of with this method include high cost, phase-behavior difficulties and lack of field experience.

Other in-situ treatment techniques have been suggested, including *radio frequency in-situ heating* or *in-situ vitrification*, using an electric current to melt the soils and waste in place. The economics for field application of these systems are unknown.

3.5.2 Biodegradation

3.5.2.1 Natural Biological Activity in the Subsurface

In-situ biorestoration of the subsurface is a relatively new technology that has recently gained considerable attention. Scarcely more than a decade ago, conventional wisdom assumed that the subsurface below the zone of plant roots was, for all practical purposes, sterile.

Recent research has indicated that the deeper subsurface is not sterile, but in fact, harbors significant populations of microorganisms. Bacterial densities around 10^6 organisms/g dry soil have been found in several noncontaminated aquifers. The water-table aquifers examined so far exhibit considerable variation in the rate of biodegradation of specific contaminants that enter the subsurface environment. Rates can vary two to three orders of magnitude between aquifers or over a vertical separation of only a few feet in the same aquifer. Although extremely variable, the rates of biodegradation are fast enough to protect ground-water quality in many aquifers.

Although they are not clearly defined, several environmental factors are known to influence the capacity of indigenous microbial populations to degrade contaminants. These factors include dissolved oxygen, pH, temperature, oxidation-reduction potential, availability of mineral nutrients, salinity, soil moisture, the concentration of specific pollutants, and the nutritional quality of dissolved organic carbon in the ground water.

Many water-table aquifers contain oxygen, which can support aerobic microorganisms that can degrade a wide variety of organic contaminants. Examples include acetone, isopropanol, methanol, ethanol, t-butanol, benzene, toluene, the xylenes, and other alkybenzenes that leak into ground water from gasoline spills or solvent spills (Novak, *et al.*, 1984; Lokke, 1984; Jhaveri and Mazzacca, 1983; Wilson *et al.*, 1986; Lee *et al.*, 1984); napthalene, the methylnaphthalenes, fluorene, acenaphthene, dibenzofuran and a variety of other polynuclear aromatic hydrocarbons released from spilled diesel oil or heating oil (Wilson, *et al.*, 1985); and many methylated phenols and heterocyclic organic compounds seen in certain industrial waste waters. Many synthetic organic compounds can also be degraded. Examples include dichlorobenzenes (Kuhn, *et al.*, 1985), the mono-, di-, and trichlorophenols (Suflita and Miller, 1985), the detergent builder nitrilotriacetic acid (NTA) (Ward, 1985), and some of the simpler chlorinated compounds such as methylene chloride (dichloromethane) (Jhaveri and Mazzacca, 1983).

The extent of biodegradation of these compounds in ground water is limited by the concentration of oxygen. For the compounds discussed above, roughly two parts of oxygen are required to completely metabolize one part of organic compound. For example, microorganisms in a well-oxygenated ground water containing 4 mg/l of molecular oxygen can degrade only 2 mg/l of benzene. The solubility of benzene (1780 mg/l) is much greater than the capacity of its aerobic degradation in ground water. Obviously, the prospects for aerobic metabolism of these compounds will depend on their concentration as well as on the concentration of other degradable organic materials in the aquifer. Concentrated plumes of organic contaminants cannot be degraded aerobically until dispersion or other processes dilute the plume with oxygenated water.

Many of the commonly encountered organic pollutants in aquifers are synthetic organic solvents that are very persistent in oxygenated waters. Examples include tetrachloroethylene (PCE), trichloroethylene (TCE), cis and trans 1,2-dichlroethylene, ethylene dichloride (1,2-dichloroethane), 1,1,1-trichloroethane (TCA), 1,1,2-trichloroethane, carbon tetrachloride, and chloroform. This important class of organic contaminants commonly enters ground water as spills from underground storage tanks. Ground-water contamination in the Santa Clara Valley of California (Silicon Valley) is a good example. Recent research has shown that this class of organic contaminants can be cometabolized by bacteria that grow on gaseous aliphatic hydrocarbons like methane or propane. The potential use of cometabolism for in situ restoration is under evaluation.

When the concentration of organic contaminants is high, oxygen in the ground water will be totally depleted and aerobic metabolism will stop. However, further biotransformations often will be mediated by a variety of anaerobic bacteria. Anaerobes that produce methane, called methanogens, are only active in highly reduced environments. Molecular oxygen is very toxic to them. Methane can be produced by the fermentation of a few simple organic compounds such as acetate, formate, methanol, or methylamines. Molecular hydrogen can also be used in the reduction of inorganic carbonate to methane. Although the microorganisms that actually produce the methane can use a very limited set of organic compounds, they can act in consort with other microorganisms which break more complex organic compounds down to substances that the methanogenic organisms can use. These partnerships or consortia can totally degrade a surprising variety of natural and synthetic organic compounds.

The rates of reaction are usually slow and often require long lag periods before active transformation begins (Wilson, 1985). Microbiologists are accustomed to microorganisms that grow to high densities in only a few days, and rarely conduct experiments that last longer than a few weeks. However, the residence time of organic pollutants in aquifers is at least months or years and is frequently decades to centuries. As a result, much of what was learned in earlier laboratory studies cannot be applied to the subsurface environment. Currently, microbiologists are re-examining the potential for biodegradation of organic contamination in ground waters that actively produce methane and are finding many expected reactions.

It was previously thought that the metabolism of benzene, toluene, the xylenes and other alkylbenzenes required molecular oxygen because oxygen is a co-substrate for the only known enzyme that can begin the metabolism of this class of compounds (Young, 1984). Thus, their metabolism would not be expected in methanogenic environments. Yet recently, the metabolism of these compounds was demonstrated in methanogenic river alluvium that has been contaminated with landfill leachate (Wilson and Rees, 1985). When radioactive toluene was added to this material at least half the carbon was metabolized completely to carbon dioxide. The same materials also metabolized several methyl and chlorophenols (Suflita and Miller, 1985). Very recently extensive anaerobic metabolism of alkylbenzenes has been demonstrated in a sandy water-table aquifer contaminated with aviation gasoline released from an undergound storage tank.

The halogenated solvents that are presistent in oxygenated ground water can be transformed in methanogenic ground water. Examples include trichloroethylene, tetrachloroethylene, the dichloroethylenes, 1,1,1-dichloroethane, carbon tetrachloride and chloroform (Parsons, *et al.*, 1984; Parsons, *et al.*, 1985; Wood *et al.*, 1985). Ethylene dibromide is also transformed (Wilson and Rees, 1985). The chlorinated ethylenes undergo a sequential reductive dehalogenation from tetrachloroethylene to trichloroethylene, then to the dichloroethylenes (primarily the CIS isomer) and finally to vinyl chloride (Wood, *et al.*, 1985). In some subsurface environments, appreciable quantities of vinyl chloride accumulate, which is unfortunate because this compound is considerably more toxic and carcinogenic than its parent compound. In other subsurface environments the vinyl chloride is further metabolized. The factors that control the fate of vinyl chloride are unknown (Wilson, 1985). The chloroalkanes follow a similar pattern (Wood, *et al.*, 1985); carbon tetrachloride is converted to chloroform, then to methylene chloride, while 1,1,1-dichloroethane is converted to 1,1-dichloroethane, which in turn goes to ethyl chloride.

These reductive dehalogenations resemble respirations. In aerobic respiration, molecular oxygen accepts an electron and is reduced to the hydrogenated compound, water. The chlorinated compounds accept electrons and are reduced to the corresponding hydrogenated compound, while chlorine is released as a chloride ion. The source of electrons can be a co-occuring contaminant, such as volatile fatty acids in landfill leachate, or it can be a geological material. Reductive dechlorination of trichloroethylene has been associated with flooded surface soil, buried soils in glaciated areas, buried layers of peat, and coal seams.

If oxygen is depleted, but conditions do not favor the methanogens, certain classes of organic compounds can be degraded by bacteria that respire nitrate or sulfate. Ground waters recharged through soils that support intensive agriculture often have high concentrations of nitrate, and ground waters with appreciable concentrations of sulfate are widespread, particularly in arid regions. Microorganisms respiring nitrate can degrade a number of phenols and cresols (methylphenols). Recently, it has been shown that nitrate respiring organisms in river alluvium could also degrade all three xylenes (dimethylbenzenes) (Kuhn, *et al.*, 1985). Nitrate-respiring microorganisms can also degrade carbon tetrachloride and a variety of brominated methanes. However, they have not been shown to degrade chloroform or those chlorinated ethylenes or ethanes which are also stable in oxygenated ground water (Bouwer and McCarty, 1983).

Like the methanogens, the sulfate-respiring bacteria can participate in consortia that degrade a wide variety of natural organic compounds. In contrast to the behavior of methanogenic subsurface material, chlorinated derivatives of naturally-occurring aromatic compounds were not degraded in river alluvium containing appreciable sulfate concentrations (200 mg/l) and exhibiting active sulfate respiration (Suflita and Miller, 1985, Suflita and Gibson, 1985). As they did under highly reduced conditions, tetrachloroethylene and trichloroethylene underwent reductive dehalogenations.

As these studies have shown, natural biorestoration does occur. Contaminants in solution in ground water as well as vapors in the unsaturated zone can be completely degraded or can be transformed to new compounds. Undoubtedly, thousands of contamination events are remediated naturally before the contamination reaches a point of detection. However, methods are needed to determine when natural biorestoration is occurring, the stage the restoration process is in, whether enhancement of the process is possible or desirable, and what will happen if natural processes are allowed to run their course. A number of researchers are presently working in this area.

3.5.2.2 Enhanced Biorestoration - Basic Principles

In subsurface situations, the populations of metabolically-capable organisms increase until they are limited by some metabolic requirement such as food or mineral nutrients, or oxygen in the case of aerobic organisms. Once this point is reached, the rate of transformation of an organic material is controlled by transport processes that supply the limiting nutrient.

The vast majority of microbes in the subsurface are firmly attached to soil particles. As a result, nutrients must be brought by advection or diffusion through the mobile phases, water and soil gas. In the simplest and perhaps most common case, the organic compound to be consumed for energy and cell synthesis is brought in aqueous solution in infiltrating water. At the same time oxygen is brought by diffusion through the soil gas. In the unsaturated zone, volatile organic compounds can also move readily as vapors in the soil gas. Below the water table all transport must be through liquid phases and as a result the prospects for aerobic metabolism is severely limited by the very low solubility of oxygen in water. In the final analysis, the rate of biological activity is controlled by:

- The stoichiometry of the metabolic process
- The concentration of the required nutrients in the mobile phases
- The advective flow of the mobile phases or the steepness of concentration gradients within the phases
- Opportunity for colonization in the subsurface by metabolically capable organisms
- Toxicity exhibited by the waste or a co-occurring material.

3.5.2.3 Enhanced Biorestoration - Current Practice

Most enhanced in-situ bioreclamation techniques available today are variations of techniques pioneered by Richard Raymond, Virginia Jameson, and co-workers at Suntech during the period 1974-1978. Suntech's process received a patent entitled "Reclamation of Hydrocarbon Contaminated Ground Waters" (Raymond, 1974). This process reduces hydrocarbon contaminants in aquifers by enhancing the indigenous hydrocarbon-utilizing microflora. Nutrients and oxygen are introduced through injection wells and circulated through the contaminated zone by pumping one or more producing wells. The increased supply of nutrients and oxygen stimulates biodegradation of the hydrocarbons. Oxygen is supplied by sparging air into the injection wells. Raymond's process has been used largely to clean up gasoline contaminated aquifers.

The first basic step in Raymond's process is usually to employ physical methods to recover as much of the gasoline product as possible. While the product is being recovered, Raymond requires a detailed investigation of the hydrogeology and the extent of contamination. A laboratory study is conducted to determine if the native microbial population can degrade the contaminants. Laboratory studies also identify the combination of minerals that promotes maximum cell growth on the contaminant in 96 hours under aerobic conditions at the ambient ground-water temperature.

Considerable variation in the nutrient requirements has been noted by Suntech. One aquifer required only the addition of nitrogen and phosphorus, while the growth of microbes in another aquifer was stimulated best by the addition of ammonium sulfate, mono- and disodium phosphate, magnesium sulfate, sodium carbonates, calcium chloride, manganese sulfate and ferrous sulfate. They found that chemical analysis of the ground water was not helpful in estimating the nutrient requirements of the system.

After the microbial laboratory investigations have established the optimal conditions for growth of the indigenous microbial population, the systems for injecting the mixture of nutrients and oxygen and for producing water to circulate them in the formation are designed and built. Controlling the ground-water flow is critical to moving oxygen and nutrients to the contaminated zone and optimizing the degradation process.

The Suntech process is reported to have met with reasonable success when applied to gasoline spills in the subsurface. Some of the sites treated by this technique have been cleaned to the point where no dissolved gasoline was present in the ground waters and state regulatory standards were satisfied. The State agencies in charge of cleaning-up other sites, however, have directed operations to continue until no trace of liquid gasoline can be detected. Most of the sites have implemented appropriate ground-water monitoring programs following clean-up. The overall percent removal of total hydrocarbons using this method has usually ranged from 70 to 80 percent.

The Suntech process does not provide for treatment of the material above the water table. Soils or geological material contaminated by leaking underground storage tanks may be physically removed during the process of removing the tank, in which case the contaminated material can be disposed of in an approved manner. However, in cases where the extent of the pollution is large or the water table extends to a depth where physical removal of contaminated material is totally impractical, alternative methods are used. One of these methods is construction of one or a series of surface infiltration galleries. These galleries are used to recirculate water, which has been treated, back through the contaminated unsaturated zone. Oxygen is generally added to the infiltrated water during an in-line stripping process for volatile organic contaminants or through aeration devices placed in the infiltration galleries. Recirculation of the water also facilitates movement of contaminants to the recovery well. The dislodged or solubilized contaminant can be treated in a surface treatment system before the water is reinjected.

In constructing an infiltration gallery, the most critical factor is the rate of water infiltration. Silty and shaley materials accept water very slowly. The site must be tested and evaluated to determine size and configuration of infiltration pits.

Whether the material is situated above or below the water table, the rate of bioreclamation in hydrocarbon contaminated zones is effectively the rate of supply of oxygen. Table 3-1 compares the number of times that water in contaminated material below the water table, or air in material above it, must be replaced to totally reclaim subsurface materials of various textures. The calculations assume typical values for the volume occupied by air, water and hydrocarbons (De Pastrovich, *et al.*, 1979; Clapp and Hornberger, 1978). The actual values at a specific site will probably be different. The calculations further assume that the oxygen content of the water was 10 mg/l, that of the air 200 mg/l, and that the hydrocarbons were completely metabolized to carbon dioxide.

It is obvious that prodigious volumes of water are needed in the finely-textured subsurface materials. This has prompted a search for some mechanism to increase the concentration of oxygen. The most obvious approach is to sparge the injection wells with oxygen instead of air. This will increase the oxygen concentration about five-fold. The water can also be supplemented with hydrogen peroxide (Brown *et al.*, 1984).

Laboratory studies have shown that hydrocarbon-degrading bacteria can adapt to tolerate hydrogen peroxide equivalent to 200 mg/l oxygen, a twenty-fold increase in oxygen supply over water sparged with air (Lee and Ward, 1985). However, the rate of decomposition of hydrogen peroxide to oxygen must be controlled. Rapid decomposition of only 100 mg/l of peroxide will exceed the solubility of oxygen in water resulting in bubble formation which could lead to gas blockage and loss of permeability. Iron catalyzes the decomposition of hydrogen peroxide in ground water. Standard practice is to add enough phosphate to the recirculated water to precipitate the iron. Some suppliers add an organic catalyst that will decompose the peroxide at a rate appropriate to the rate of infiltration, so that the oxygen demand of the bacteria attached to the solids is balanced by the oxygen supplied by decomposing peroxide in the recirculated water.

Obviously, successful use of hydrogen peroxide requires careful control of the geochemistry and hydrology of the site. In addition to the factors mentioned already, hydrogen peroxide can mobilize metals such as lead and antimony; and if the water is hard, magnesium and calcium phosphates can precipitate and plug up the injection well or infiltration gallery.

3.6 Treatment Trains

In most contaminated hydrogeologic systems a remediation process may be so complex in terms of contaminant behavior and site characteristics that no one system or unit is going to meet all requirements. Very often, it is necessary to combine several unit operations, in series and sometimes in parallel into

Table 3-1 Estimated Volumes of Water or Air Required to Completely Renovate Subsurface Material that Contained Hydrocarbons at Residual Saturation.

	Proportion of Total Subsurface Volume Occupied by:			Volumes Required to meet Hydrocarbons Oxygen Demand	
Texture	Hydrocarbons (when drained)	Air (when drained)	Water (when flooded)	Air	Water
Stone to Coarse Gravel	0.005	0.4	0.4	250	5,000
Gravel to Coarse Sand	0.008	0.3	0.4	530	8,000
Coarse to Medium Sand	0.015	0.2	0.4	1,500	15,000
Medium to Fine Sand	0.025	0.2	0.4	2,500	25,000
Fine Sand to Silt	0.040	0.2	0.5	4,000	32,000

one treatment process train in order to effectively restore ground-water quality to a required level. Barriers and hydrodynamic controls alone merely serve as temporary plume control measures. However, hydrodynamic processes must also be integral parts of any withdrawal and treatment or in-situ treatment measures.

Most remediation projects, where enhanced biorestoration has been applied, have started by removing heavily contaminated soils. This was usually followed by installing pumping systems, to remove free product floating on the ground water, before biorestoration enhancement measures were initiated to degrade the more diluted portions of the plume.

As noted earlier, there are numerous proven surface treatment processes available for treating a variety of organic and inorganic wastewaters. However, regardless of the source of ground-water contamination and the remediation measures anticipated, the limiting factor is getting the contaminated subsurface material to the treatment unit or units, or in the case of in-situ processes, getting the treatment process to the contaminated material. The key to success is a thorough understanding of the hydrogeologic and geochemical characteristics of the area. Such an understanding will permit full optimization of all possible remedial actions, maximum predictability of remediation effectiveness, minimum remediation costs, and more reliable cost estimates.

3.7 Institutional Limitations on Controlling Ground-Water Pollution

The principal criteria for selecting remediation procedures should be the water quality level to which an aquifer should be restored and the most economic technology available to reach that quality level. Unfortunately, there are numerous institutional limitations that sometimes override these criteria in determining if, when, what, and how remediation will be selected and carried out.

Response to a ground-water contamination problem is likely to require compliance with several local, state and federal pollution control laws and regulations. If the response involves handling hazardous wastes, discharging substances into the air or surface waters, or the underground injection of wastes, federal pollution laws apply. These laws do not exempt the activities of federal, state, or local officials or other parties attempting to remediate contamination events. They apply to generators and responding parties alike, and it is not unusual for these pollution control laws to conflict. For example, a hazardous waste remediation project may be slowed, altered or abandoned by the imposition, upon the party undertaking the effort, of elaborate RCRA permit requirements governing the transport and disposal of hazardous wastes.

In-situ remediation procedures may be subject to permitting or other requirements of federal or state underground injection control programs. Withdrawal and treatment approaches may be subject to regulation under federal or state air pollution control programs or to pretreatment requirements if contaminated ground water will be discharged to a municipal wastewater treatment system. Also, pumping from an aquifer may involve a state's ground-water regulations on well construction standards and well spacing requirements as well as interfere with various competing legal rights to pump ground water.

Other factors influencing remediation decisions are the availability of alternate sources of water supply, the political and judicial pressure, and the availability of funds. If alternate water supplies are plentiful and economical, there may be little incentive for more than cosmetic remediation, if any. Conversely, if there is great pressure from the public, press and/or courts to "do something", there is a tendency to overreact--to install remediation measures that offer more in appearance than in substance. In the final analysis, responsible agencies can pursue only those remediation measures for which they have resources.

3.8 References

Borden, R.C., M.D. Lee, J.T. Wilson, C.H. Ward and P.B. Bedient. 1984. Modeling the Migration and Biodegradation of Hydrocarbons Derived from a Wood-Creosoting Process Waste. Proceedings of Petroleum Hydrocarbons and Organic Chemicals in Ground Water: Prevention, Detection, and Restoration, Conference, November 5-7, 1984, Houston, TX.

Bouwer, E.J., and P.L. McCarty. 1983. Transformation of Halogenated Organic Compounds Under Dentrification Conditions. Applied and Environmental Microbiology 45(4)1295-1299.

Clapp, R.B., and G.M. Horn berger. 1978. Empirical Equations for Some Soil Hydraulic Properties. Water Resources Research 14:601-604.

Cooper, D.G. 1982. Biosurfactants and Enhanced Oil Recovery. Proceedings of the 1982 International Conference on Microbiological Enhancement of Oil Recovery, May 16-21, 1982, Shangri-La, Afton, OK.

Cooper, D.G., and J.T. Zajic. 1980. Surface-Active Compounds from Microorganisms. Advanced Applied Microbiology 26:229-253.

De Pastrorich, T.L., Y. Baradat, R. Barthal, A. Chiarelli, and D.R. Fussel. 1979. Protection of Ground Water from Oil Pollution. CONCAWE Report No. 3179, The Oil Companies' International Study Group, Den Haag, The Netherlands.

Ehrenfeld, J., and J. Bass. 1984. Evaluation of Remedial Action Unit Operations of Hazardous Waste Disposal Sites. Pollution Technology Review No. 110. Noyes Publications, Park Ridge, NJ.

Ehrlich, G.G., R.A. Schroeder, and P. Martin. 1985. Microbial Populations in a Jet-Fuel Contaminated Shallow Aquifer at Tustin, California. U.S. Geological Survey Open File Report 85-335.

Henson, R.W. and R.E. Kallio. 1957. Inability of Nitrate to Serve as a Terminal Oxidant for Hydrocarbons. Science 125:1198-1199.

Jones, J.N., R.M. Bricka, T.E. Myers, and D.W. Thompson. 1985. Factors Affecting Solidification/Stabilization of Hazardous Waste. Proceedings of the Eleventh Annual Research Symposium for Land Disposal of Hazardous Waste. EPA-600/9-85-013, U.S. Environmental Protection Agency, Hazardous Waste Environmental Research Laboratory, Cincinnati, OH.

Knox, R.C., L.W. Canter, D.F. Kincannon, E.L. Stover and C.H. Ward. 1984. State-of-the-Art of Aquifer Restoration. EPA-600/2-84/182a and b, U.S. Environmental Protection Agency, Robert S. Kerr Environmental Research Laboratory, Ada, OK.

Kuhn, E.P., P.J. Colberg, J.L. Schoor, D. Wanner, A.J.B. Zehnder, and R.P. Schwarzenbach. 1985. Environmental Science and Technology 19:961-968.

Lee, M.D., and C.H. Ward. 1985. Restoration Techniques for Aquifers Contaminated with Hazardous Waste. Journal of Hazardous Materials (In Press).

Lee, M.D., and C.H. Ward. 1984. Reclamation of Contaminated Aquifers: Biological Techniques. Proceedings of the 1984 Hazardous Material Spills Conference. April 9-12, 1984, Nashville, TN.

Overcash, M.R., and D. Pal. 1979. Design of Land Treatment Systems for Industrial Waste -- Theory and Practice. Ann Arbor Science. Ann Arbor, MI.

Parsons, F., G.B. Lage, and R. Rice. 1985. Biotransformation of Chlorinated Organic Solvents in Static Microcosms. Environmental Toxicology and Chemistry 4:739-742.

Parsons, F., P.R. Wood, and J. DeMarco. 1984. Transformations of Tetrachloroethene and Trichlonethene in Microcosms and Ground Water. Journal American Water Works Association 76(2):56-59.

Perry, J.J. 1979. Microbial Cooxidations Involving Hydrocarbons. Microbiology Review 43:59-72.

Raymond, R.L. 1974. Reclamation of Hydrocarbon Contaminated Ground Waters. U.S. Patent Office, 3,846,290. Patented November 5, 1974.

Suflita, J.M., and S.A. Gibson. 1985. Biodegradation of Haloaromatic Substrates in a Shallow Anoxic Ground Water Aquifer. Proceedings of the Second International Conference on Ground Water Quality Research, March 26-29, 1984, Tulsa, OK.

Suflita, J.M., and G.D. Miller. 1985. Microbial Metabolism of Chlorophenolic Compounds in Ground Water Aquifers. Environmental Toxicology and Chemistry 4:751-758.

U.S. Army Engineers. 1985. Guidelines for Preliminary Selection of Remedial Actions for Hazardous Waste Sites. EM 1110-2-505, DA-USAE, Washington, DC.

U.S. Environmental Protection Agency. 1985. Remedial Action at Waste Disposal Sites. EPA/625/6-82-006, U.S. Environmental Protection Agency, Hazardous Waste Environmental Research Laboratory, Cincinnati, OH.

van der Waarden, M., L.A. Bridie, and W.M. Groenewoud. 1977. Transport of Mineral Oil Components to Ground Water II. Water Research 11:359-365.

Wilson, B. 1985. Behavior of Trichloroethylene, 1,1-Dichloroethylene in Anoxic Subsurface Environments. M.S. Thesis, University of Oklahoma.

Wilson, B.H., and J.F. Rees. 1985. Biotransformation of Gasoline Hydrocarbons in Methanogenic Aquifer Material. Proceedings of the NWWA/API Conference on Petroleum Hydrocarbons and Organic Chemicals in Ground Water, November 13-15, 1985, Houston, TX.

Wilson, J.L. and S.H. Conrad. 1984. Is Physical Displacement of Residual Hydrocarbons a Realistic Possibility in Aquifer Restoration? Proceedings of the Petroleum Hydrocarbons and Organic Chemicals in Ground Water: Prevention, Detection, and Restoration Conference, November 5-7, 1984, Houston, TX.

Wood, P.R., R.F. Lang, and I.L. Payan. 1985. Anaerobic Transformation, Transport, and Removal of Volatile Chlorinated Organics in Ground Water. In: Ground Water Quality, edited by C.H. Ward, W. Giger and P.L. McCarty, John Wiley & Sons, New York, NY.

Young, L.Y. 1984. Anaerobic Degradation of Aromatic Compounds. In: Microbial Degradation of Aromatic Compounds, edited by D.R. Gibson, Marcel Dekker, New York, NY.

Part II

Scientific and Technical Background for Assessing and Protecting the Quality of Ground-Water Resources

4. Basic Hydrogeology

Hydrogeology is the study of ground water -- its origin, occurrence, movement, and quality. Ground water is a part of the hydrologic cycle and, in order to understand the influence of the hydrologic cycle on ground water, it is essential to have some basic knowledge of precipitation, infiltration, the relationship between ground water and streams, and the impact of the geologic framework on water resources. This chapter provides a brief outline of these topics.

4.1 Precipitation

Much precipitation never reaches the ground; it evaporates in the air and from trees and buildings. That which reaches the land surface is variable in time, areal extent, and intensity. The variability has a direct impact on streamflow, evaporation, transpiration, soil moisture, ground-water recharge, ground water, and ground-water quality. Therefore, precipitation should be examined first in any type of hydrogeologic study in order to determine how much is available, its probable distribution, and when and under what conditions it is most likely to occur. In addition, a determination of the amount of precipitation is the first step in a water balance calculation.

4.1.1 Seasonal Variations in Precipitation

Throughout much of the United States, the spring months are most likely to be the wettest months. This is because low intensity rains often continue for several days at a time. The rain, in combination with springtime snowmelt, will saturate the soil and streamflow is generally at its peak over a period of several weeks or months. Because the soil is saturated, this is the major period of ground-water recharge. In addition, because all of the surface runoff consists of precipitation and snowmelt, surface waters most likely will contain less dissolved mineral matter than at any other time during the year.

Not uncommonly, the fall is also a wet period although precipitation is not as great or prolonged as in the spring. Because ground-water recharge can occur over wide areas during spring and fall, one should expect some natural changes in the chemical quality of ground water in surficial or shallow aquifers.

During the winter in northern states the ground is frozen, largely prohibiting infiltration and ground-water recharge. An early spring flow coupled with widespread precipitation may lead to severe flooding over large areas.

Summer precipitation is more likely to be convective in nature and the result is high intensity rainfall that occurs during a short time interval in a small area. Most of the rain does not infiltrate, there is a soil-moisture deficiency, and ground-water recharge over wide areas is not to be expected. On the other hand, these typically small, local showers can have a significant impact on shallow ground-water quality because some of the water flows quickly through fractures or other macropores in the unsaturated zone, carrying water-soluble compounds leached from the dry soil to the water table. In this case, certain chemical constituents, and perhaps microbes as well, may increase dramatically.

4.1.2 Types of Precipitation

Precipitation is classified by the conditions that produce the rising column of unsaturated air which is antecedent to precipitation.

Convectional precipitation is the result of uneven heating of the ground, which causes the air to rise and expand, vapor to condense, and precipitation to occur. This is the major type of precipitation during the summer, producing high intensity, short duration storms usually of small areal extent. They often cause flash floods in small basins. Ground-water recharge caused by convective storms is likely to be of a local nature.

Orographic precipitation is caused by topographic barriers that force the moisture laden air to rise and cool. This occurs, for example, in the Pacific Northwest where precipitation exceeds 100 inches per year, and in Bangladesh, which receives more than 425 inches per year nearly all of which falls during the monsoon season. In this vast alluvial plain, rainfall commonly averages 106 inches during June for a daily average exceeding 3.5 inches.

Cyclonic precipitation is related to large low pressure systems that require 5 or 6 days to cross the United States from the northwest or Gulf of Mexico. These systems are the major source of winter precipitation. During the spring, summer, and fall they lead to rainy periods that may last 2 or 3 days or more. They are characterized by low intensity and long duration, and cover a wide area. They probably have a major impact on natural recharge to ground-water systems during the summer and fall and impact ground-water quality as well.

4.1.3 Recording Precipitation

The amount of precipitation is measured by recording and nonrecording rain gages. Many are located throughout the country but, because of the inadequate density of gages, our estimate of annual, and particularly summer, precipitation is too low. Records can be obtained from the Climatological Data, which are published annually by The National Oceanic and Atmospheric Administration (NOAA). Precipitation is highly variable, both in time and space. The areal extent of precipitation is evaluated by means of contour or isohyet maps (Figure 4-1).

A rain gage should be installed in the vicinity of a site under investigation in order to know exactly when precipitation occurred, how much fell, and its intensity. Data such as these are essential to the interpretation of hydrographs of both wells and streams, and they provide considerable insight into the causes of fluctuations in ground-water quality.

4.2 Infiltration

The variability of streamflow depends on the source of the supply. If the source of streamflow is from surface runoff, the stream will be characterized by short periods of high flow with long periods of low flow or no flow at all. Streams of this type are known as "flashy." If the basin is permeable, there will be little surface runoff and ground water will provide the stream with a high sustained uniform flow. These streams are known as "steady." Whether a stream is steady or flashy depends on the infiltration of precipitation and snowmelt.

When it rains, some of the water is intercepted by trees or buildings, some is held in low places on the ground (this is known as depression storage), some flows over the land surface without infiltrating and eventually reaches a stream (surface runoff), some is evaporated, and some infiltrates. Of the water that infiltrates, some replenishes the soil-moisture deficiency, if any, while the remainder percolates deeper, perhaps becoming ground water. The depletion of the soil-moisture begins immediately after a rain due to evaporation and transpiration.

Infiltration capacity (f) is the maximum rate at which a soil is capable of absorbing water in a given condition. Several factors control infiltration capacity:

- o Antecedent rainfall and soil-moisture conditions. Soil moisture fluctuates seasonally, usually being high during winter and spring and low during the summer and fall. If the soil is dry, wetting the top of it will create a strong capillary potential just under the surface, supplementing gravity. When wetted, the clays forming the soil swell, which reduces the infiltration capacity shortly after a rain starts.
- o Compaction of the soil due to rain.
- o Inwash of fine material into soil openings, which reduces infiltration capacity. This is especially important if the soil is dry.
- o Compaction of the soil due to animals, roads, trails, urban development, etc.

Figure 4-1 Distribution of annual average precipitation in Oklahoma, 1970-79 (from Pettyjohn and others).

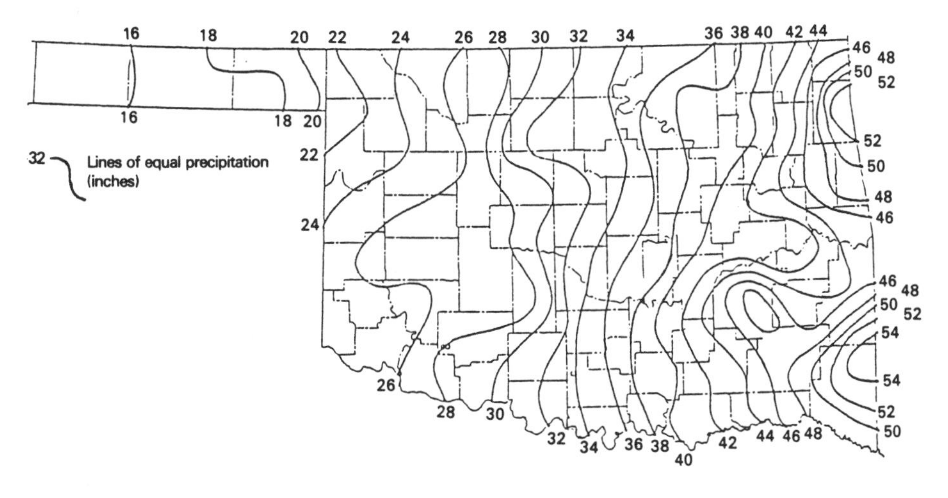

- Certain microstructures in the soil will promote infiltration, such as openings caused by burrowing animals, insects, decaying roots and other vegetative matter, frost heaving, dessication cracks, and other macropores.
- Vegetative cover, which tends to increase infiltration because it promotes populations of burrowing organisms and retards surface runoff, erosion, and compaction by raindrops.
- Decreasing temperature, which increases water viscosity, reducing infiltration.
- Entrapped air in the unsaturated zone, which tends to reduce infiltration.
- Surface gradient.

Infiltration capacity is usually greater at the start of a rain that follows a dry period, but it decreases rapidly (Figure 4-2). After several hours it is nearly constant because the soil becomes clogged by particles and swelling clays. Thus a sandy soil, as opposed to a clay-rich soil, may maintain a high infiltration capacity for a considerable time.

Figure 4-2 Infiltration capacity decreases with time during a rainfall event.

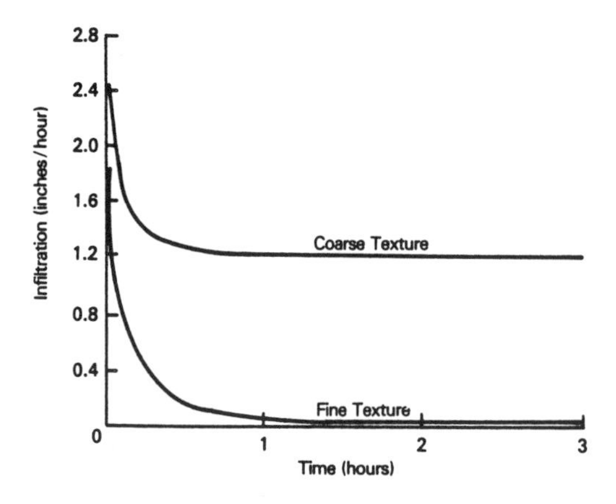

As the duration of rainfall increases, infiltration capacity continues to decrease. This is partly due to the increasing resistance to flow as the moisture front moves downward; that resistance is a result of frictional increases due to the increasing length of flow channels and the general decrease in permeability owing to swelling clays. If precipitation is greater than the infiltration capacity, surface runoff occurs. If precipitation is less than the infiltration capacity, all moisture is absorbed.

When a soil has been saturated by water, then allowed to drain by gravity, the soil is said to be holding its field capacity of water. Drainage generally requires no more than two or three days and most occurs within one day. A sandy soil has a low field capacity that is reached quickly; clay-rich soils are characterized by a high field capacity that is reached slowly (Figure 4-3).

Figure 4-3 Relation between grain size and field capacity and wilting point (from Smith and Ruhe, 1955).

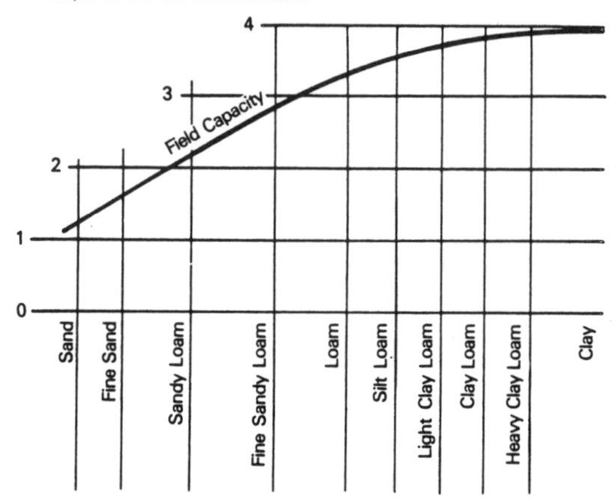

The water that moves down becomes ground-water recharge. Since recharge occurs even when field capacity is not reached, there must be a rapid transfer of water through the unsaturated zone. This probably occurs through macropores (Pettyjohn, 1983).

4.3 Surface Water

Streamflow, runoff, discharge, and yield of drainage basin are all nearly synonymous terms. Channel storage refers to all of the water contained at any instant within the permanent stream channel. Runoff includes all of the water in a stream channel flowing past a cross section; this water may consist of precipitation that falls directly into the channel, surface runoff, ground-water runoff, and effluent.

Although the total quantity of precipitation that falls directly into the channel may be large, it is quite small in comparison to the total flow. Surface runoff, including interflow or stormflow, is the only source of water in ephemeral streams and intermittent streams during part of the year. It is the major cause of flooding. During dry weather ground-water runoff may account for a stream's entire flow. It is the major source of water to streams from late summer to winter; at this time streams are also most highly mineralized under natural conditions. Ground water moves slowly to the stream, depending on the hydraulic gradient and permeability; the contribution is slow but the supply is steady. When ground-water runoff provides a stream's entire discharge, the flow is called dry-weather flow. Other sources of runoff include the discharge of industrial or municipal effluent, or irrigation return flow.

4.3.1 Stream Types

Streams are generally classified on the basis of their length, size of the drainage basin, or discharge; the latter is probably the most significant index of a stream's utility in a productive society. Rates of flow are generally reported as cubic feet per second (cfs), millions of gallons per day (mgd), acre-feet per day, month, or year, cfs per square mile of drainage basin (cfs/mi^2), or inches depth on drainage basin per day, month, or year. In the United States, the most common unit of measurement for rate of flow is cubic feet per second (cfs). The discharge (Q) is determined by measuring the cross-sectional area of the channel (A), in square feet, and the average velocity of the water (v), in feet per second, so that:

$$Q = vA \tag{4-1}$$

From a hydrogeologic point of view, there are three major stream types -- ephemeral, intermittent, and perennial. They are determined by the relation between the water table and the stream channel.

An ephemeral stream owes its entire flow to surface runoff, it may have no well-defined channel, and the water table consistently remains below the bottom of the channel (Figure 4-4). Water leaks from the channel into the ground, recharging the underlying strata.

Intermittent streams flow only part of the year, generally from spring to midsummer, as well as during wet periods. During dry weather these streams flow only because of the ground water that discharges into them. This is possible because the water table is then above the base of the channel (Figure 4-4). Eventually sufficient ground water has discharged throughout the basin to lower the water table below the channel, which then becomes dry. This reflects a decrease in the quantity of ground water in storage. During late summer or fall, a wet period may temporarily cause the water table to rise enough for ground water to again discharge into the stream. Thus during part of the year the flood plain materials are full to overflowing, which causes the discharge to increase in a downstream direction, but at other times water will leak into the ground, causing a reduction of the discharge.

Many streams, particularly those in humid and semiarid regions, flow throughout the year. These are called perennial streams. In these cases, the water table annually remains above the stream bottom, ground water is discharged, and streamflow increases downstream (Figure 4-4). A stream in which the discharge increases downstream is called a gaining stream. When the discharge of a stream decreases downstream due to leakage, it is called a losing stream.

Figure 4-4 Relation between water table and stream type.

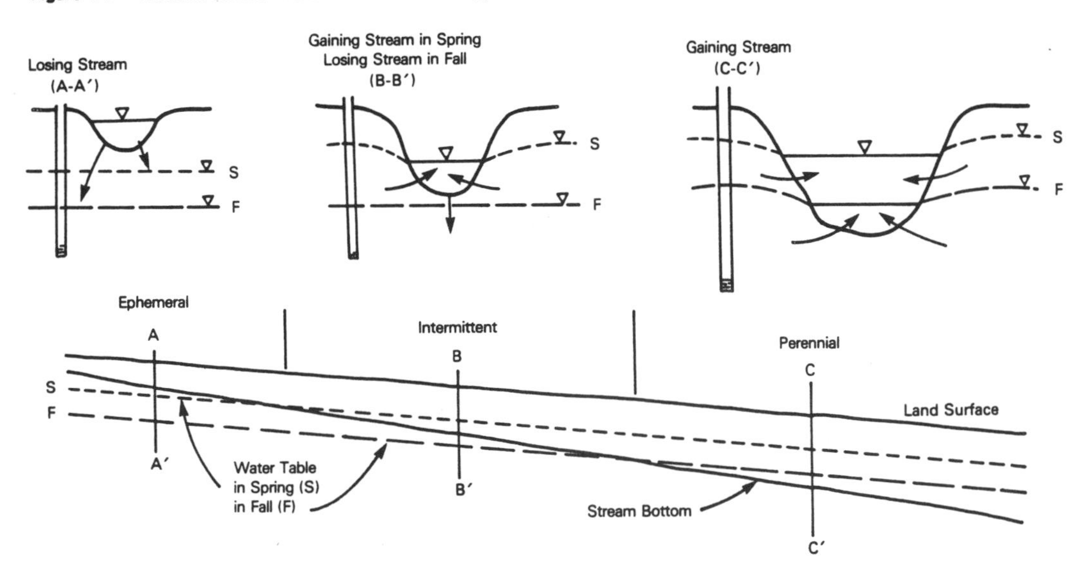

Figure 4-5 Water quality data for Cottonwood Creek near Navina, Oklahoma (from U.S. Geological Survey Water Resources Data for Oklahoma).

ARKANSAS RIVER BASIN

07159720 COTTONWOOD CREEK NEAR NAVINA, OK--Continued

WATER-QUALITY RECORDS

PERIOD OF RECORD.--Water years 1978 to current year.

PERIOD OF DAILY RECORD.--
SPECIFIC CONDUCTANCE: October 1977 to November 1980.
WATER TEMPERATURE: October 1977 to November 1980.

REMARKS.--Samples were collected monthly and specific conductance, pH, water temperature, and dissolved oxygen were determined in the field.

WATER QUALITY DATA, WATER YEAR OCTOBER 1982 TO SEPTEMBER 1983

DATE	TIME	AGENCY ANA- LYZING SAMPLE (CODE NUMBER)	STREAM- FLOW, INSTAN- TANEOUS (CFS)	SPE- CIFIC CON- DUCT- ANCE (UMHOS)	PH (STAND- ARD UNITS)	TEMPER- ATURE (DEG C)	OXYGEN, DIS- SOLVED (MG/L)	OXYGEN, DIS- SOLVED (PER- CENT SATUR- ATION)	HARD- NESS (MG/L AS CACO3)	HARD- NESS NONCAR- BONATE (MG/L AS CACO3)	CALCIUM DIS- SOLVED (MG/L AS CA)
OCT											
27...	1330	80020	15	1400	7.7	14.5	6.8	66	360	127	93
NOV											
29...	1300	80020	55	935	8.0	7.0	8.2	71	280	108	73
DEC											
15...	1320	80020	28	1300	7.8	6.0	8.0	66	400	157	100
JAN											
18...	1300	80020	26	1430	7.9	5.0	6.8	55	410	127	100
FEB											
23...	1345	80020	100	955	7.7	10.0	7.4	69	310	123	76
MAR											
29...	1430	80020	134	1100	7.6	8.0	8.2	72	380	170	94
APR											
27...	1430	80020	76	1290	7.8	19.0	4.9	55	410	161	97
MAY											
24...	1330	80020	266	850	7.8	20.0	6.9	78	270	87	67
JUN											
21...	1245	80020	57	1300	7.6	24.0	5.4	67	460	168	110
SEP											
15...	1030	80020	24	1320	7.7	21.5	6.3	75	350	141	91

WATER QUALITY DATA, WATER YEAR OCTOBER 1982 TO SEPTEMBER 1983

DATE	MAGNE- SIUM, DIS- SOLVED (MG/L AS MG)	SODIUM, DIS- SOLVED (MG/L AS NA)	PERCENT SODIUM	SODIUM AD- SORP- TION RATIO	POTAS- SIUM, DIS- SOLVED (MG/L AS K)	ALKA- LINITY LAB (MG/L AS CACO3)	SULFATE DIS- SOLVED (MG/L AS SO4)	CHLO- RIDE, DIS- SOLVED (MG/L AS CL)	SILICA, DIS- SOLVED (MG/L AS SIO2)	SOLIDS, RESIDUE AT 180 DEG. C DIS- SOLVED (MG/L)	SOLIDS, SUM OF CONSTI- TUENTS, DIS- SOLVED (MG/L)
OCT											
27...	30	140	45	3	9.5	229	220	170	13	850	810
NOV											
29...	24	84	39	2	5.8	173	160	95	10	574	560
DEC											
15...	36	130	41	3	7.7	241	230	150	14	865	810
JAN											
18...	39	150	44	3	8.9	284	250	170	11	895	900
FEB											
23...	30	77	34	2	4.8	191	170	95	12	591	580
MAR											
29...	36	92	34	2	4.4	213	220	100	12	718	690
APR											
27...	40	110	37	2	5.2	247	230	140	12	789	780
MAY											
24...	26	62	33	2	4.3	188	140	71	12	508	500
JUN											
21...	44	120	36	3	5.9	289	240	130	17	870	840
SEP											
15...	29	140	46	3	11	206	240	170	12	852	820

Nearly all water courses have headwater regions characterized by ephemeral streams. Farther downbasin, intermittent streams predominate and, even farther, the water courses are perennial. Some streams fed by springs or glacial meltwater are perennial throughout their entire length.

The natural gradation from one stream type to another may be interrupted by either natural or man-made causes. Irrigation may provide enough recharge to cause the water table to rise sufficiently to increase ground-water runoff, while pumping from wells may have the opposite effect.

Streams flowing through saturated permeable deposits, such as sand and gravel, are normally gaining streams, but streams flowing through karst regions may be losing in one reach and gaining in another. High dry-weather flow may reflect the discharge of water from mine workings.

From a hydraulic perspective, a stream is similar to an exceedingly long, very shallow, horizontal well. Consequently, the chemical quality of water in the stream during dry weather reflects the quality of ground water in the zone of active circulation within the basin if the stream is not contaminated by some surface source. During wet weather, the chemical quality of water in a stream varies largely because of the mixing of dilute surface runoff with the more highly mineralized ground-water runoff (Figure 4-5). The sediment load, reflecting erosion in the basin and stream channel, also affects the quality of the stream. The loading of a stream with either sediment or dissolved constituents is commonly reported in units of tons per day (Tons per day = Discharge x Concentration x .0027).

4.3.2 Stream Discharge Measurements and Records

At a stream gaging site the discharge is measured periodically at different rates of flow, which are plotted against the elevation of the water level in the stream (stage or gage-height). This forms a rating curve (Figure 4-6). At a gaging station the stage is continuously measured and this record is converted, by means of the rating curve, into a discharge hydrograph. The terminology used to describe the various parts of a stream hydrograph are shown in Figure 4-7.

Discharge, water quality, and ground-water level records are published each year by the U.S. Geological Survey for each state. An example of the annual record of a stream is shown in Figure 4-8. Notice that these data are reported in "water years." The water year is designated by the calendar year in which it ends, which includes 9 of the 12 months. Thus, water year 1985 extends from October 1, 1984 to September 30, 1985.

Figure 4-6 A generalized stream stage vs. discharge rating curve.

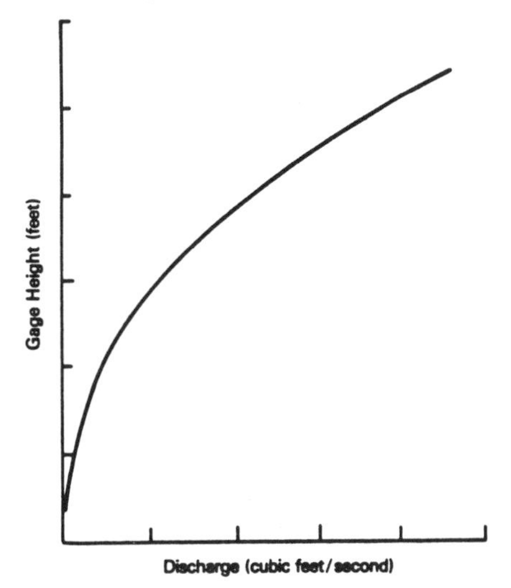

Figure 4-7 Stream hydrograph showing definition of terms.

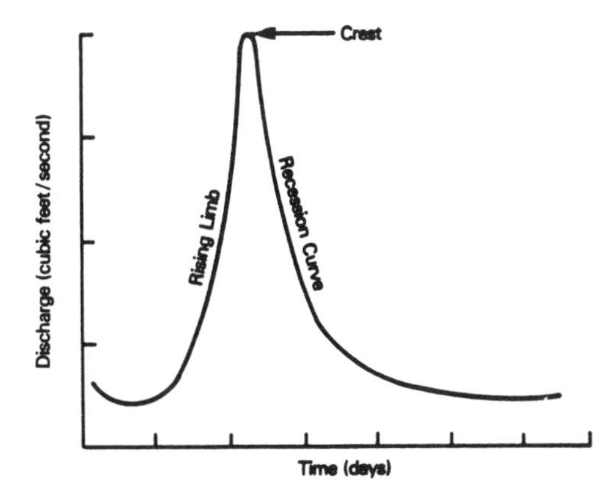

4.4 The Relation Between Surface Water and Ground Water

There are many tools for learning about ground water without basing estimates on the ground-water system itself -- that is, one can use streamflow data. Analyses of streamflow data permit an evaluation of the basin geology, permeability, the amount of ground-water contribution, and the major areas of discharge. In addition, if chemical quality data are available or collected for a specific stream, they can be used to determine background

Figure 4-8 **Stream discharge record for Cottonwood Creek near Navina, Oklahoma (from U.S. Geological Survey Water Resources Data for Oklahoma).**

ARKANSAS RIVER BASIN

07159720 COTTONWOOD CREEK NEAR NAVINA, OK

LOCATION.--Lat 35°46'36", long 97°32'45", SW 1/4 NW 1/4 sec. 17, T.15 N., R.4 W., Logan County, Hydrologic Unit 11050002 on downstream right bank, 0.5 mi (0.8 km) downstream from Deer Creek, 1.7 mi (2.7 km) southeast of Navina, 10.7 mi (17.2 km) southwest of Guthrie, and at mile 25.0 (40.2 km).

DRAINAGE AREA.--247 mi^2 (640 km^2).

WATER-DISCHARGE RECORDS

PERIOD OF RECORD.--October 1977 to September 1980, March 1982 to current year.

GAGE.--Water-stage recorder. Datum of gage is 962.10 ft (293.248 m) National Geodetic Vertical Datum of 1929.

REMARKS.--Records poor. Low flow sustained by part of sewage effluent from Oklahoma City.

EXTREMES FOR PERIOD OF RECORD.--Maximum discharge, 12,300 ft^3/s (348 m^3/s) May 30, 1980, gage height, 22.43 ft (6.837 m); minimum daily, 8.0 ft^3/s (0.23 m^3/s) Oct. 14, 15, 1977.

EXTREMES FOR CURRENT YEAR.--Maximum discharge, 3,600 ft^3/s (102 m^3/s) at 0645 May 14, gage height, 20.87 ft (6.361 m), no other peak above base of 2,000 ft^3/s (56.6 m^3/s); minimum daily discharge, 15 ft^3/s (0.42 m^3/s) Oct. 22, 23, 26.

Discharge, in Cubic Feet per Second, Water Year October 1982 to September 1983

Mean Values

DAY	OCT	NOV	DEC	JAN	FEB	MAR	APR	MAY	JUN	JUL	AUG	SEP
1	31	32	27	27	199	47	105	64	217	86	21	24
2	31	36	25	26	186	48	103	62	159	63	20	24
3	36	34	24	24	93	51	94	58	138	53	19	22
4	31	37	32	24	72	52	92	56	118	45	19	22
5	30	39	42	25	57	67	196	53	107	42	19	21
6	28	41	34	24	57	66	302	52	108	40	19	22
7	32	44	28	24	52	58	164	50	94	38	21	20
8	26	46	24	22	48	51	126	46	88	37	23	20
9	24	49	23	24	50	45	117	46	82	35	24	19
10	21	47	21	28	80	43	150	48	78	35	22	20
11	21	48	30	28	68	41	142	447	329	32	21	20
12	22	55	29	28	52	41	114	146	259	32	20	19
13	22	46	26	20	49	44	101	942	243	31	20	21
14	20	41	26	16	48	48	93	3330	199	29	21	26
15	33	46	27	22	47	48	85	1980	96	30	20	22
16	34	40	22	23	45	48	82	467	80	29	19	19
17	24	40	22	26	47	48	82	294	71	30	19	18
18	18	38	23	26	44	56	80	326	67	31	19	19
19	31	38	25	21	42	48	77	273	64	30	19	18
20	19	39	23	29	68	44	77	205	60	28	178	25
21	18	43	21	28	99	48	75	723	54	27	426	48
22	15	39	22	28	210	46	82	550	52	26	142	22
23	15	38	21	31	101	45	274	389	48	25	81	20
24	16	37	25	34	68	44	205	267	47	24	55	19
25	17	36	40	30	58	47	105	189	49	23	46	21
26	15	40	31	32	56	493	85	156	63	23	40	20
27	16	68	29	65	51	478	75	140	81	21	36	19
28	24	73	58	41	48	197	68	131	139	20	33	20
29	26	52	45	36	---	143	63	123	465	20	30	18
30	21	38	35	33	---	123	63	134	150	23	28	18
31	27	---	32	34	---	110	---	393	---	21	26	---
TOTAL	744	1300	892	879	2095	2768	3477	12140	3805	1029	1506	646
MEAN	24.0	43.3	28.8	28.4	74.8	89.3	116	392	127	33.2	48.6	21.5
MAX	36	73	58	65	210	493	302	3330	465	86	426	48
MIN	15	32	21	16	42	41	63	46	47	20	19	18
AC-FT	1480	2580	1770	1740	4160	5490	6900	24080	7550	2040	2990	1280

WTR YR 1983 TOTAL 31281 MEAN 85.7 MAX 3330 MIN 15 AC-FT 62050 concentrations of various parameters and locate areas of ground-water contamination as well.

The interrelationship between surface and ground water is of great importance in both regional and local hydrologic situations, and a wide variety of information can be obtained by analyzing stream flow data. Evaluation of the ground-water component of streamflow can provide important and useful information regarding regional recharge rates, aquifer characteristics, ground-water quality, and indicate areas of high potential yield to wells. To determine the ground-water component of runoff a stream hydrograph must first be separated into its component parts. There are many ways in which this can be accomplished, although all are quite subjective. Several methods will be briefly described.

4.4.1 The Regional System

In order to better appreciate the origin and significance of ground-water runoff and its quality, one should briefly examine the regional ground-water flow system. In humid and semiarid regions, in particular, the water table generally conforms to the surface topography. The hydraulic gradient or water table slopes away from divides and topographically high areas toward adjacent low areas, such as streams and rivers. The high areas serve as ground-water recharge areas, while the low places are ground-water discharge zones (Figure 4-9).

As water infiltrates in a recharge area, the mineral content is relatively low. The quality changes, however, along the flow path and dissolved solids as well as other constituents increase with increasing distances traveled in the ground. The water eventually flows into a stream or body of surface water and, due to the different lengths of flow paths and rock solubility, even streams and small lakes in close proximity may have large differences in both flow and quality.

4.4.2 Bank Storage

As a flood wave passes a particular stream cross section, the water table may rise in the adjacent stream side deposits. This occurs because the elevation of the water level in the stream, or the "stage," rises quickly and soon becomes higher than the water table. This blocks the ground water that would normally flow into the stream and causes the water table to rise in the flood plain. In addition, because of the higher stage, water will flow from the stream into the ground. Once the stage begins to fall, the water, which was recently added to the ground

Figure 4-9 The chemical quality of ground water commonly changes along a flow path in the regional system as water flows from areas of recharge to areas of discharge.

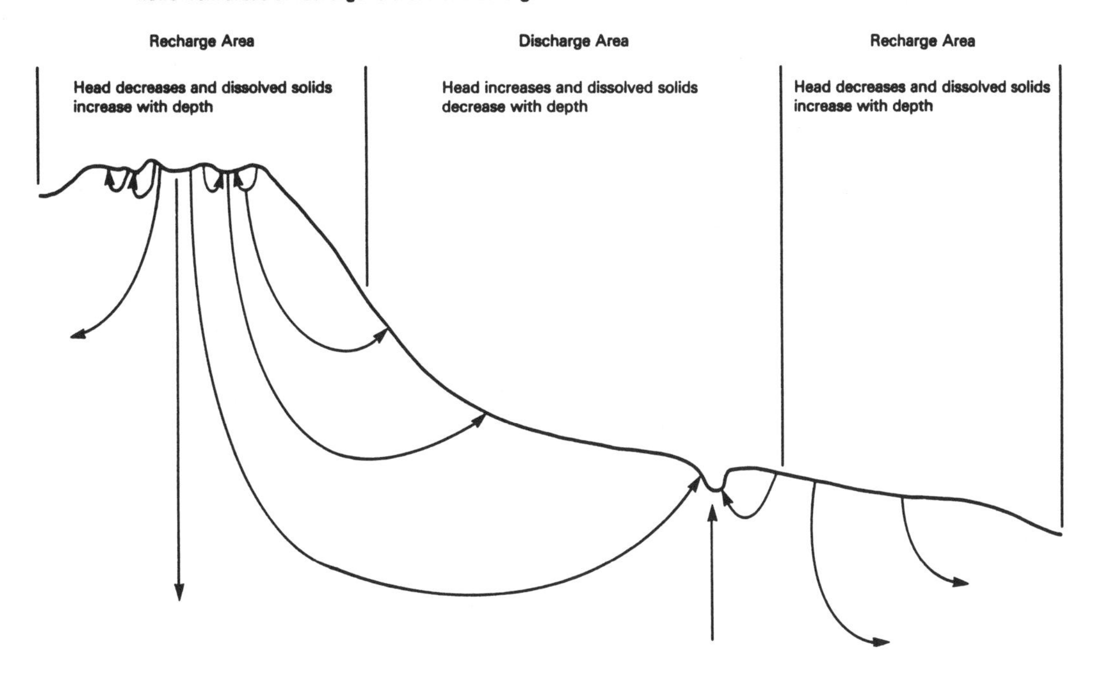

water, will begin to flow back into the stream, rapidly at first and then more slowly as the water-table gradient declines. This temporary storage of water in the near vicinity of the stream channel is called bank storage (Figure 4-10).

As the drainage from bank storage progresses, the recession segment of the hydrograph gradually tapers off into what is called a depletion curve, the shape of which is controlled by the permeability of the stream side deposits. A master depletion curve is used to separate a stream hydrograph.

4.4.3 Master Depletion Curve

Intervals between surface runoff events are generally short and, therefore, depletion curves must be constructed from a combination of several arcs of the hydrograph with the arcs overlapping in their lower parts (Figure 4-11). To plot a depletion curve, tracing paper is placed over a hydrograph of daily flows and, using the same horizontal and vertical scales, the lower arcs are traced, working backward in time from the lowest discharge to a period of surface runoff. The tracing paper is moved horizontally until the arc of another runoff event coincides in its lower part with the arc already traced. The process is continued until all the available arcs are plotted on top of one another. The upward curving parts of individual arcs are disregarded. The resulting continuous arc is a mean or normal depletion curve that presumably represents the hydrograph that would result from the ground-water runoff alone during a prolonged dry period.

4.4.4 Separating a Hydrograph by Graphical Methods

A hydrograph can be separated in the following ways. A depletion curve is positioned on the lower part of the recession limb of a runoff hydrograph, as shown in Figure 4-12. Notice that it departs from the actual recession curve at point D, which should reflect the end of surface runoff. The master curve is extended backward to its intersection at C with a vertical line drawn through the peak. A second line originating at A, which is the start of surface runoff, is drawn to C. The area or discharge below the line ACD is ground-water runoff.

It may be difficult to locate D with the depletion curve and a second method is to estimate its position with the equation:

$$N = A^2 \qquad (4\text{-}2)$$

where:

N = number of days after a peak when surface runoff ceases

A = drainage basin area, in square miles.

The distance N can be measured directly on the hydrograph.

Another method for separating a hydrograph consists of extending a line from point A, the start of surface runoff, to point D (Figure 4-12). A third method consists of extending the presurface runoff depletion trend to a point directly under the hydrograph peak, B, and then from B to D. This reflects a stream that is influenced by bank storage.

4.4.5 Separating a Hydrograph by Chemical Methods

Hydrographs also can be separated by chemical means. During baseflow the natural quality of a stream is at or near its maximum concentration of dissolved solids but, as surface runoff reaches the channel and provides an increasing percentage of the flow, the mineral concentration decreases. After the peak, ground-water runoff increases, surface runoff decreases, and the mineral content increases.

Several investigators (including Toler, 1965; Kunkle, 1965; Pinder and Jones, 1969; Visocky, 1970; and LaSala, 1968) have used the relation between runoff and water quality to calculate ground-water runoff from one or more aquifers or to measure streamflow. This method is based on the concentration of a selected chemical parameter that is characteristic of ground-water and surface runoff. The basic equation, which can take several forms, is as follows:

$$Qg = Q\,(C - Cs)/(Cg - Cs)$$

where:

Qg = quantity of ground-water runoff

C = concentration of the specific chemical parameter on conductance of runoff

Q = runoff

Cs = concentration of the specific chemical parameter or conductance of surface runoff

Qs = surface runoff

Cg = concentration of the specific chemical parameter or conductance of ground water.

Specific conductance is most often used because of the ease in obtaining it. Cg is measured in a well or series of wells and it should be about the same as C in a stream during baseflow. Cs is measured from a sample collected from the surface of the ground before the water reaches the stream. It is assumed that Cs and Cg are constant. Q and C are measured directly in the stream.

Toler (1965) used this method during baseflow to determine the quantity of water discharging from a surficial sand aquifer (Q_1) and an underlying artesian limestone aquifer (Q_2) in Florida. In this case, as shown in Figure 4-13, the dissolved solids in water from the limestone (C_2) averaged 50 mg/l, while that from the sand (C_1) averaged 10. When the stream had a discharge of 18 cfs (Q) and corresponding

Figure 4-10 Movement of water into and out of bank storage along a stream in Indiana. (from Daniels *et al*, 1970),

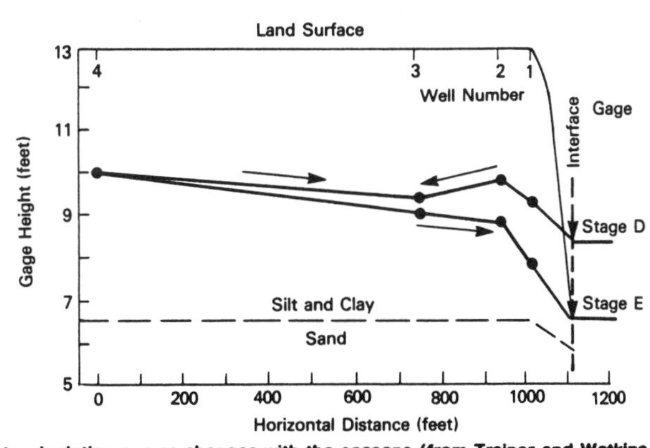

Figure 4-11 The shape of ground-water depletion curves changes with the seasons (from Trainer and Watkins, 1975).

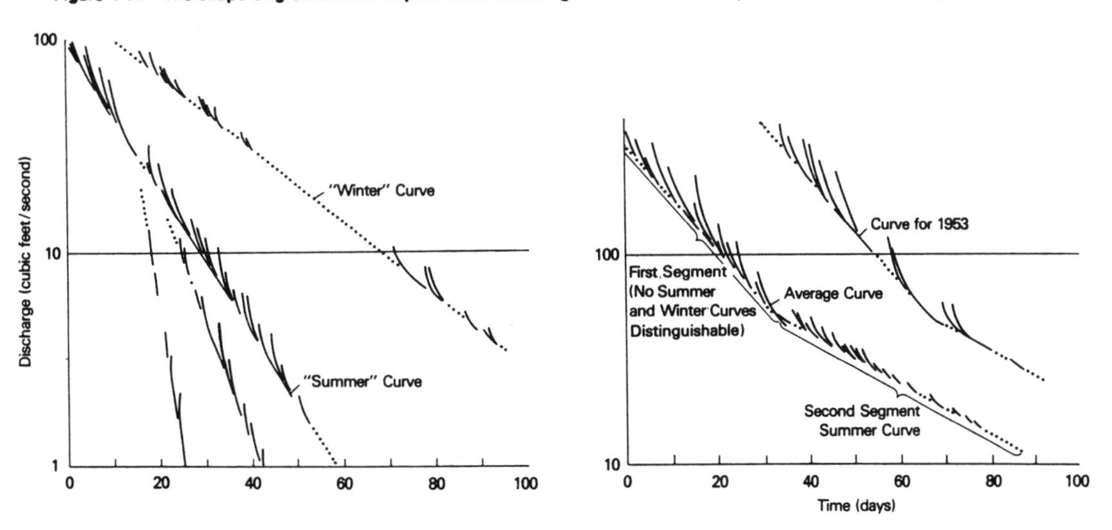

Figure 4-12 **A stream hydrograph can be separated by three different methods.**

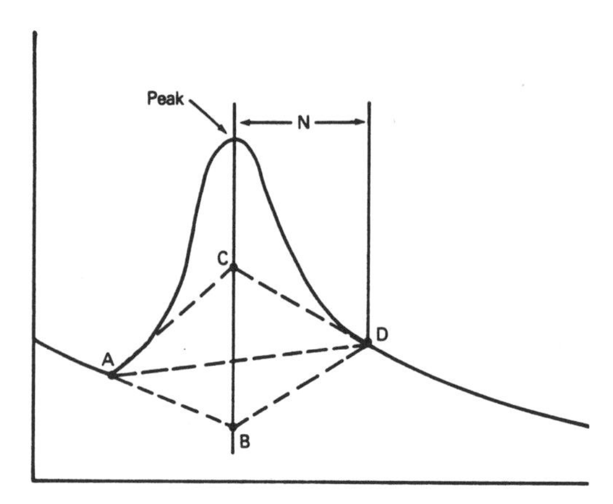

Figure 4-13 **Schematic showing the contribution of water from different aquifers to Econfina Creek, Florida.**

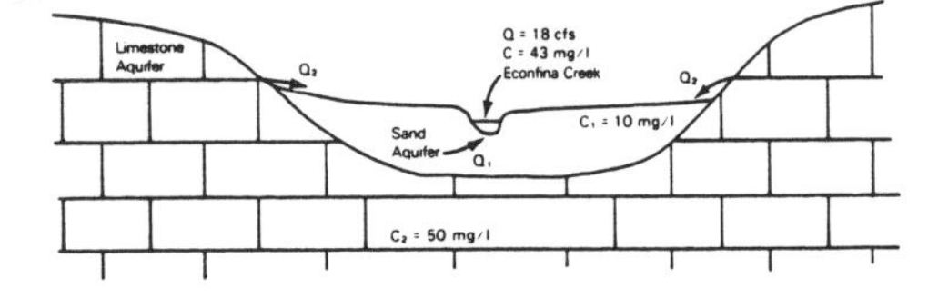

dissolved solids of 43 mg/l (C), ground water from the limestone was discharging through a series of springs at a rate of about 14.85 cfs:

$$Q_2 = Q\ (C - C_1)/(C_2 - C_1)$$

$$= 18\ (43 - 10)/(50 - 10) = 14.85 \text{ cfs}$$

Kunkle (1965) used specific conductance measurements to separate a runoff event hydrograph of Iowa's Four Mile Creek. In this case, continuous recordings of discharge and conductance were available. Specific conductance of the ground water and the stream at low flow averaged 520 micromhos (C_1), while surface runoff averaged 160 (C_2). Instantaneous ground-water runoff during the event was calculated for several points under the hydrograph (Figure 4-14). For example, when the discharge and conductivity of Four Mile Creek was 2.3 cfs (Q) and 410 micromhos (C), respectively, ground-water runoff (Q_1) was 1.6 cfs:

$$Q_1 = Q\ (C - C_2)/(C_1 - C_2)$$

$$= 2.3\ (410 - 160)/(520 - 160) = 1.6 \text{ cfs}$$

4.4.6 Ground-Water Rating Curves

A ground-water rating curve shows the relation between the water table and streamflow. Water levels are measured in one or more wells that are not influenced by pumping. At the same time, the discharge is determined during periods of baseflow. Selected water level and discharge measurements are then plotted on a graph and a smooth curve is drawn through the points as illustrated in Figure 4-15. The rating curve shows what the discharge should be relative to some particular ground-water level; the difference, if any, is surface runoff. For example, in Figure 4-15 when the ground-water level is 46.4 feet, baseflow should be 20 cfs. If the stream discharge happened to be 35 cfs, for example, then the difference, 15 cfs, would have been caused by surface runoff.

Olmsted and Hely (1962) used a ground-water rating curve to evaluate the water-bearing properties of folded igneous and metamorphic rocks in the Piedmont Upland of the Delaware River Basin. Here the average depth of water in all of the observation wells averaged about 17.5 feet and the annual fluctuation was 5.75 feet; precipitation averages about 44 inches per year. Hydrographs of runoff and ground-water runoff for Brandywine Creek are shown in Figure 4-16. The study found that ground-water runoff accounted for 67 percent of the total flow over a 6-year period. This compares favorably with the 64 percent determined for North Branch Rancocas Creek in the coastal plain of New Jersey; 74 percent for Beaverdam Creek in the coastal plain of Maryland (Rasmussen and Andreasen, 1959); 42 percent for Perkiomen Creek, a flashy stream in the Triassic Lowland of Pennsylvania; and 44 percent for the Pomperaug River Basin, a

Figure 4-14 Hydrographs showing the discharge, specific conductance, and computed ground-water runoff in Four Mile Creek, Iowa (from Kunkle, 1965).

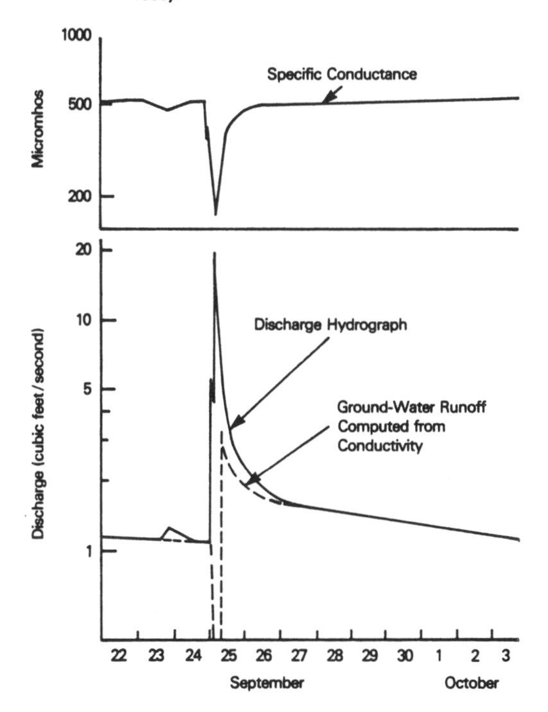

small stream in Connecticut (Meinzer and Stearns, 1929).

A single rating curve cannot be used with much accuracy during certain times of the year when the water table lies at a shallow depth because of significant losses of ground water to evapotranspiration. In their study of Panther Creek in Illinois, Schicht and Walton (1961) developed two rating curves, one for use when evapotranspiration is high, the other when it is low (Figure 4-17). Double rating curves also can be used to estimate evapotranspiration losses. For example, in Figure 4-17 a ground-water level stage for stream of 6 feet below land surface would indicate about 24 cfs of ground-water runoff when evapotranspiration is high and about 48 cfs when it is low. Therefore, streamflow is depleted by 24 cfs during periods of high evapotranspiration; this can be converted to losses per square mile of drainage basin above the gage.·

Various methods of hydrograph separation are available, all of which are laborious, time consuming, quite subjective, and open to questions of accuracy and interpretation. In each case a technique is used to provide a number of points on a hydrograph through which a line can be drawn to separate ground-water runoff from surface runoff. Once this line is drawn, one must then determine, directly on the hydrograph, the daily value of each of the separated components and then sum the results.

4.4.7 Determining Regional Ground-Water Recharge Rates

Annual ground-water runoff divided by total discharge provides the percentage of stream flow that consists of ground water. Effective ground-water recharge is that quantity of precipitation that infiltrates, is not removed by evapotranspiration, and eventually discharges into a stream. It is equivalent to ground-water runoff.

Effective ground-water recharge rates can be easily estimated with a computer program (Pettyjohn and Henning, 1978). This program separates a hydrograph by three different methods, provides monthly recharge rates, an annual rate, and produces a flow-duration curve. The results compare favorably with those obtained by other means. The data base is obtained from annual U.S. Geological Survey streamflow records.

4.4.8 Seepage Measurements

Seepage or dry-weather measurements consist of discharge determinations made at several locations along a stream during a short time interval when runoff is comprised entirely of baseflow. Rather than actually measuring the discharge, published records of a single day can be used by merely plotting on a map the daily mean flow of all the gages in the basin. Measurements such as these permit an evaluation of the basin geology, permeability, the amount of ground-water contribution, and the major areas of discharge. In addition, if chemical quality data are collected in the same manner, they can be used to determine background concentrations of various parameters and locate areas of ground-water contamination as well.

The flow of some streams increases substantially within short distances. Under natural conditions, this increase probably indicates the presence of deposits or zones of high permeability adjacent to the stream channel. These zones may consist of deposits of sand and gravel, fractures or faults, solution openings in limestone, or merely local changes in grain size and increased permeability. In gaining stretches,

Figure 4-15 Rating curve of mean ground-water level compared with base flow of Beaverdam Creek, Maryland (from Rasmussen and Andreason, 1959).

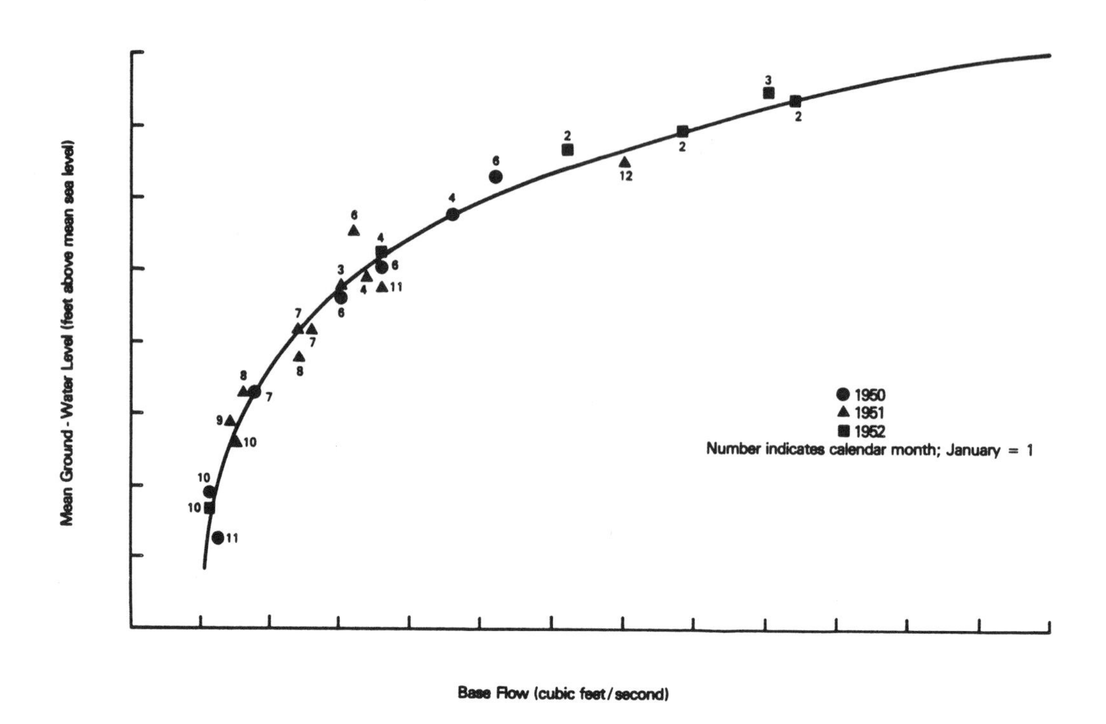

Figure 4-16 Hydrograph of Brandywine Creek, Chadd's Ford, Pennsylvania, 1952-1953.

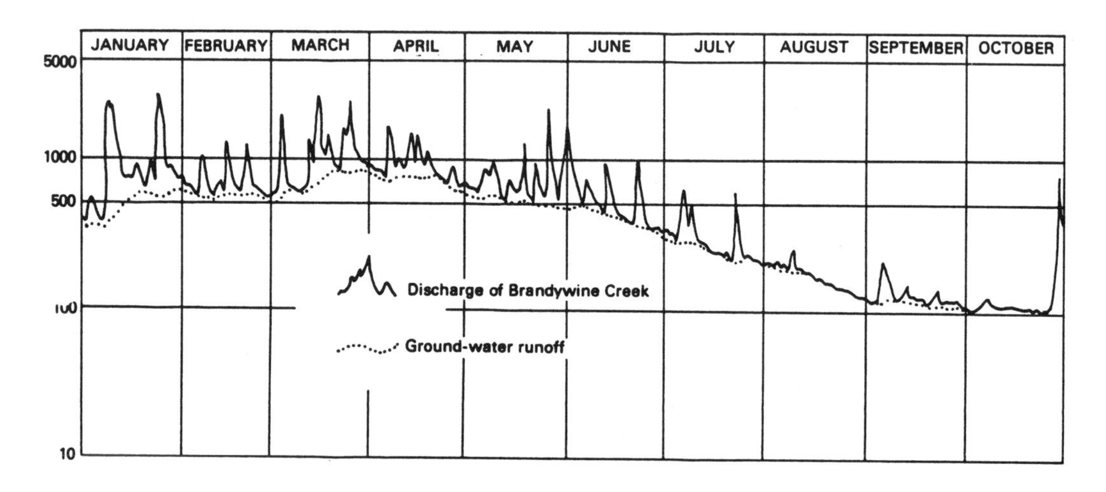

Figure 4-17 Rating curve of mean ground-water level and base flow in the Panther Creek basin, Illinois (from Schicht and Walton, 1961).

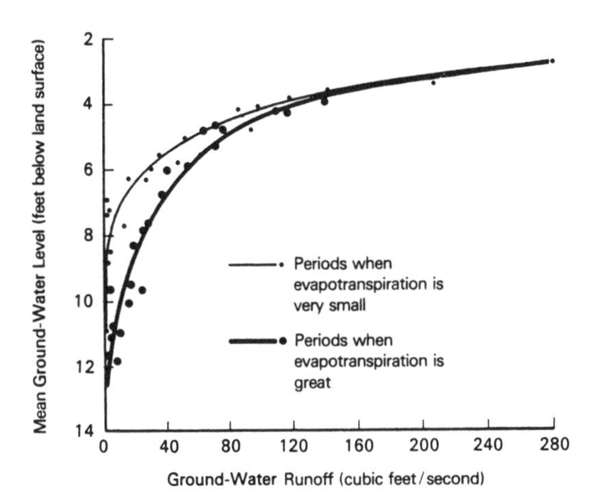

ground water may discharge through a series of springs or seeps along valley walls or the stream channel, or it may seep upward directly into the channel. During certain periods, particularly springtime and after heavy rainfalls, ground water may discharge with such a high velocity that the sand grains are partly suspended and lose all strength, that is, they become quicksand.

A number of discharge measurements were made in the Scioto River Basin, which lies in a glaciated part of central Ohio. The flow measurements themselves are important in that they show the actual discharge, in this case at about the 90 percent flow, which is the discharge that is equaled or exceeded 90 percent of the time. In this case the discharge was reported in units of millions of gallons per day (mgd) since it was a municipal water-supply study, rather than the usual cfs. As Figure 4-18 shows, the discharge at succeeding downstream sites on the Scioto River, as well as its tributaries, is greater than that immediately upstream. This shows that the river is gaining and that water is being added to it by ground-water runoff from the adjacent deposits.

4.4.9 Maps of Potential Ground-Water Yield

A particularly useful method for evaluating the hydrogeology of a basin consists of relating the discharge to the size of the drainage basin (cfs or mgd per square mile of basin). One can use this method to examine Figure 4-18 and then relate the flow index (cfs or mgd/mi^2) to the geology and hydrology of the area. A cursory examination of the data shows that the flow indices can be conveniently separated into three distinct but arbitrary groups: where the flow index is between 0.01 and 0.020 mgd/mi^2; between 0.021 and 0.035 mgd/mi^2; and greater than 0.036 mgd/mi^2. Notice that even though several watercourses fall into the larger flow index group, the actual discharge ranges from 3.07 to 1.81 mgd.

Logs of wells drilled along the streams with a flow index in the first group show a preponderance of fine grained material that contains only a few layers of sand and gravel; these wells generally yield less than 10 gpm (gallons per minute). For the stream segments in the second group, however, logs of wells and test holes indicate that several feet of sand and gravel underlie fine-grained alluvial material, the latter of which ranges from 5 to about 25 feet in thickness. Adequately designed and constructed wells that tap the outwash deposits produce as much as 500 gpm. Glacial outwash, much of it very coarse-grained, forms an extensive, very permeable deposit through which flow the streams and river of the third group. The outwash extends from the surface to depths that in places exceed 200 feet. Here ground-water recharge rates range from 200,000 gpd/mi^2 during dry months to more than 500,000 in spring. Industrial wells tapping these deposits can produce more than 1,000 gpm.

The above example shows that by combining dry-weather discharge data and well yields with a map showing the areal extent of the deposits that are characteristic of the stream valleys, a map can be developed that indicates the potential yield of the area. The map of potential ground-water yield relies heavily on streamflow measurements and some geologic data, but it provides a good first cut approximation of ground-water availability.

4.4.10 Quality as an Indicator

Stream reaches typified by significant increases in ground-water runoff may also have unusual quality. In northern Ohio the discharge of a small stream that drains into Lake Erie increases over a 3-mile stretch from less than 1 to more than 28 cfs and remains relatively constant afterward. The increase begins at an area of springs where limestone, which has an abundance of solution openings, crops out in and near the stream. The glacial till-limestone surface dips downstream, eventually exceeding 90 feet in depth.

In the upper reaches of the stream, baseflow is provided by ground-water runoff from the adjacent thin covering of till, which has a low permeability. Because this water has been in the ground but a short time, the mineral content is low, as indicated by the specific conductance of 583 to 638 microohms (Figure 4-19). Where streamflow begins to significantly increase, the limestone aquifer provides the largest increment. Moreover, the bedrock water contains excessive concentrations of dissolved solids,

Figure 4-18 Discharge and low flow indices of the Scioto River in central Ohio are strongly influenced by local geologic conditions. These data allow the development of a potential ground-water yield map (from Pettyjohn and Henning, 1979).

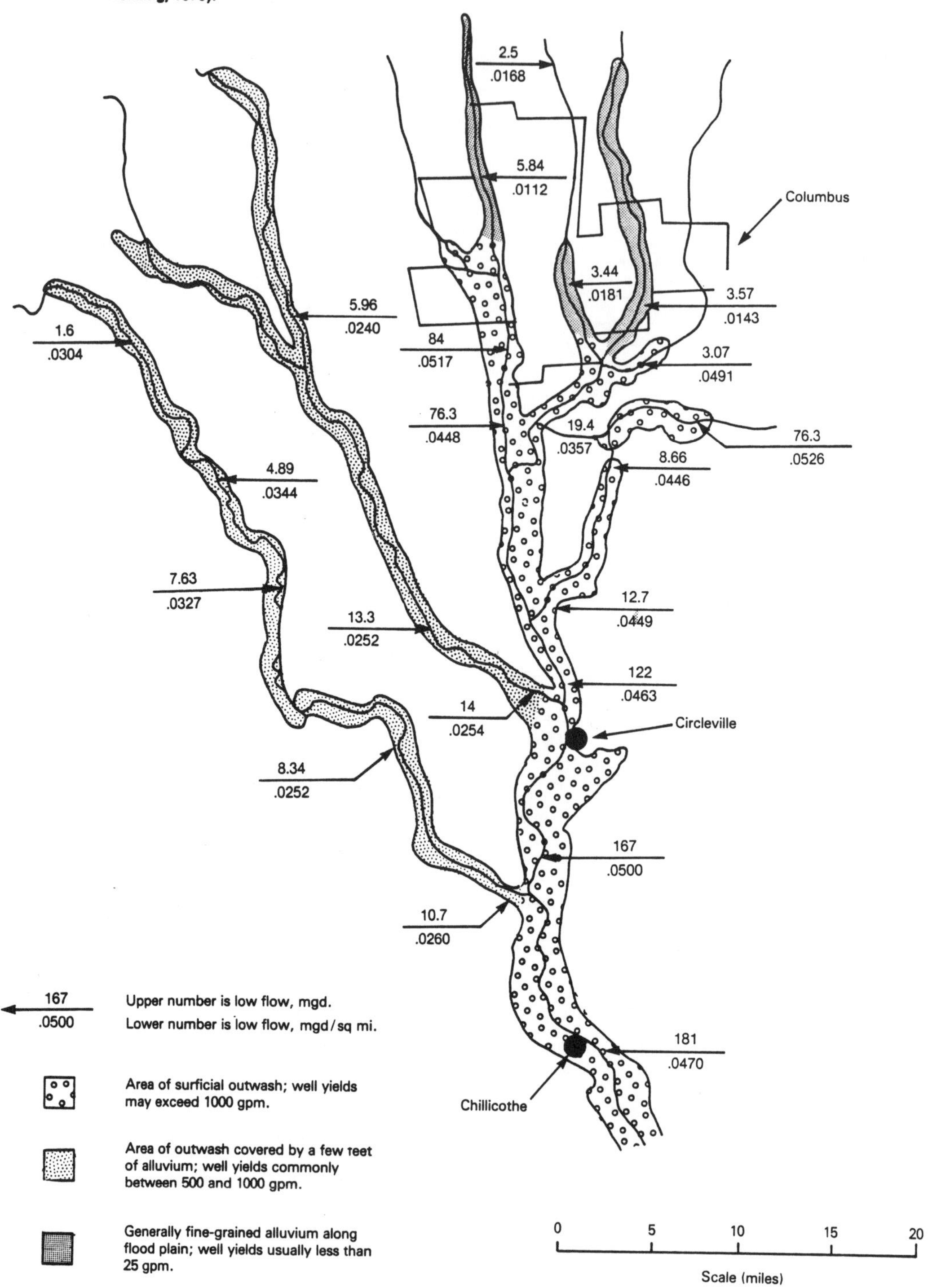

Figure 4-19 Fish populations are controlled by discharge of mineralized water from an underlying carbonate aquifer in Green Creek, in northeastern Ohio.

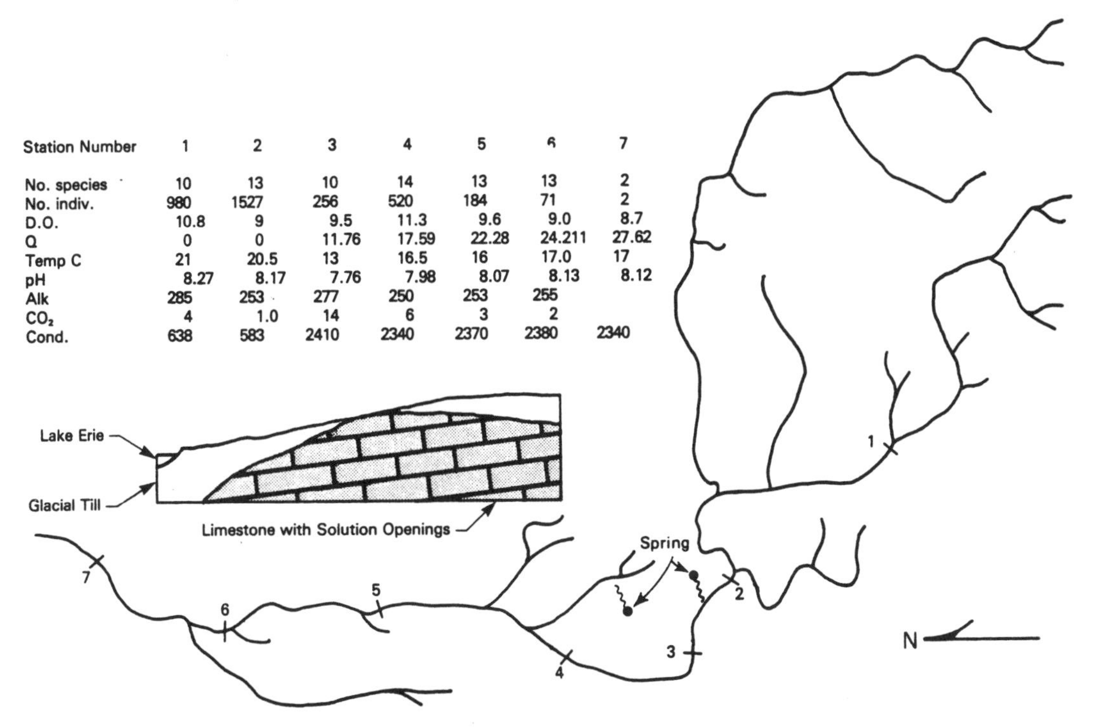

Station Number	1	2	3	4	5	6	7
No. species	10	13	10	14	13	13	2
No. indiv.	980	1527	256	520	184	71	2
D.O.	10.8	9	9.5	11.3	9.6	9.0	8.7
Q	0	0	11.76	17.59	22.28	24.211	27.62
Temp C	21	20.5	13	16.5	16	17.0	17
pH	8.27	8.17	7.76	7.98	8.07	8.13	8.12
Alk	285	253	277	250	253	255	
CO_2	4	1.0	14	6	3	2	
Cond.	638	583	2410	2340	2370	2380	2340

hardness, and sulfate and in this stretch calcite precipitates on rocks in the stream channel. In the upper reaches of this stream, the fish population is exceedingly abundant, but in the vicinity of the springs it diminishes quickly and remains in a reduced state throughout the remaining length. No doubt the reduction in fish population is directly related to the natural quality of the water that flows from the limestone.

A method to locate relatively small areas of ground-water contamination by means of stream quality was described by Pettyjohn (1975, 1985). In this case, the municipal water supply at the central Ohio city of Westerville periodically contained excessive concentrations of chloride, producing a salty taste. The water was obtained from Alum Creek, which was being contaminated by oil-field brines from scores of wells in the 189 square mile upstream part of the basin. The contamination was largely the result of leakage of brine from "evaporation pits" to the water table and, eventually, the contaminated ground water reached a water course.

In order to locate specific areas of contamination, water samples were collected from Alum Creek and many of its small tributaries during a single day in which the streamflow consisted entirely of ground-water runoff. The background concentration of chloride (less than 25 mg/l) was established on the basis of its concentration in uncontaminated small tributaries. Concentrations exceeding background were assumed to be the result of contamination (Figure 4-20).

The chloride concentrations were plotted on a base map showing the location of all oil and gas wells and tests, both operating and abandoned. All contaminated tributaries contained oil wells and "evaporation pits" within their subbasins, some of which were the source of the chloride. Next, the configuration of each small contaminated basin was delineated on a topographic map. The well location provided some control on the point source of contamination. It was then possible to estimate the general size of each contaminated site because it had to lie in the vicinity of a well within that small basin and the plume had to trend downgradient toward the stream (Figure 4-21).

The approach described above allows an investigator to minimize drilling costs for monitoring wells because uncontaminated areas are readily evident and the investigator can then key on selected sites. Once

Figure 4-20 Distribution of chloride and oil and gas wells in Alum Creek basin, Ohio.

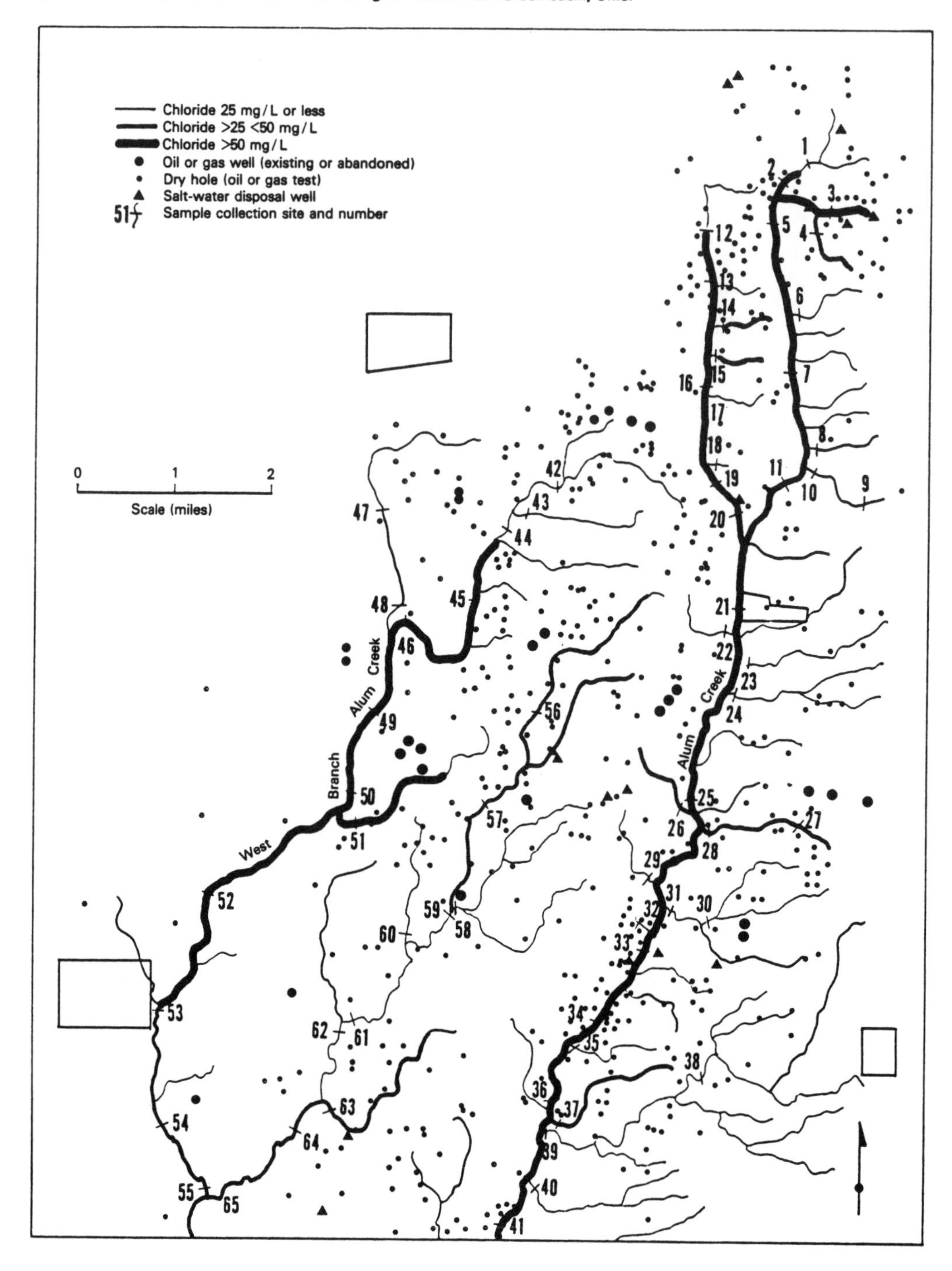

Figure 4-21 Areas of ground-water pollution in Alum Creek basin, Ohio..

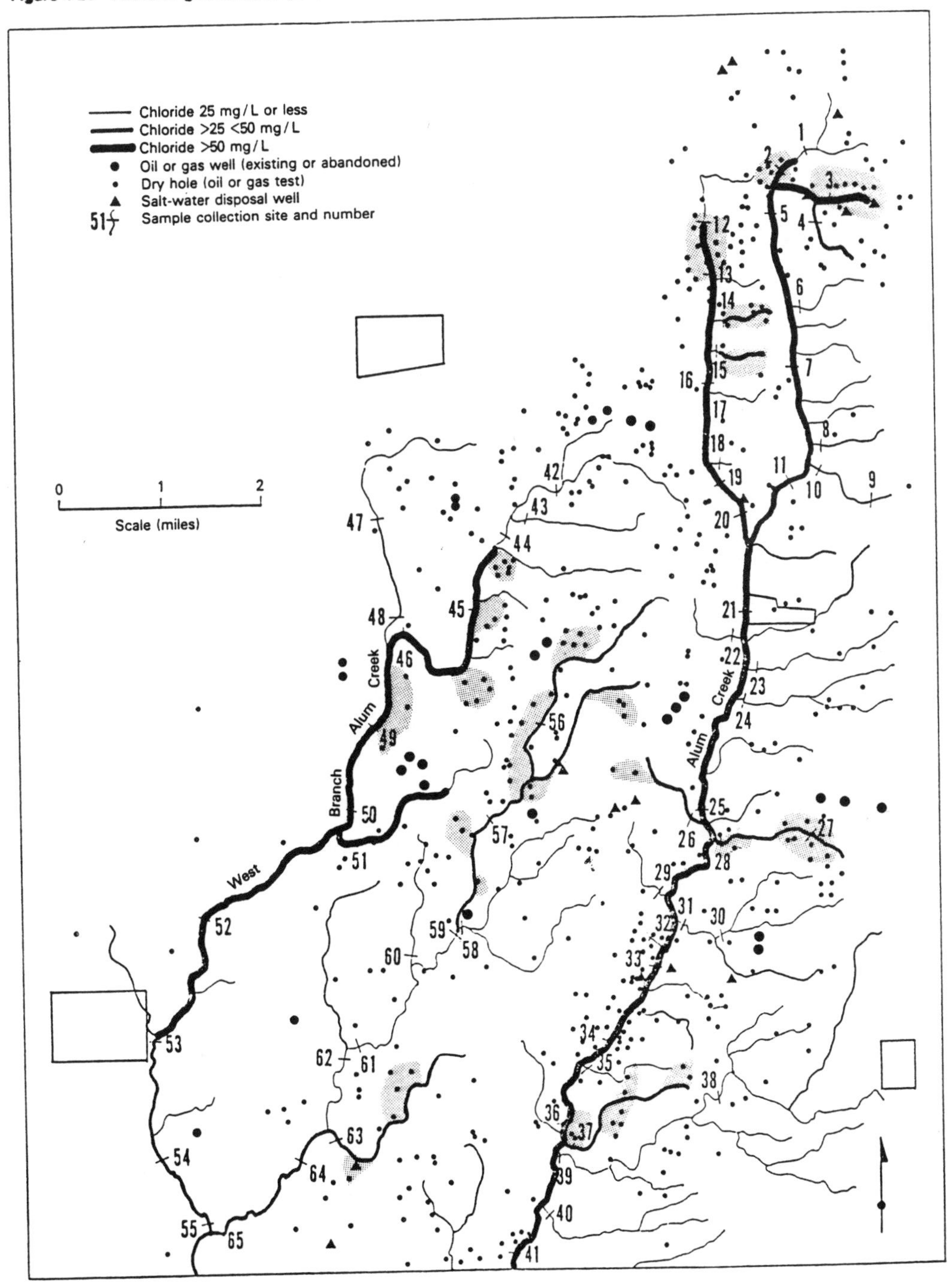

contaminated areas have been located, additional surface water samples can be collected from the small basins to permit a more detailed assessment.

4.4.11 Temperature as an Indicator

The temperature of shallow ground water is nearly uniform, reflecting the mean annual temperature of the region. It ranges from a low of about 37°F in the north-central part of the United States to more than 77°F in southern Florida (Figure 4-22). Surface water temperatures, however, range within wide extremes, freezing in the winter in northern regions and exceeding 100°F during hot summer days in the south. Mean monthly stream temperatures during July and August range from a low of 55°F in the northwest to more than 85°F in the southeast (Figure 4-23).

During the summer when ground water provides a significant increment of flow, the temperature of water in a stream's gaining reach will decline. Conversely, during winter the ground water will be warmer than that on the surface and, although ice will normally form, parts of a stream may remain open. In central Iowa, for example, winter air temperatures commonly drop below zero and ice quickly forms on streams, ponds, and lakes. Here ground-water temperatures are about 52°F and, if a sufficient amount is discharging into a surface water body, ice may not form. In summer the relatively cold ground water (52°F) mixes with the warm (more than 79°F) surface water to produce a mixture colder than that in nongaining reaches.

The point to be made here is that the evaluation of stream temperature provides clues to changes in permeability and perhaps even chemical quality.

4.4.12 Flow Duration Curves

A flow-duration curve shows the frequency of occurrence of various rates of flow. They are useful for regional evaluations of hydrogeologic conditions.

When used in conjunction with some of the other methods described above, the investigator can readily determine areas that are subject to ground-water contamination. That is, areas and zones that provide, relatively speaking, large amounts of ground-water runoff reflect permeable zones that are most sensitive to contamination.

The flow-duration curve is a cumulative frequency curve prepared by arranging all discharges of record in order of magnitude and subdividing them according to the percentages of time during which specific flows are equaled or exceeded. All chronologic order is lost (Cross and Hedges, 1959). Flow-duration curves may be plotted on either probability or semilog paper. In either case, the shape of the curve is an index of the natural storage in a basin, including ground water. Since dry- weather flow consists entirely of ground-water runoff, the lower end of the curve indicates the general hydrogeologic characteristics of shallow aquifers.

Several flow-duration curves for Ohio streams are shown in Figure 4-24. During low-flow conditions, the curves for several of the streams, such as the Mad, Hocking, and Scioto Rivers as well as Little Beaver Creek trend toward the horizontal, whereas Grand River, White Oak and Home Creeks all remain very steep. The former contain permeable deposits.

Mad River flows through a broad valley that is filled with very permeable sand and gravel and, as expected, the river maintains a high sustained flow. The Hocking River locally contains outwash in and along its flood plain, which provides a considerable amount of ground-water runoff. Above Columbus, the Scioto River flows across thin layers of limestone that crop out along the stream valley; the adjacent uplands are covered with glacial till. In this reach, ground-water runoff is relatively small. Immediately south of Columbus, however, the valley widens considerably and is filled with coarse, permeable outwash. Mad River has a higher low-flow index than the Scioto River at Chillicothe because the Mad receives ground-water runoff throughout its entire length, while the flow of the Scioto increases significantly only in the area of outwash.

White Oak and Home Creeks originate in bedrock areas where relatively thin alternating layers of sandstone, shale, and limestone crop out along the steep hillsides. The greater relief in these basins promotes surface runoff and the rocks are distinguished by moderately low permeability. As the flow-duration curves indicate, ground-water runoff from these basins is far less than those that contain outwash.

The above examples and techniques can be used mainly for regional hydrogeologic evaluations. Increases in dry-weather flow, excluding inflow from tributaries, are usually caused by an increase in permeability. This, in turn, implies the presence of an aquifer or zone that might serve as a major source of water supply and it therefore should be protected. Abrupt changes in a stream's chemical quality during dry-weather flow probably will indicate zones of permeability that are greater than the predominant strata. The change in quality should indicate the presence of discharge areas of contaminated ground-water runoff or the natural chemical quality of underlying aquifers.

The major purpose of stream hydrograph separation is to develop an estimate of the amount of ground-water runoff. If the percentage of ground-water runoff if large, such as 60 percent or more, then the rocks within the basin are permeable, infiltration and ground-water recharge are large, and the basin has a good potential for the development of ground-water sources of supply. Consequently, the basin or

Figure 4-22 Typical ground-water temperatures (°F) (from Johnson, 1966).

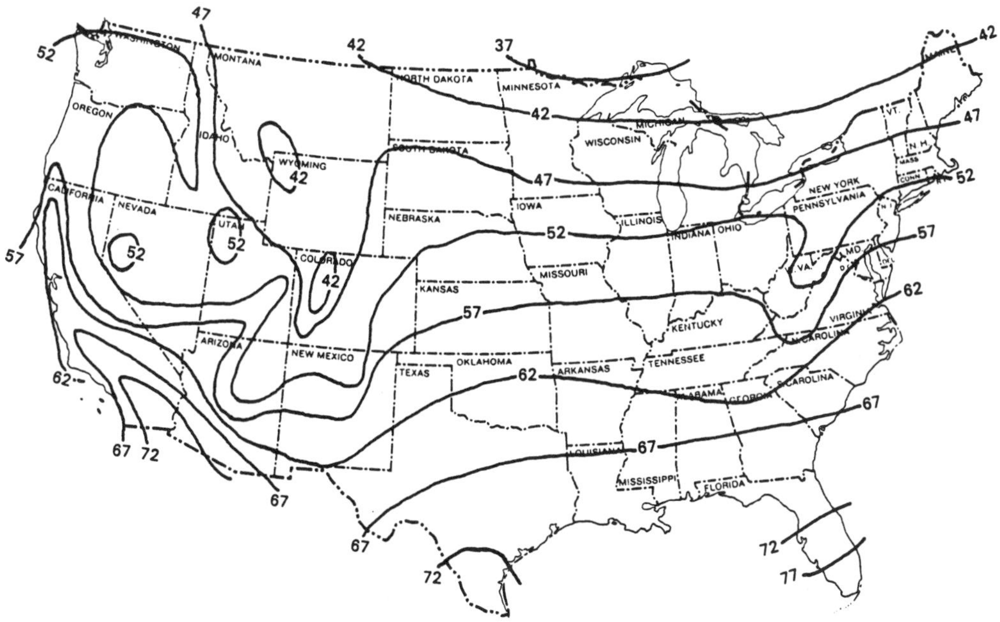

Figure 4-23 Summer stream temperatures (°F).

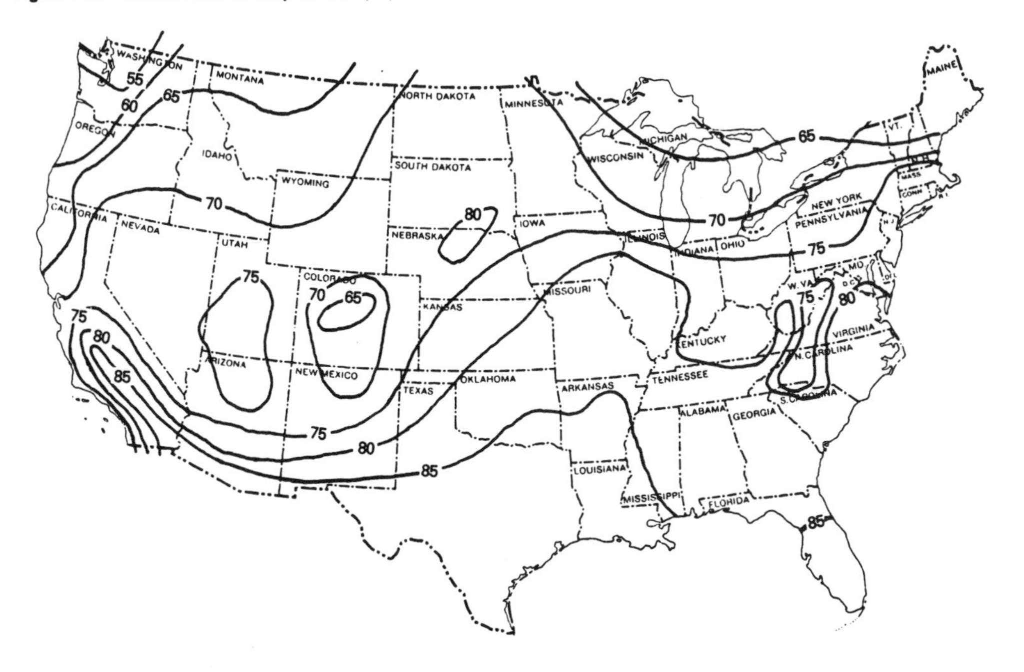

Figure 4-24 Flow-duration curves for selected Ohio streams (from Cross and Hedges, 1959).

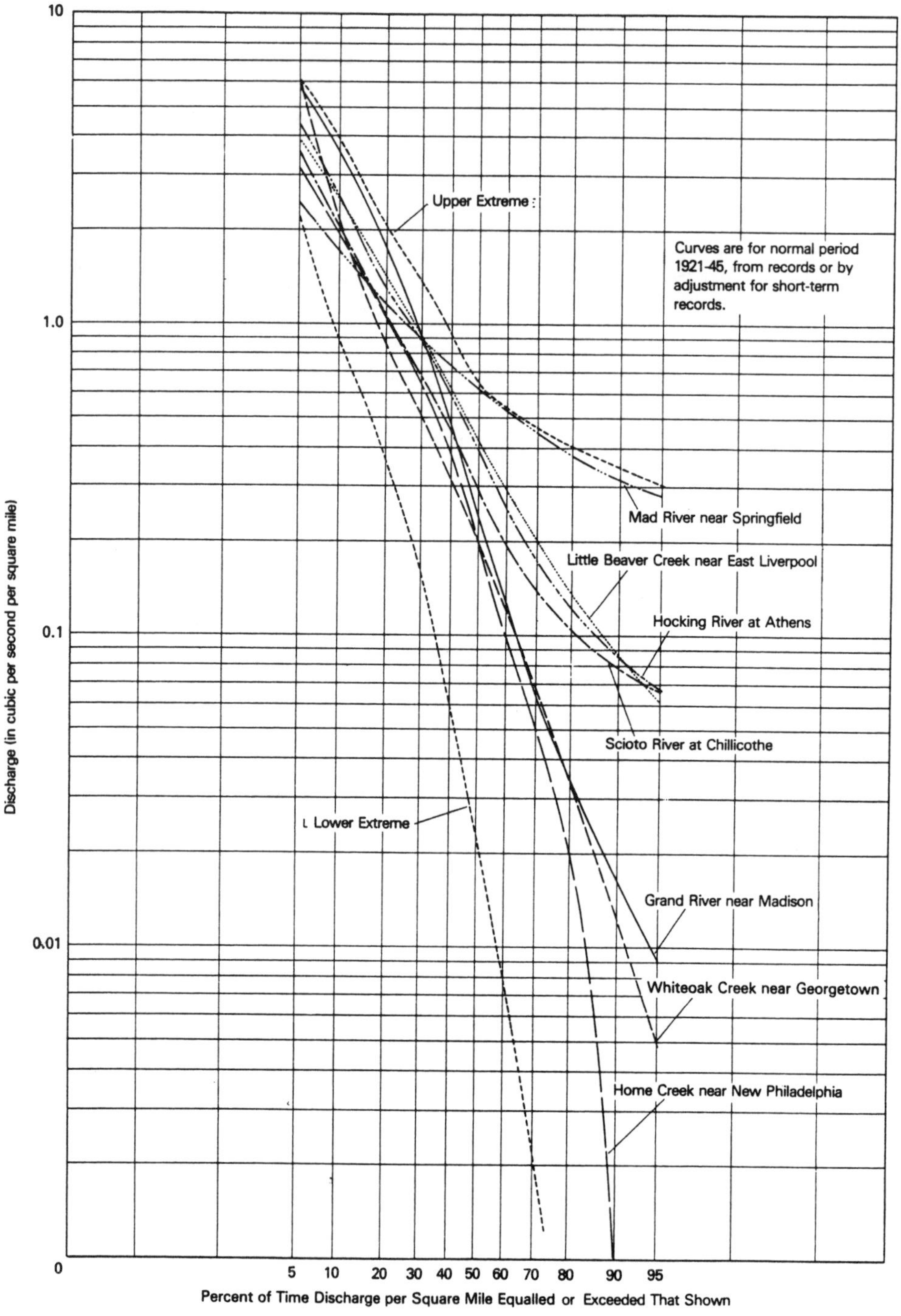

Discharge (in cubic per second per square mile)

10
1.0
0.1
0.01
0

Upper Extreme

Curves are for normal period 1921-45, from records or by adjustment for short-term records.

Mad River near Springfield
Little Beaver Creek near East Liverpool
Hocking River at Athens
Scioto River at Chillicothe
Lower Extreme
Grand River near Madison
Whiteoak Creek near Georgetown
Home Creek near New Philadelphia

5 10 20 30 40 50 60 70 80 90 95

Percent of Time Discharge per Square Mile Equalled or Exceeded That Shown

Figure 4-25 The water table generally conforms to the surface topography.

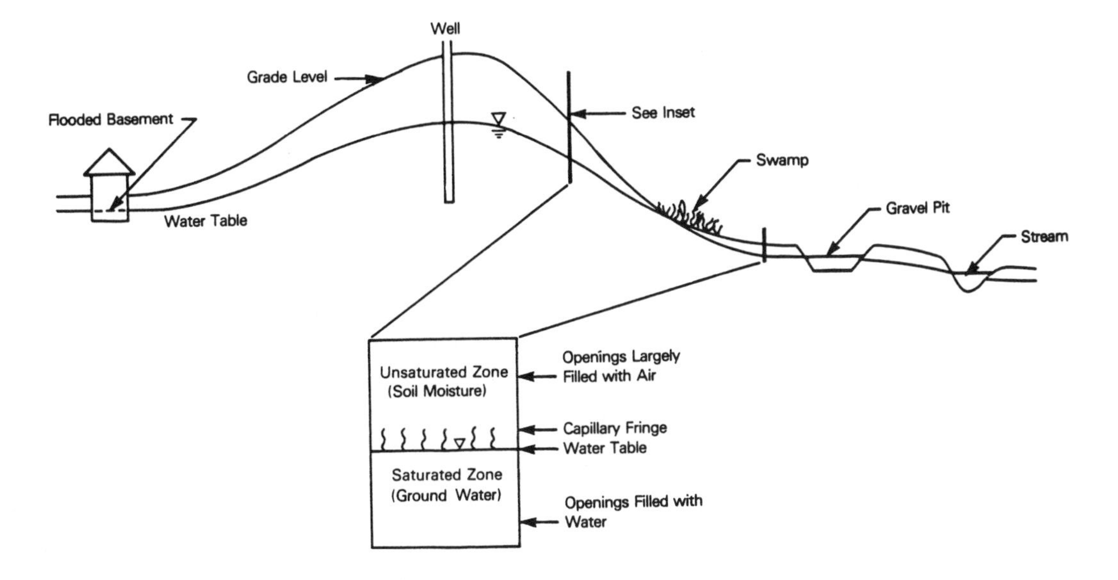

parts of its should be protected because it is readily subject to contamination.

4.5 Ground Water

The greatest difficulty in working with ground water is that it is hidden from view, cannot be adequately tested, and occurs in a complex environment. On the other hand, the general principles governing ground-water occurrence, movement, and quality are quite well known, which permits the investigator to develop a reasonable degree of confidence in his predictions. The experienced investigator is well aware, however, that these predictions are only an estimate of the way the system functions. Ground-water hydrology is not an exact science, but it is possible to develop a good understanding of a particular system if one pays attention to fundamental principles.

4.5.1 The Water Table

Water under the surface of the ground occurs in two zones, an upper unsaturated zone and the deeper saturated zone (Figure 4-25). The boundary between the two zones is the water table. In the unsaturated zone, most of the pore space is filled with air, but water occurs as soil moisture and in a capillary fringe that extends upward from the water table. The water in this zone is under a negative hydraulic pressure; that is, it is less than atmospheric. Ground water occurs below the water table and all of the pores are filled with fluid that is under pressure greater than atmospheric.

The water table conforms to the surface topography, but it lies at a greater depth under hills than it does under valleys (Figure 4-25). In general, the water table lies at depths ranging from 0 to about 20 feet or so in humid and semiarid regions, but its depth exceeds hundreds of feet in some desert environments.

The elevation of the water table must be determined with care, and many such measurements have been incorrectly taken. The position of the water table can be determined from the water level in swamps, flooded excavations (abandoned gravel pits, highway borrow pits, etc.), sumps in basements, lakes, ponds, streams, and shallow dug wells. In some cases there may be no water table at all or it may be seasonal. Measurement of the water level in drilled wells, particularly if they are of various depths, will more likely reflect the pressure head of one or more aquifers that are confined than the actual water table.

Accurately determining the position of the water table is important because the thickness, permeability, and composition of the unsaturated zone exert a major control on ground-water recharge and the movement of contaminants, particularly organic compounds, from land surface to an underlying aquifer. Attempting to determine the position of the water table by measuring the water level in drilled wells nearly always will indicate an unsaturated zone that is substantially thicker than it actually is and thus provides a false sense of security.

Ground water has different origins; however, all fresh ground water originated from precipitation that infiltrated. Magmatic or juvenile water is "new" water that has been released from molten igneous rocks. The steam that is so commonly given off during volcanic eruptions is probably not magmatic, but rather shallow ground water heated by the molten magma. Connate ground water is defined as that entrapped within sediments when they were deposited. Ground water, however, is dynamic and there is probably no connate water that meets this definition. Rather, the brines that underlie all or nearly all fresh ground water have changed substantially through time because of chemical reactions with the geologic framework.

4.5.2 Aquifers and Aquitards

In the subsurface, rocks serve either as confining units or as aquifers. A confining unit or aquitard is characterized by low permeability that does not readily permit water to pass through it despite the fact that it stores large quantities of water. Examples include shale, clay, and silt. An aquifer has sufficient permeability to permit water to flow through it with relative ease and, therefore, it will provide a usable quantity to a well or spring.

Water occurs in aquifers under two different conditions -- unconfined and confined (Figure 4-26). An unconfined or water-table aquifer has a free water surface that rises and falls in response to differences between recharge and discharge. A confined or artesian aquifer is overlain by an aquitard and the water is under sufficient pressure to rise above the base of the confining bed if it is perforated. In some cases, the water is under enough pressure to rise to some point above land surface. This is called a flowing artesian well. The water level in an unconfined aquifer is referred to as the water table; with confined aquifers the water level is called the potentiometric surface.

Water will arrive at some point in an aquifer through one of several means. The major source is direct infiltration of precipitation, which occurs nearly everywhere. Where the water table lies below a stream or canal, water will infiltrate. This source is important part of the year in some places and is a continuous source in others. Interaquifer leakage, or flow from one aquifer to another, is probably the most significant source in deeper, confined aquifers. Likewise, leakage from aquitards is very important where pumping from adjacent aquifers has lowered the head sufficiently for leakage to occur. Underflow, which is the normal movement of water through an aquifer, will also transport ground water to a specific point. Additionally, water can reach an aquifer through

Figure 4-26 Aquifer A is unconfined and aquifers B and C are confined, but water may leak through confining units to recharge adjacent water-

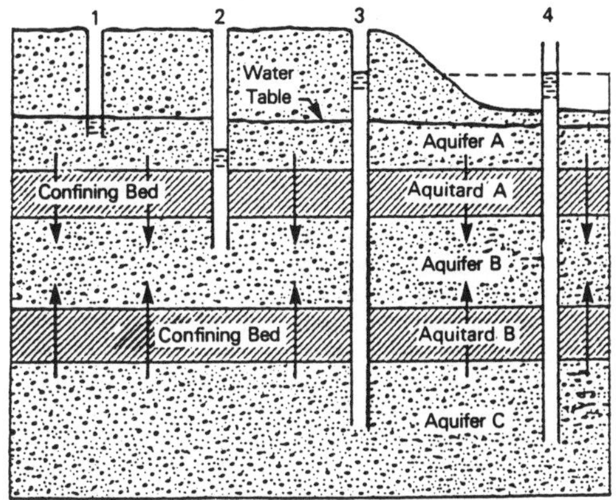

Aquifer A is unconfined and aquifers B and C are confined, but water may leak through confining units to recharge adjacent water-bearing zones.

artificial means, such as leakage through ponds, pits, and lagoons.

An aquifer serves two functions; one as a conduit through which flow occurs, and the other as a storage reservoir. This is accomplished by means of openings in the rock. The openings include those between individual grains and those present in joints, fractures, tunnels, and solution openings. There are also artificial openings, such as engineering works, abandoned wells, and mines. The openings are primary if they were formed at the time the rock was emplaced; they are secondary if they developed after lithification. Examples of the latter include fractures and solution openings.

4.5.3 Porosity and Hydraulic Conductivity

Porosity, expressed as a percentage or decimal fraction, is the ratio between the openings and the total rock volume. It defines the amount of water a saturated rock volume can contain. If a unit volume of saturated rock is allowed to drain by gravity, not all of the water it contains will be released. The volume drained is the specific yield, a percentage, and the volume retained is the specific retention. It is the specific yield that is available to wells. Therefore, porosity is equal to specific yield plus specific retention. Typical values for various rock types are listed in Table 4-1.

Permeability (P) is used in a qualitative sense, while hydraulic conductivity (K) is a quantitative term. They are often expressed in units of gpd/ft^2 (gallons per day per square foot) and refer to the ease with which water can pass through a rock unit. It is the hydraulic conductivity that allows an aquifer to serve as a conduit. Hydraulic conductivity values range widely from one rock type to another and even within the

Table 4-1 Selected Values of Porosity, Specific Yield, and Specific Retention.

Material	Porosity	Specific Yield (% by vol)	Specific Retention
Soil	55	40	15
Clay	50	2	48
Sand	25	22	3
Gravel	20	19	1
Limestone	20	18	2
Sandstone, semiconsolidated	11	6	5
Granite	0.1	0.09	0.01
Basalt, young	11	8	3

same rock. Those rocks or aquifers in which the hydraulic conductivity is nearly uniform are called homogeneous and those in which it is variable are heterogeneous or nonhomogeneous. Hydraulic conductivity can also vary horizontally, in which case the aquifer is anisotropic. If uniform in all directions, which is rare, it is isotropic. The fact that both unconsolidated and consolidated sedimentary strata are deposited in horizontal units is the reason that hydraulic conductivity is generally greater horizontally than vertically by at least an order of magnitude. Typical ranges in values of hydraulic conductivity for most common water-bearing rocks are shown in Table 1-3.

4.5.4 Hydraulic Gradient

The hydraulic gradient (I) is the slope of the water table or potentiometric surface and is the change in water level per unit of distance along the direction of maximum head decrease. It is determined by measuring the water level in several wells. The water level in a well, usually expressed as feet above sea level, is the total head (h_t), which consists of elevation head (z) and pressure head (h_p).

$$h_t = z + h_p \qquad (4\text{-}6)$$

The hydraulic gradient is the driving force that causes ground water to move in the direction of decreasing total head. It is generally expressed in consistent units such as feet per foot. For example, if the difference in water level in two wells 1,000 feet apart is 8 feet, the gradient is 8/1,000 or 0.008. The direction of ground-water movement and the hydraulic gradient can be determined by information from three wells (Figure 4-27).

4.5.5 Potentiometric Surface Map

A potentiometric surface or water-level map is a graphical representation of the gradient. One can be prepared by plotting water-level measurements on a base map and then drawing contours. The map should be drawn so that it actually reflects the hydrogeological conditions. An example is shown in Figure 4-28.

Figure 4-27 The generalized direction of ground-water movement can be determined by means of the water level in three wells of similar depth (from Heath and Trainer, 1981).

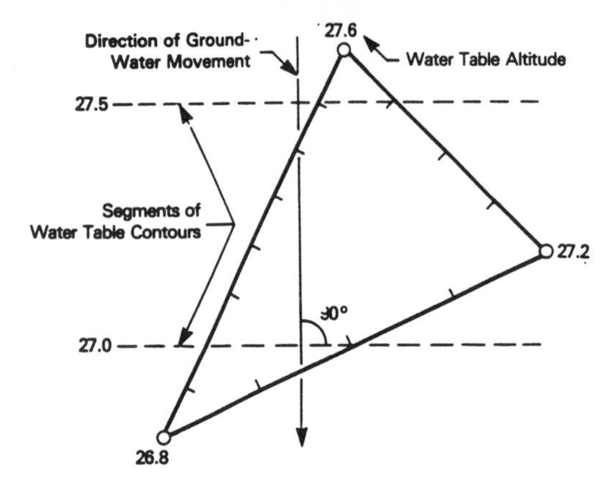

The contours are called equipotential lines, indicating that the water has the potential to rise to that elevation. In the case of a confined aquifer, however, the water may have the potential to rise to a certain elevation, but it cannot actually do so until the confining unit is perforated by a well. Potentiometric surface maps are an essential part of any ground-water investigation because they indicate the direction in which ground water is moving and provide an estimate of the gradient, which controls velocity.

A potentiometric surface map can be developed into a flow net by constructing flow lines that intersect the equipotential lines or contour lines at right angles. Flow lines are imaginary paths that would be followed by particles of water as they flow through the aquifer. Although there are an infinite number of both equipotential and flow lines, the former are constructed with uniform differences in elevation between them and the latter so that they form, in combination with equipotential lines, a series of squares. A carefully prepared flow net in conjunction with Darcy's Law (discussed below) can be used to estimate the quantity of water flowing through an area.

Figure 4-28 A potentiometric surface map representing the hydraulic gradient.

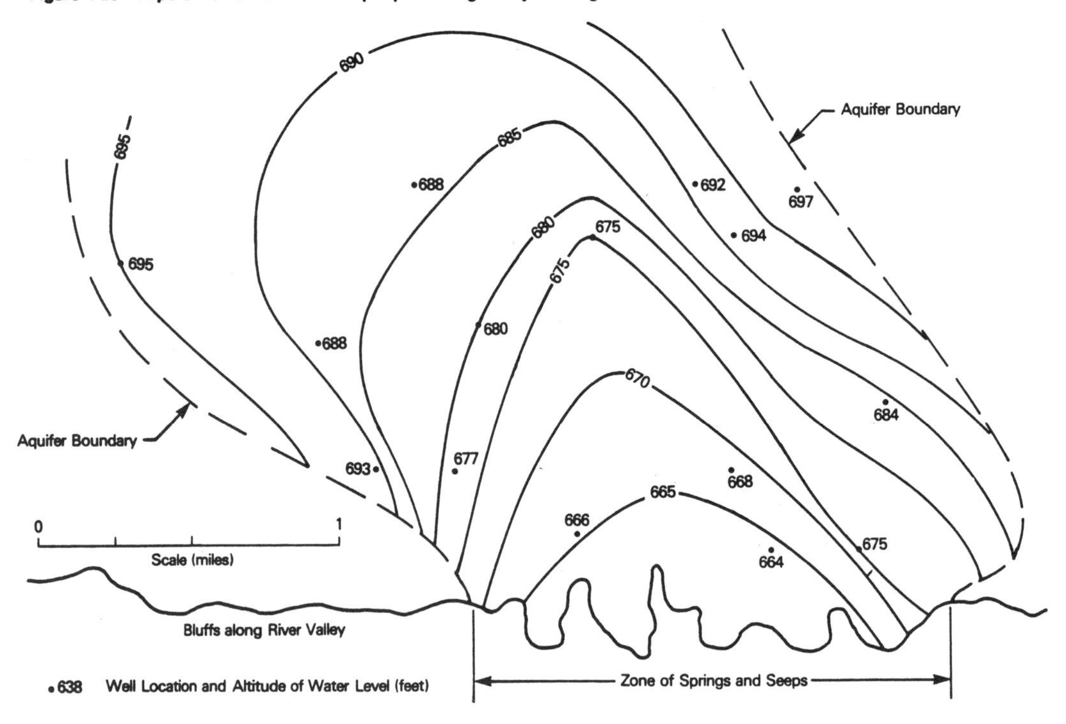

4.5.6 Calculating Ground-Water Flow

Darcy's Law, expressed in many different forms, is used to calculate the quantity of underflow or vertical leakage. One means of expressing it is:

$$Q = KIA \qquad (4\text{-}7)$$

where:

- Q = quantity of flow per unit of time, in gpd
- K = hydraulic conductivity, in gpd/ft^2
- I = hydraulic gradient, in ft/ft
- A = cross-sectional area through which the flow occurs, in ft^2

The flow rate is directly proportional to the gradient and therefore the flow is laminar, which means the water will follow distinct flow lines rather than mix with other flow lines. Where this does not occur, as in the case of unusually high velocity which might be found in fractures, solution openings, or adjacent to some pumping wells, the flow is turbulent.

Notice in Figure 4-29a that a certain quantity of fluid (Q) enters the sand-filled tube and the same amount exits. The water level declines along the length of the flow path (L) and the head is higher in the manometer at the beginning of the flow path than it is at the other end. The difference in head (H) along the flow path (L) is the hydraulic gradient (H/L or I). The head loss reflects the energy required to move the fluid this distance. If Q and A remain constant but K is increased, then the head loss decreases. It is particularly important to keep in mind that the head loss occurs in the direction of flow.

In Figure 4-29b, the flow tube has been inverted and the water is flowing from bottom to top or top to bottom. Q, K, A, and I all remain the same. This illustrates an important concept when the manometers are considered as wells. Notice that the deeper well has a head that is higher than the shallow well when the water is moving upward, whereas the opposite is the case when the flow is downward. Where this occurs in the field, it clearly shows the existence of recharge and discharge areas. In recharge areas, shallow wells have a higher head than deeper wells; the difference indicates the energy required to vertically move the water the distance between the screens of the two wells. Where the flow is horizontal, there should be no difference in head. Along stream valleys, which are regional discharge areas, the deeper well will have the higher head (Figure 4-29c). The location of waste disposal sites in recharge areas might lead to the vertical migration of leachate to deeper aquifers, and from this perspective, disposal sites should be locatedin discharge areas.

An example of the use of Darcy's Law is shown in Figure 4-30. In this case, a sand aquifer about 30 feet thick lies within the flood plain of a river that is about a mile wide. The aquifer is covered by a confining unit of glacial till, the bottom of which is about 45 feet below land surface. The difference in water level in two wells a mile apart is 10 feet. The hydraulic conductivity of the sand is 500 gpd/ft^2. The quantity of underflow passing through cross section A-A' (Figure 4-30) is:

$$Q = KIA$$

$$= 500 \text{ gpd/ft}^2 \times (10 \text{ ft}/5280 \text{ ft}) \times (5280 \times 30)$$

$$= 150{,}000 \text{ gpd}$$

Ground water moves both through aquifers and confining units. Because the difference in hydraulic conductivity between aquifers and confining units commonly differs by several orders of magnitude, the head loss per unit of distance in an aquifer is far less than in a confining unit. Consequently, lateral flow in confining units is small compared to aquifers, but vertical leakage through them can be significant. Owing to the large differences in hydraulic conductivity, flow lines in aquifers tend to parallel the boundaries but in confining units they are much less dense (Figure 4-31). The flow lines are refracted at the boundaries in order to produce the shortest flow path in the confining unit. The angles of refraction are proportional to the differences in hydraulic conductivity.

If one is concerned about the flow from one aquifer to another via a confining unit, a slightly modified form of Darcy's Law can be used:

$$Q_L = (p/m)AH \qquad (4\text{-}8)$$

where:

- Q_L = quantity of leakage, in gpd
- p = vertical hydraulic conductivity of the confining unit, in gpd/ft^2
- m = thickness of the confining unit, in ft
- A = cross sectional area, in ft^2
- H = difference in head between the two wells

As illustrated in Figure 4-32, assume two aquifers are separated by a layer of silt. The silty confining unit is 10 feet thick and has a vertical permeability of 2 gpd/ft^2. The difference in water level in wells tapping the upper and lower aquifers is 15 feet. Let us also assume that these hydrogeologic conditions exist in an area of 1 square mile. The daily quantity of leakage that occurs within this area from the shallower aquifer to the deeper one is:

$$Q_L = (2 \text{ gpd/ft}^2/10 \text{ ft}) \times 5280^2 \times 15 \text{ ft}$$

$$= 83{,}635{,}200 \text{ gpd}$$

This calculation clearly shows that the quantity of leakage, either upward or downward, can be highly

Figure 4-29 Graphical explanation of Darcy's Law. Notice that the flow in a tube can be horizontal or vertical in the direction of decreasing head.

A. Horizontal sand-filled tube.

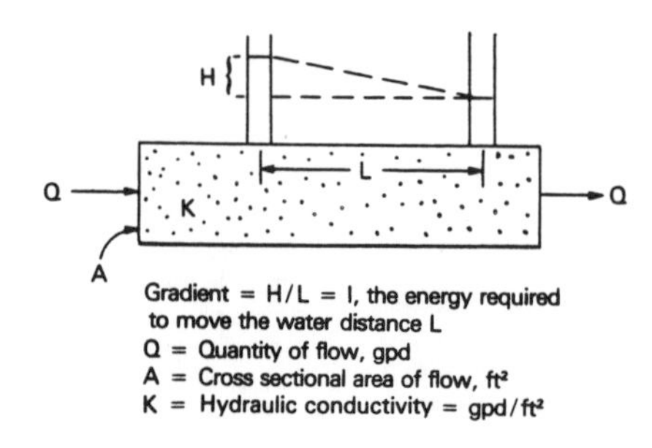

B. Vertical tube with flow from bottom to top.

Vertical tube with flow from top to bottom.

H
Q
L
A
K

C. Field conditions.

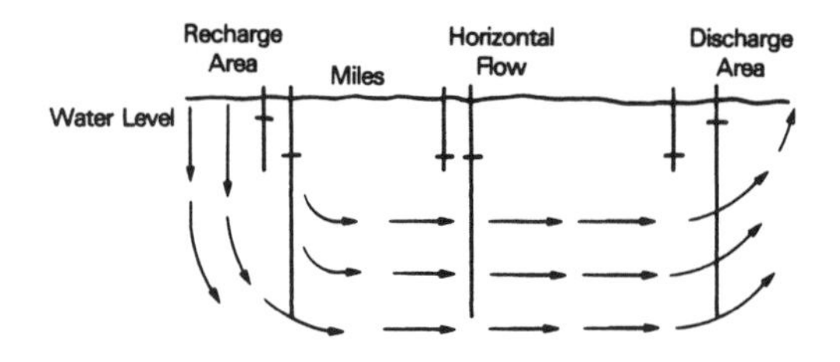

Figure 4-30 Using Darcy's Law to estimate underflow in an aquifer.

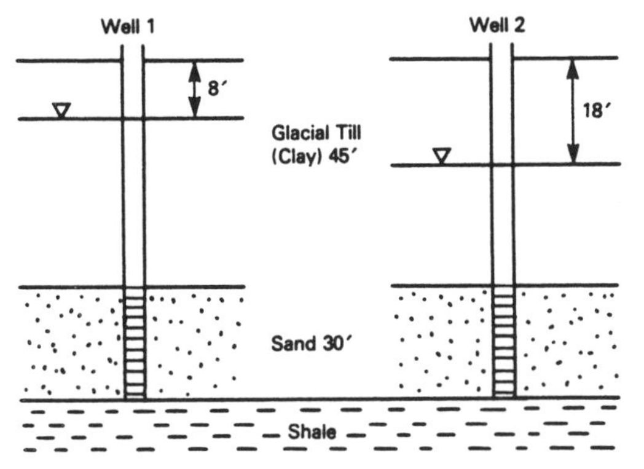

significant even if the hydraulic conductivity of the confining unit is small.

4.5.7 Interstitial Velocity

The interstitial velocity of ground water is of particular importance in contamination studies. It can be estimated by the following equation:

$$v = KI/7.48n \qquad (4\text{-}9)$$

where:

v = average velocity, in ft/d

n = effective porosity

Other terms are as previously defined.

As an example, assume there is a spill that consists of a conservative substance such as chloride. The liquid waste infiltrates through the unsaturated zone and quickly reaches a water-table aquifer that consists of sand and gravel with hydraulic

Figure 4-31 Long-term ground-water hydrographs show that the water level fluctuates in response to differences between recharge and discharge.

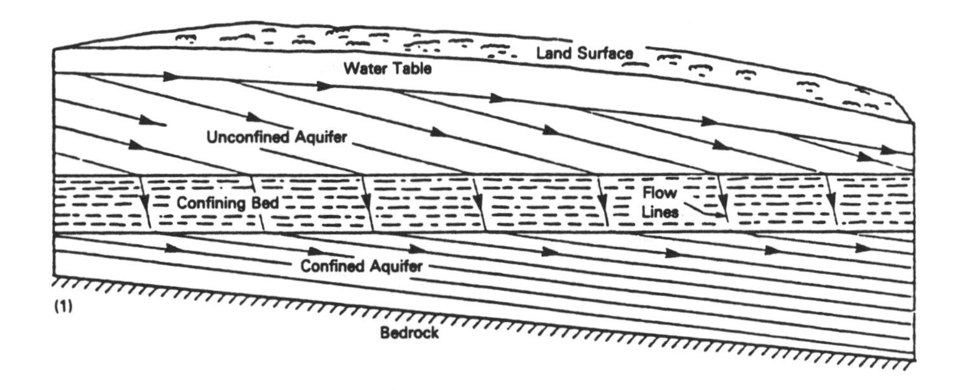

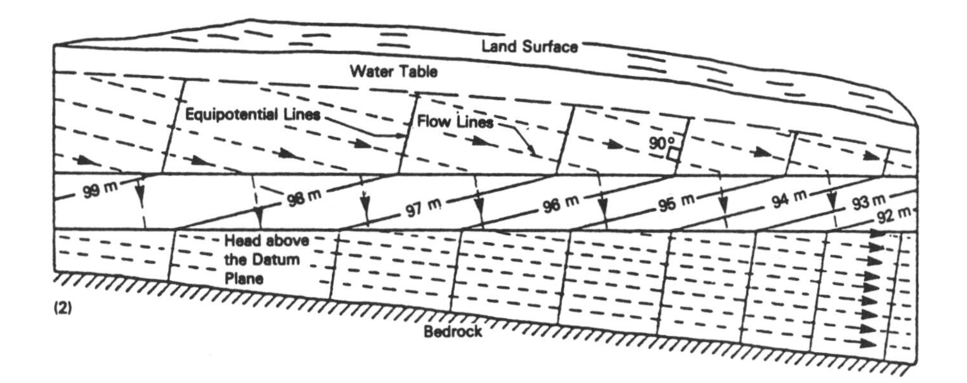

Figure 4-32 Using Darcy's Law to calculate the quantity of leakage from one aquifer to another.

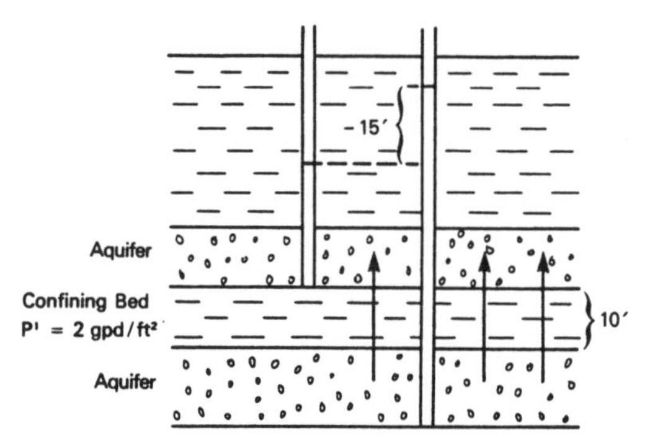

Area of leakage = 1 mi²
$P' = 2$ gpd/ft²
$m' = 10$ ft
$\Delta h = 15$ ft

$$Q = PIA = \frac{P'}{m'} A\Delta h$$

$$Q = \frac{2}{10} \times (5280 \times 5280) \times 15 = 83{,}635{,}200 \text{ gpd}$$

Figure 4-33 Using ground-water velocity calculations, it would require nearly six years for a contaminant to reach the downgradient well under the stated conditions.

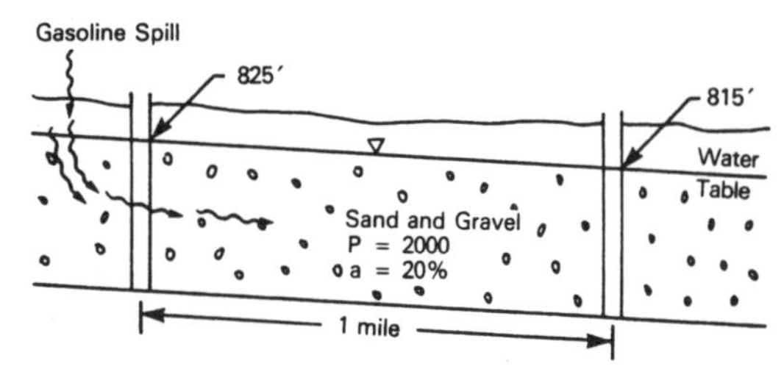

$$V = \frac{PI}{7.48a} = \frac{2000 \times \frac{10}{5280}}{7.48 \times 0.2} = \frac{3.8}{1.5} = 2.5 \text{ ft/day}$$

$$\text{Time} = \frac{5280'}{2.5 \text{ ft/day}} = 2112 \text{ days or } 5.8 \text{ years}$$

It would require 5.8 years for gasoline to reach downgradient under existing conditions.

conductivity of 2,000 gpd/ft² and effective porosity of 0.20. The water level in a well at the spill lies at an altitude of 1,525 feet and, at a well a mile directly downgradient, it is at 1,515 feet (Figure 4-33). What is the velocity of the water and contaminant and how long will it be before the second well is contaminated by chloride?

$$v = (2000 \text{ gpd/ft}^2) \times (10 \text{ ft}/5280 \text{ ft})/7.48 \times .20$$

$$= 2.5 \text{ ft/d}$$

$$\text{Time} = 5280 \text{ ft}/2.5 \text{ ft/d}$$

$$= 2112 \text{ days or } 5.8 \text{ yr}$$

This velocity value is crude at best and can only be used as an estimate. Hydrodynamic dispersion, for example, is not considered in the equation. This phenomenon causes particles of water to spread in a direction that is transverse to the major direction of flow and to move downgradient at a rate faster than expected. It is caused by an intermingling of streamlines due to differences in interstitial velocity brought about by the irregular pore space and interconnections.

Furthermore, most chemical species are retarded in their movement by reactions with the geologic framework, particularly with certain clays, soil-organic matter, and certain hydroxides. Only conservative substances such as the chloride ion will move unaffected by retardation.

In addition, it is not only the water below the water table that is moving, but also fluids within the capillary fringe. Here the velocity diminishes rapidly upward from the water table. Movement in the capillary fringe is important where the contaminant is gasoline or other substances less dense than water.

4.5.8 Transmissivity and Storativity

Hydrogeologists commonly use the term transmissivity (T) to describe the capacity of an aquifer to transmit water. Transmissivity is equal to the product of the aquifer thickness (m) and hydraulic conductivity (K) and is measured in units of gpd/ft of aquifer thickness:

$$T = Km \tag{4-11}$$

Another important term is storativity (S), which describes the quantity of water that an aquifer will release from or take into storage per unit surface area of the aquifer per unit change in head. In unconfined aquifers the storagtivity is, for all practical purposes, equal to the specific yield and, therefore, should range between 0.1 and 0.3. The storativity of confined aquifers is substantially smaller because the water which is released from storage when the head declines comes from the expansion of water and compression of the aquifer, both of which are very small. For confined aquifers, storativity generally ranges between 0.0001 and 0.00001; for leaky confined aquifers it is in the range of 0.001. The small storativity for confined aquifers means that to obtain a sufficient supply from a well there must be a large pressure change throughout a wide area. This is not the case with unconfined aquifers because the water derived is not related to expansion and compression but comes instead from gravity drainage and dewatering of the aquifer.

Hydrogeologists have found it necessary to use transmissivity and storativity to calculate the response of an aquifer to stresses and to predict future water level trends. These terms are also required as input for most flow and transport computer models.

4.5.9 Water-Level Fluctuations

Ground-water levels fluctuate throughout the year in response to natural changes in recharge and discharge (or storage), to changes in pressure, and to artificial stresses. Fluctuations brought about by changes in pressure are limited to confined aquifers. Most of these changes are short term and are caused by loading, such as a passing train compressing the aquifer, or by an increase in discharge from an overlying stream. Others are related to changes in barometric pressure, tides, earthtides, and earthquakes. None of these fluctuations reflect a change in the volume of water in storage.

Fluctuations that involve changes in storage are generally more long lived (Figure 4-34). Most ground-water recharge takes place during the spring, which causes the water level to rise. Following this period, which is a month or two long, the water level declines in response to natural discharge, which is largely to streams. Although the major period of recharge occurs in the spring, minor events can happen any time there is a rain.

Evapotranspiration effects on a surficial or shallow aquifer are both seasonal and daily. Plants, each serving as a minute pump, remove water from the capillary fringe or even from beneath the water table during hours of daylight in the growing season. This results in a diurnal fluctuation in the water table and stream flow.

4.5.10 Cone of Depression

When a well is pumped, the water level in its vicinity declines to provide a gradient to drive water toward the discharge point. The gradient becomes steeper as the well is approached because the flow is converging from all directions and the area through which the flow is occurring gets smaller. This results in a cone of depression around the well (Figure 4-35). Relatively speaking, the cone of depression around a well tapping an unconfined aquifer is small if compared to that around a well in a confined system. The former may be a few tens to a few hundred feet in diameter, while the latter may extend outward for miles.

Cones of depression from several pumping wells may overlap and, since their drawdown effects are additive, the water-level decline throughout the area of influence is greater than from a single cone (Figure 4-36). In ground-water studies and particularly contamination problems, evaluation of the cone or cones of depression can be critical because they represent an increase in the hydraulic gradient, which in turn controls ground-water velocity and direction of flow. In fact, properly spaced and pumped wells provide a mechanism to control the migration of leachate plumes. Discharging and recharging well schemes are commonly used in attempts to restore contaminated aquifers.

4.5.11 Specific Capacity

The decline of the water level in a pumping well, or any well for that matter, is called the drawdown and the prepumping level is the static water level.

(Figure 4-37). The discharge rate of the well divided by the difference between the static and the pumping level is the specific capacity. The specific capacity indicates how much water the well will produce per foot of drawdown:

$$\text{Specific capacity} = Q/s \qquad (4\text{-}11)$$

where:

Q = discharge rate, in gpm

s = drawdown, in ft

If a well produces 100 gpm and the drawdown is 8 ft, the well will produce 12.5 gpm for each foot of available drawdown. One can rather crudely estimate transmissivity by multiplying specific capacity by 2,000.

Figure 4-34 Long-term ground-water hydrographs show that the water level fluctuates in response to differences between recharge and discharge.

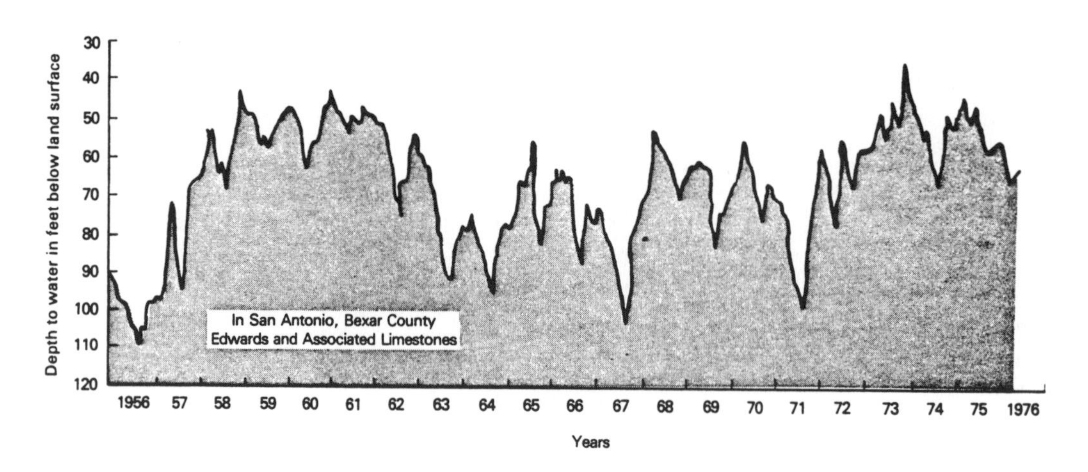

Figure 4-35 Cones of depression in unconfined and confined aquifers (from Heath, 1983).

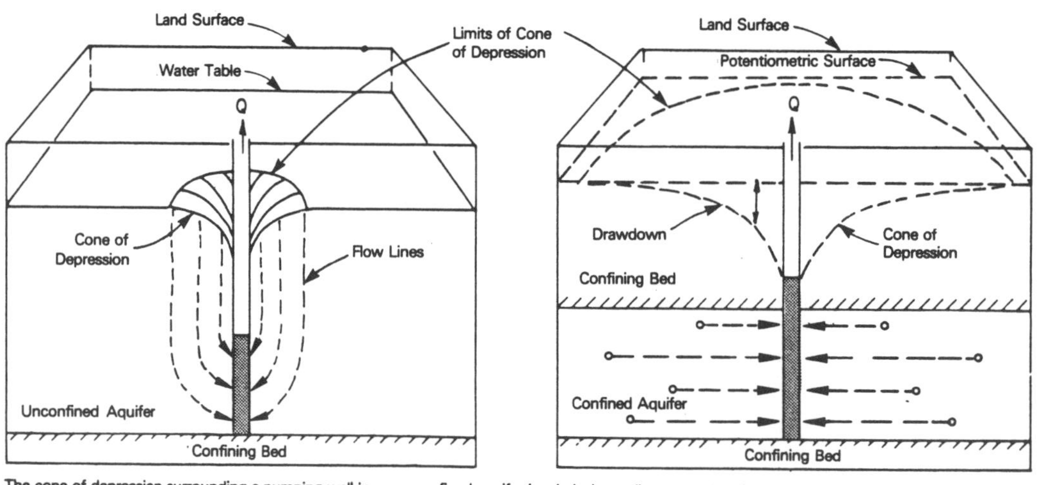

The cone of depression surrounding a pumping well in an unconfined aquifer is relatively small compared to that in a confined system.

Figure 4-36 Overlapping cones of depression result in more drawdown than would be the case for a single well (from Heath, 1983).

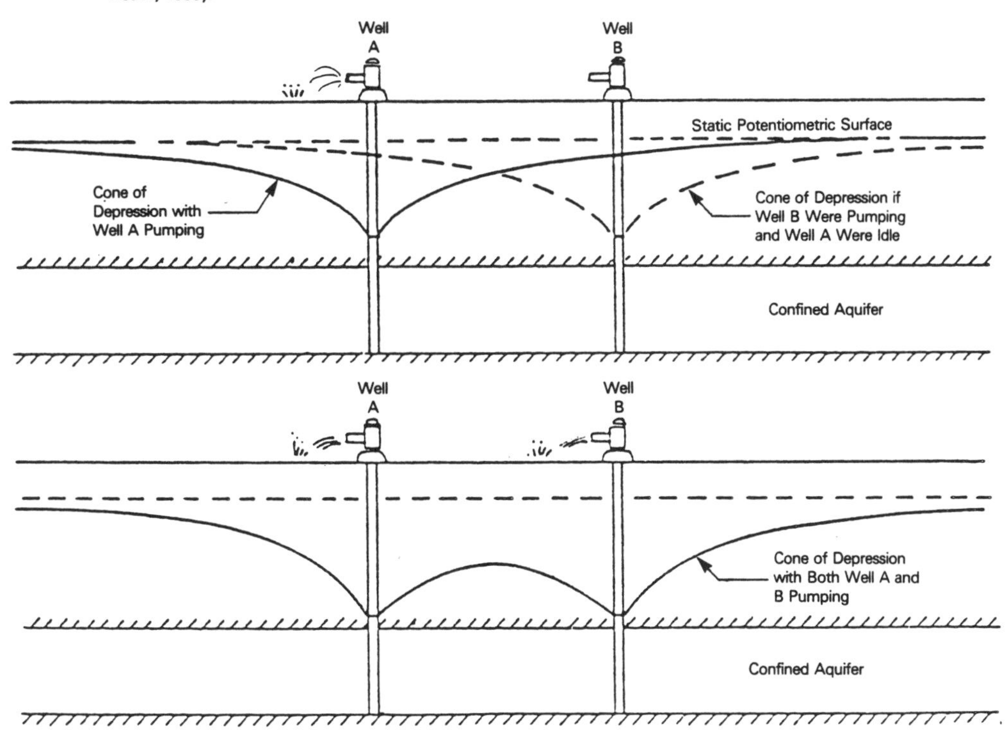

4.6 References

Cross, W.P., and R.E. Hedges. 1959. Flow Duration of Ohio Streams. Ohio Division of Water Bulletin 31.

Daniel, J.F., L.W. Cable, and R.J. Wolf. 1970. Ground Water - Surface Water Relation During Periods of Overland Flow. U.S. Geological Survey Professional Paper 700-B, U.S. Government Printing Office, Washington, D.C.

Durfor, C.N., and E. Becker. 1962. Public Water Supplies of the 100 Largest Cities in the United States. U.S. Geological Survey Water-Supply Paper 1812. U.S. Government Printing Office, Washington, D.C.

Freeze , R.A., and J.A. Cherry. 1979. Groundwater. Prentice-Hall Publishing Co., Inc., Englewood Cliffs, NJ.

Heath, R.C. 1984. Ground-Water Regions of the United States. U.S. Geological Survey Water-Supply Paper 2242, U.S. Government Printing Office, Washington, D.C.

Heath, R.C. 1983. Basic Ground-Water Hydrology. U.S. Geological Survey Water-Supply Paper 2220, U.S. Government Printing Office, Washington, D.C.

Heath, R.C., and F.W. Trainer. 1981. Introduction to Ground Water Hydrology. Water Well Journal Publishing Co., Worthington, OH.

Johnson, E.E. 1966. Ground-Water and Wells. Edward E. Johnson, Inc., Saint Paul, MN.

Kunkle, G.R. 1965. Computation of Ground-Water Discharge to Streams During Floods, or to Individual Reaches During Base Flow, by Use of Specific Conductance. U.S. Geological Survey Professional Paper 525-D, U.S. Government Printing Office, Washington, D.C.

LaSala, A.M. 1967. New Approaches to Water-Resources Investigations in Upstate New York. Ground Water 5(4).

Meinzer, O.E., and N.D. Stearns. 1928. A Study of Ground Water in the Pomerang Basin. U.S.

Figure 4-37 Values of transmissivity based on specific capacity are commonly too small because of well construction details (from Heath, 1983).

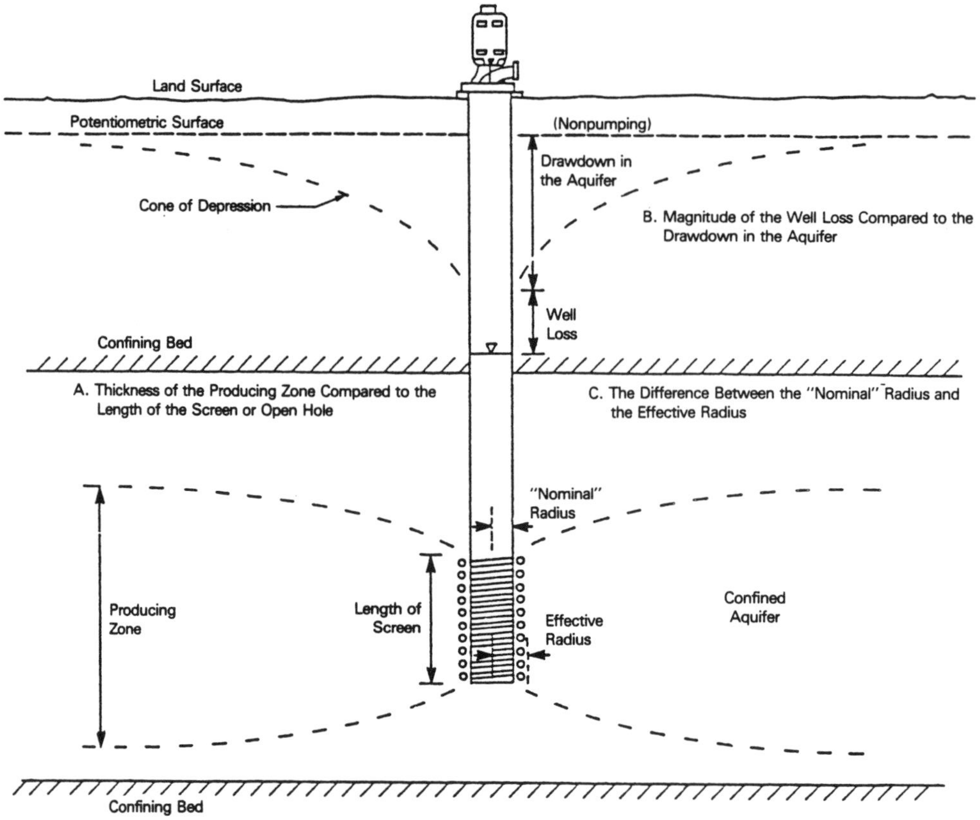

Geological Survey Water-Supply Paper 597-B, U.S. Government Printing Office, Washington, D.C.

Olmsted, F.H., and A.G. Hely. 1962. Relation Between Ground Water and Surface Water in Brandywine Creek Basin, Pennsylvania. U.S. Geological Survey Professional Paper 417-A, U.S. Government Printing Office, Washington, D.C.

Pettyjohn, W.A. 1985. Regional Approach to Ground-Water Investigations. In: Ground Water Quality, edited by C.H. Ward, W. Giger, and P.L. McCarty, John Wiley & Sons, New York, NY

Pettyjohn, W.A., H. White, and S. Dunn. 1983. Water Atlas of Oklahoma. University Center for Water Research, Oklahoma State University, OK

Pettyjohn, W.A. 1982. Cause and Effect of Cyclic Changes in Ground-Water Quality. Ground-Water Monitoring Review 2(1).

Pettyjohn, W.A. and R.J. Henning. 1979. Preliminary Estimate of Ground-Water Recharge Rates, Related Streamflow and Water Quality in Ohio. Project Completion Report 552, Ohio State University Water Resources Center, OH.

Pettyjohn, W.A. 1975. Chloride Contamination in Alum Creek, Central Ohio. Ground Water 13(4).

Rasmussen, W.C., and G.E. Andreason. 1959. Hydrologic Budget of the Beaver Dam Creek Basin, Maryland. U.S. Geological Survey Water-Supply Paper 1472, U.S. Government Printing Office, Washington, D.C.

Seaber, P.R. 1965. Variations in Chemical Character of Water in the Englishtown Formation of New Jersey. U.S. Geological Survey Professional Paper 498-B, U.S. Government Printing Office, Washington, D.C.

Schicht, R.J., and W.C. Walton. 1961. Hydrologic Budgets for Three Small Watersheds in Illinois. Illinois State Water Survey Report of Investigation 40.

Stefferud, Alfred. 1955. Water, the Yearbook of Agriculture. U.S. Department of Agriculture.

Todd, D.K. 1980. Groundwater Hydrology. John Wiley & Sons, New York, NY.

Toler, L.G. 1965. Use of Specific Conductance to Distinguish Two Base-Flow Components in Econfina Creek, Florida. U.S. Geological Survey Professional Paper 525-C, U.S. Government Printing Office, Washington, D.C.

Trainer, F.W., and F.A. Watkins. 1975. Geohydrologic Reconnaissance of the Upper Potomac River Basin. U.S. Geological Survey Water-Supply Paper 2035, U.S. Government Printing Office, Washington, D.C.

U.S. Environmental Protection Agency. 1985. Protection of Public Water Supplies from Ground-Water Contamination. EPA-625/4-85-016, Center for Environmental Research Information, Cincinnati, OH.

U.S. Geological Survey. 1985. Water Resources Data, Oklahoma, Water Year 1983. U.S. Geological Survey Water-Data Report OK-83-1, U.S. Government Printing Office, Washington, D.C.

Viscoky, A.P. 1970. Estimating the Ground-Water Contribution to Storm Runoff by Electrical Conductance Method. Ground Water 8(2).

5. Monitoring Well Design and Construction

The principal objective of constructing monitoring wells is to provide access to an otherwise inaccessible environment. Monitoring wells are used to evaluate topics within various disciplines, including geology, hydrology, chemistry, and biology. In ground-water quality monitoring, wells are used for collecting ground-water samples, which upon analysis, may allow description of a contaminant plume, or the movement of a particular chemical (or biological) constituent, or ensure that potential contaminants are not moving past a particular point.

5.1 Ground-Water Monitoring Program Goals

Each purpose for ground-water monitoring, ambient monitoring, source monitoring, case preparation monitoring, and research monitoring, (Barcelona *et al.*, 1983) must satisfy somewhat different requirements, and may require different strategies for well design and construction. At the outset, it must be clearly understood what the intended monitoring program is to accomplish and the potential future use of the wells in other, possibly different, monitoring programs.

Regional investigations of ground-water quality fall into the *ambient* monitoring category. Such investigations seek to establish an overall picture of the quality of water within all or portions of an aquifer. Generally, sample collection is conducted routinely over a period of many years to determine changes in quality over time. Often, changes in quality are related to long-term changes in land use (e.g., the effects of urbanization). Monitoring conducted for Safe Drinking Water Act compliance generally falls in this category.

Samples are often collected from a variety of public and private water supply wells for ambient quality investigations. Because of this, the data obtained through some ambient monitoring programs may not meet the strict well design and construction requirements imposed by the three other types of monitoring. However, such programs are important for detecting significant changes in aquifer water quality over time and space and protecting public health.

Regulatory monitoring at potential contaminant sources is considered *source* monitoring. Under this type of program, monitoring wells are located and designed to detect the movement of specific pollutants outside the boundaries of a particular facility (e.g., treatment, storage, or disposal). Ground-water sampling to define contaminant plume extent and geometry would fall into this classification of monitoring. Monitoring well design and construction are tailored to the site geology and contaminant chemistry. Quantitative aspects of analytical results become most important because the level of contaminant concentration may require specific regulatory action.

Monitoring for *case preparation*, such as for legal proceedings in environmental enforcement, requires a level of detail similar to source monitoring. Source monitoring, in fact, often becomes part of legal proceedings to establish whether or not environmental damage has occurred and identify the responsible party. This is a prime example of one type of monitoring program evolving into another. The appropriateness and integrity of monitoring well design and construction methods will come under much scrutiny. In such cases, the course of action taken during the monitoring investigation, the decisions that were made concerning well design and construction, and the reasons why those decisions were made must be clearly established and documented.

Monitoring for *research* generally requires a level of sophistication beyond that required of any other type of monitoring (this, of course, depends upon the types and concentrations of constituents being sought and the overall objectives of the research). Detailed information is often needed to support the basic concepts and expand the levels of understanding of the complex mechanisms of ground-water movement and solute/contaminant transport.

The goals of any proposed ground-water monitoring program should be clearly stated and understood before decisions are made on the types and numbers of wells needed, where they should be located, how deep they should be, what constituents are of interest, and how samples should be collected, stored, transported, and analyzed.

As each of these decisions is made, consideration must be given to the costs involved in each step of the monitoring program and how compromises in one step may affect the integrity and outcome of the other steps. For example, cost savings in well construction materials may so severely limit the usefulness of a well that another well may need to be constructed at the same location for the reliable addition of a single chemical parameter.

5.2 Monitoring Well Design Components

Monitoring well design and construction methods follow production well design and construction techniques; however, it must be remembered that a monitoring well is built specifically to give access to the ground water so a "representative" sample of water can be withdrawn and analyzed. While it is important to pay attention to well efficiency and yield, the ability to produce large amounts of water for supply purposes is not the primary objective.

Emphasis, then, is placed on constructing a well that will provide easily obtainable ground-water samples that will give reliable, meaningful information. It follows from this emphasis that the materials and techniques used for constructing a monitoring well must not materially alter the quality of the water being sampled. An understanding of the chemistry of suspected pollutants and the geologic setting in which the monitoring well is to be constructed play a major role in the drilling technique and well construction materials used.

There are several components to be considered in monitoring well design. These include: location (and number of wells), diameter, casing and screen material, screen length and depth of placement, sealing material, *well development*, and *well security*. Often, discussion of one component will impinge upon other components.

5.2.1 Location and Number

Locating monitoring wells spatially and vertically to ensure that the ground- water flow regime of concern is being monitored is obviously one of the most important components in ground-water quality monitoring design. It is impossible to divorce prescribing monitoring well locations (sites) and the number of wells in the monitoring program. The number of wells and their location are principally determined by the purpose of the monitoring program. In most monitoring situations, the goal is to determine the effect some surface or near-surface activity has had on nearby ground-water quality. Most dissolved constituents will descend vertically through the unsaturated zone beneath the area of activity and then, upon reaching the saturated zone, move horizontally in the direction of ground-water flow. Therefore, monitoring wells are normally completed downgradient in the first permeable water-bearing unit encountered. Consideration should be given to natural (seasonal) and artificial fluctuations in water table elevation. Artificial fluctuations include pumpage, which will cause water levels to fall, and lagoon operation, which can cause a rise or "mound" in the water table.

Preliminary boreholes and/or monitoring wells can be constructed for the collection and analysis of geologic material samples, ground-water levels, and water quality samples to guide the placement of additional wells. Accurate water level information must be established to determine if local ground-water flow paths and gradients differ significantly from the regional appraisal.

The analysis of water quality samples from the preliminary wells can also direct the placement of additional wells. Such wells are particularly helpful in the vertical arrangement of sampling points (especially for a contaminant that is denser than water). Without some preliminary chemical data, it is usually very difficult to know where the most contaminated zone is.

A number of factors will govern where and how many wells should be constructed. These factors include: site geology, site hydrology, source characteristics, contaminant characteristics, and the size of the area under investigation. Certainly, the more complicated the geology and hydrology, the more complex the contaminant and source, and the larger the area being investigated, the greater the number of monitoring wells that will be required. Details of some of these factors are discussed in Chapters 1, 2, 4, and 9 and in the following sections.

5.2.2 Diameter

In the past, the diameter of a monitoring well was based primarily on the size of the device (bailer, pump, etc.) being used to withdraw the water samples. This practice was similar to that followed for water supply well design. For example, a domestic water well is commonly 4 to 6 inches in diameter to accommodate a submersible pump capable of delivering from 5 to 20 gallons per minute. Municipal, industrial, and irrigation wells have greater diameters to handle larger pumps and to increase the available screen open area so the well can produce water efficiently.

This practice worked well in very permeable formations, where an aquifer capable of furnishing large volumes of water was present. However, unlike most water supply wells, monitoring wells are quite often completed in very marginal water-producing zones. Pumping one or more well volumes of water (the amount of water stored in the well casing under nonpumping conditions) from a well built in low-yielding materials (Gibb *et al.*, 1981) may present a serious problem if the well has a large diameter.

Figure 5-1 illustrates the amount of water in storage per foot of casing for different well casing diameters. Well casings with diameters of 2 to 6 inches will contain 0.16 and 1.47 gallons of water per foot of casing, respectively. Purging four well volumes from a well containing 10 feet of water would require removal of 6.4 gallons of water from a 2-inch well and 58.8 gallons of water from a 6-inch well. Under low-yielding conditions, it can take considerable lengths of time to recover enough water in the well to collect a sample (Figure 5-2).

In addition, when hazardous constituents are present in the ground water, proper disposal of the purged water will be necessary. This amount of water should be kept to a minimum, for safety's sake as well as disposal cost. Cost of well construction is also a consideration. Small diameter wells (less than 4 inches) are much less expensive than large diameter wells in terms of both cost of materials and cost of drilling.

For these reasons and with the advent of numerous commercially available small-diameter pumps (less than 2 inches OD) capable of lifting water over 100 feet, 2-inch ID monitoring wells have become the standard in monitoring well technology.

Large diameter wells can be useful in situations where monitoring may be followed by remedial actions involving reclamation and treatment of the contaminated ground water. In some instances, the "monitoring" well may become a "supply" well to remove contaminated water from the ground for treatment. Larger diameter wells also merit consideration when monitoring is required at depths of hundreds of feet and in other situations where the additional strength of large diameter casing is needed. For sampling at several depths beneath one location, several monitoring wells have been nested in a single borehole (Johnson, 1983). A technique such as this will require drilling a larger diameter hole to accommodate the multiple well casings. Again, the use of smaller diameter casing provides advantages by allowing more wells to be nested in the borehole, thus easing construction and saving costs in drilling expenses.

5.2.3 Casing and Screen Material

The type of material used for a monitoring well can have a distinct effect on the quality of the water sample to be collected (Barcelona *et al.*, 1983; Gillham *et al.*, 1983 and Miller, 1982). The materials of choice should retain their structural integrity for the duration of the monitoring program under actual subsurface conditions. They should neither adsorb nor leach chemical constituents which would bias the representativeness of the samples collected.

Galvanized steel casing can impart iron, manganese, zinc, and cadmium to many waters. Steel casing may impart iron and manganese to a sample. PVC pipe

Figure 5-1 Volume of water stored per foot of well casing for different diameter casings (from Rinaldo-Lee, 1983).

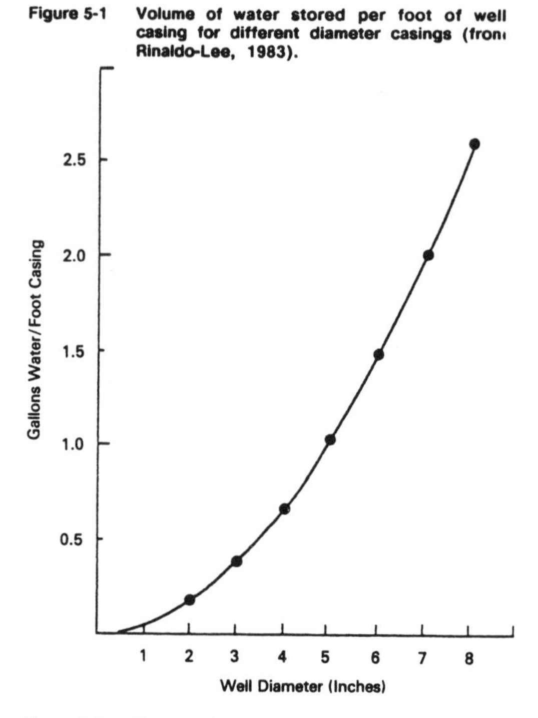

Figure 5-2 Time required for recovery after slug of water removed (from Rinaldo-Lee, 1983).

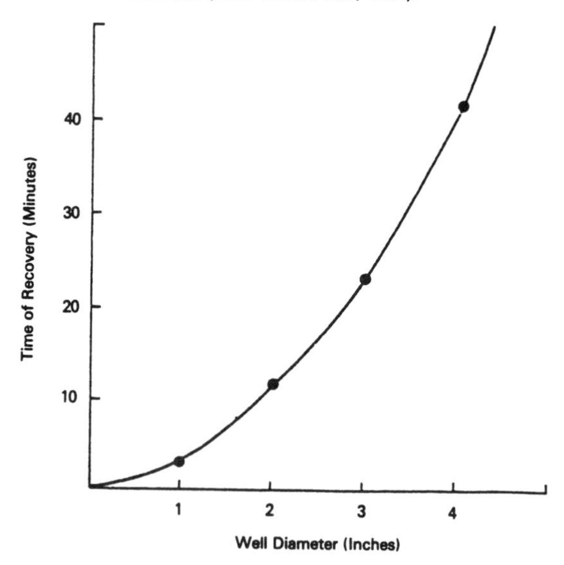

Assumptions: $K = 1 \times 10^{-5}$ cm/sec, well screen = 10′, 10′ of water above screen, 6′ of water instantaneously removed

has been shown to release and adsorb trace amounts of various organic constituents to water after prolonged exposure (Miller, 1982). PVC solvent cements used to attach sections of PVC pipe have also been shown to release significant quantities of organic compounds.

Teflon^R and glass are among the most inert materials that have been considered for monitoring well construction. Glass, however, is difficult and expensive to use under most field conditions. Teflon^R is also very expensive; with technological advances, Teflon^R-coated casings and screens may become available. Stainless steel also offers desirable properties from a monitoring perspective, but it too is expensive.

A reasoned strategy for ground-water monitoring must consider the effects of contaminated water on well construction materials. Unfortunately, there is limited published information on the performance of specific materials in varied hydrogeologic settings (Pettyjohn *et al.*, 1981). A preliminary ranking of commonly used materials exposed to different solutions representing the principal soluble species present in hazardous waste site investigations produced the following list, in order of best to worst (Barcelona *et al.*, 1983):

Teflon^R
Stainless Steel 316
Stainless Steel 304
PVC Type I
Lo-Carbon Steel
Galvanized Steel
Carbon Steel

Polyvinyl chloride (PVC Type I) has very good chemical resistance except to low molecular weight ketones, aldehydes, and chlorinated solvents. As the organic content of a solution increases, direct attack on the polymer matrix or solvent absorption, adsorption, or leaching, may occur. The only exception to this observation is Teflon. Provided that sound construction practices are followed, Teflon^R can be expected to out perform all other casing and sampling materials (Barcelona *et al.*, 1983).

Stainless steels are the most chemically resistant of the ferrous materials. Stainless steel may be sensitive to the chloride ion, which can cause pitting corrosion, especially over long term exposures under acidic conditions. Given the similarity in price, workability, and performance, the remaining ferrous materials (lo-carbon, galvanized steel, and carbon) provide little advantage over one another for casing/screen construction.

Significant levels of organic components found in PVC primers and adhesives (such as tetrahydrofuran, methylethylketone, cyclohexanone, and methylisobutylketone) were detected in well water several months after installation (Sosebee, *et al.*, 1982). The presence of compounds such as these can mask the presence of other similar volatile compounds (Miller, 1982). Therefore, when using PVC and other similar materials (e.g., ABS, polypropylene, or polyethylene) for well construction, threaded joints are the preferred means for connecting sections together.

In many situations, it may be possible to compromise accuracy or precision for initial cost, depending on the objectives of the monitoring program. For example, if the contaminants of interest are already defined and they do not include substances which might bleed or sorb, it may be reasonable to use wells cased with a less expensive material.

Wells constructed of less than optimum materials might be used for sampling if identically constructed wells are constructed in uncontaminated parts of the monitored aquifer to provide ground-water samples for use as "blanks" (Pettyjohn *et al.*, 1981). However, such blanks may not adequately address problems of adsorption on or leaching from the casing material induced by contaminants in the ground water. It may be feasible to use two or more kinds of casing materials in the saturated zone and above the seasonal high water tables, such as Teflon^R or stainless steel, and use a more appropriate material, such as PVC or galvanized steel casing, above static water level.

It must be remembered, however, that trying to save money by compromising on material quality or suitability may eventually increase program cost by causing reanalysis, or worse, monitoring well reconstruction. Careful consideration is required in each case, and the analytical laboratory should be fully aware of the construction materials used.

Care must also be given to the preparation of the casing and well screen materials prior to installation. At a minimum, materials should be washed with detergent and rinsed thoroughly with clean water. Steam-cleaning and high pressure, hot water cleaners provide excellent cleaning of cutting oils and lubricants left on casings and screens after manufacture (this is particularly true for metal casing and screen materials). To ensure that these and other sampling materials are protected from contamination prior to placement down-hole, materials should be covered (with plastic sheeting or other material) and kept off the ground.

All wells should allow free entry of water. The water produced should be as clear and silt-free as possible. For drinking water supplies, sediment in the raw water can create additional pumping and treatment costs and lead to the general unpalatability of the water. With monitoring wells sediment-laden water can greatly lengthen filtering time and create chemical interference in sample analyses.

Commercially manufactured well screens are preferred for monitoring well design given that the proper screen slot size is chosen. Sawed or torch-cut casing may be appropriate in deposits where medium to coarse sand or gravel predominates. In formations where fine sand, silt, and clay predominate, sawed or torch-cut slots will not be small (or uniform) enough to retain the materials and the well may clog. The practice of sawing slots in PVC pipe should be avoided in monitoring situations where organic chemicals are of concern because this procedure exposes fresh surfaces of PVC, increasing the possibility of releasing compound ingredients or reaction products.

It may be helpful to have several slot-sized well screens on-site so the correct manufactured screen can be placed in the hole *after* the materials within the zone of interest have been inspected. Gravel pack compatible with the selected screen slot size will further help retain the finer fractions of material and allow freer entry of water into the well by creating a zone of higher permeability around the well screen.

For natural-packed wells (no gravel pack), where relatively homogeneous, coarse materials predominate, a slot size should be selected that will retain from 40 to 50 percent of the screened material. In cases where adequate well development procedures may be difficult to follow, a screen that will retain about 70 percent of the screened formation should be selected. If an artificial pack is used, a uniform gravel-pack size that is from three to five times the 50 percent size of the formation and a screen size that will retain at least 90 percent of the pack material should be selected (Walker, 1974). The gravel-pack should be composed of clean, uniform quartz sand.

Placement of the gravel-pack should be done carefully to avoid bridging in the hole and to allow uniform settling around the screen. A tremie pipe can be used to guide the sand to the bottom of the hole and around the screen. The pipe should be slowly lifted as the annulus between the screen and borehole as the borehole fills. If the depth of water standing in the annulus is not great, the sand can be simply poured from the surface. Calculations should be made to determine what volume of sand will be required to fill the annulus to the desired depth (usually about one foot above the top of the screen). Field measurements should be taken to confirm the pack has reached this level before backfilling or sealing procedures start.

5.2.4 Screen Length and Depth of Placement

The length of screen chosen and the depth at which it is placed in monitoring well design are dependent on, to a large degree, the behavior of the contaminant as it moves through the unsaturated and saturated zones and, again, the goal of the monitoring program.

When monitoring a potable water supply aquifer, the entire thickness of the water-bearing formation could be screened (just as a production well would be). For regional aquifer studies, production wells are commonly used for sampling. Such samples would provide water integrated over the depth of the water-bearing zone(s) and would provide a sample similar in quality to what would be found in a drinking water supply.

When sampling specific depth intervals at one location is necessary, vertical nesting of wells is common. This technique is often necessary when the saturated zone is too thick to adequately monitor with one long screened section (causing dilution of the collected sample). Contaminants tend to stratify within the saturated zone; collection of a sample integrated over a thick zone will give little information on the depth and concentration that a contaminant may have reached.

Screen lengths of one to two feet are common in detailed plume geometry investigations. Thick aquifers would require that several wells be completed at different depth intervals. In such situations (and depending on the magnitude of the aquifer saturated thickness), screen lengths of no more than 5 to 10 feet are used. Monitoring wells can be constructed in separate holes placed closely together or in one larger diameter hole, as in Figure 5-3. Prevention of the vertical movement of contaminants in the well bore before and after well completion may be difficult to achieve since multiple wells in one hole are difficult to seal. Thus, the drilling of multiple holes may be required to insure well integrity. Specially constructed installations have been developed to sample a large number of points vertically over short intervals (Morrison, 1981; Pickens, 1981; and Torstensson, 1984; Figures 5-4 and 5-5).

In other situations, only the first water-bearing zone encountered will require monitoring (for example, when monitoring near a potential contaminant source in a relatively impermeable glacial till). Here, the "aquifer" or zone of interest may be only 6 inches to a few feet thick. Screen length should be limited to 1 to 2 feet in these cases to minimize siltation problems from surrounding fine-grained materials and possible dilution effects from water contributed by uncontaminated zones.

Because of the chemical reactions which occur when ground water contacts the atmosphere, particularly for volatile compounds, aeration of the screened section should be avoided. Well depth should assure that the screened section is always fully submerged. Fluctuations in the elevation of the top of the saturated zone caused by seasonal variations or man-induced changes must be considered.

Figure 5-3 Typical multiwell installations (from Johnson, 1983).

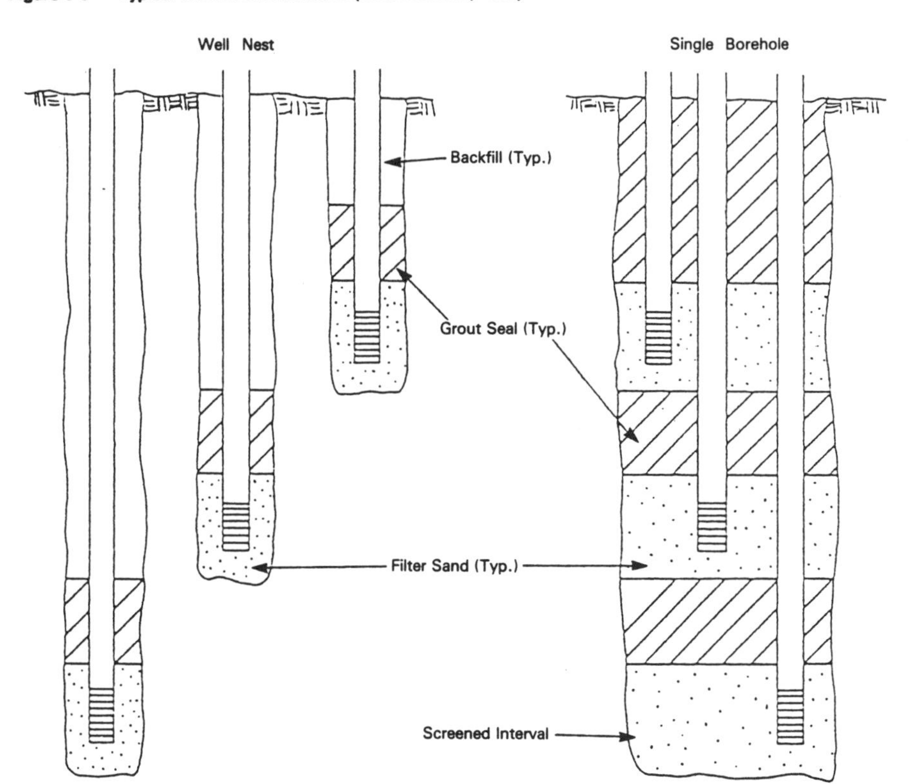

Monitoring for contaminants with densities different than water calls for special attention. In particular, low density organic compounds such as gasoline will float on the ground-water surface (Gillham *et al.*, 1983). Monitoring wells constructed for floating contaminants should contain screens which extend above the zone of saturation so that these lighter substances can enter the well. The screen length and position must accommodate the magnitude and depth of variations in water table elevation. However, the thickness of floating products in the well does not necessarily indicate the thickness of the product in the aquifer.

5.2.5 Sealing Materials and Procedures

It is critical that the screened portion of each monitoring well access ground water from a specific depth interval. Vertical movement of ground water in the vicinity of the well can greatly influence sample quality (Keith *et al.*, 1982). Rainwater can infiltrate backfill, potentially diluting or contaminating samples; vertical seepage of leachate along the well casing will also produce unrepresentative samples (this is particularly important in multilevel installations such as in Figures 5-3, 5-4, and 5-5). Even more importantly, the creation of a conduit in the annulus of the monitoring well that could contribute to or hasten the spread of contamination is to be strictly avoided. Several methods have been employeed successfully to isolate contaminated zones during the drilling process (Burkland and Raber, 1983; Perry and Hart, 1985).

Monitoring wells are usually sealed with neat cement grout, dry benonite (powdered, granulated, and pelletized), or bentonite slurry. Well seals usually occur at two places within the annulus created by the drilling operation. One area is within or near the saturated zone to isolate the screened interval for sampling. The other is at the ground surface to inhibit downward leakage of surface contaminants.

The use of bentonite traditionally has been considered to provide a much better seal than cement. However, recent investigations on the use of clay liners for hazardous waste disposal have shown that some organic compounds migrate through bentonite with little or no attenuation (K.W. Brown, *et*

Figure 5-4 Schematic diagram of a multilevel sampling device (from Pickens, 1981).

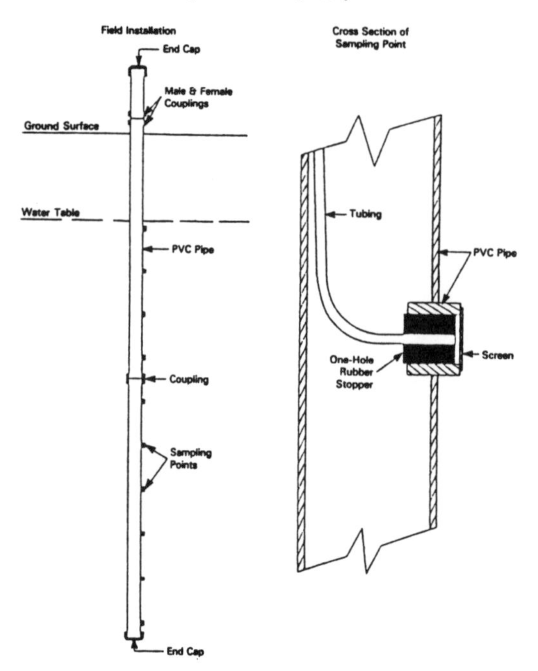

al., 1983). Therefore, cement may offer some benefits over bentonite.

Bentonite is most often used as a down-hole seal to prevent vertical migration within the well annulus. When bentonite must be placed below the water table (or where water has risen in the bore hole), it is recommended that a bentonite slurry be tremied down the annulus to fill the hole from the bottom upward. In collapsible material conditions, where the borehole has collapsed to a point just above the water table, dry bentonite (granulated or pelletized works best) can be poured down the hole.

Bentonite clay has appreciable ion exchange capacity which may interfere with the chemistry of collected samples when the seal is proximate to the screen or well intake. Cement grout has been known to seriously affect the pH of sampled water when improperly placed. Therefore, special attention and care should be exercised during placement of a down-hole seal. Approximately one foot (at a minimum) of gravel pack or naturally collapsed material should extend above the top of the well intake to ensure that the sealing materials do not migrate downward into the well screen. If the sealing material is too watery before it is placed down the hole, settling or migration of sealing materials into the gravel-pack or screened area may occur and the fine materials in the seal may penetrate the natural or artificial pack.

Figure 5-5 Single (a) and multiple (b) installation configurations for an air-lift sampler (from Morrison, 1981).

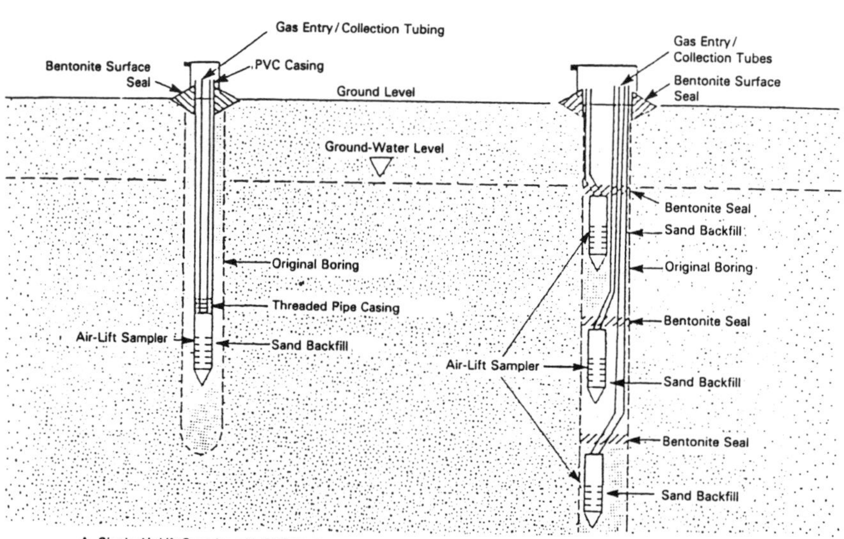

While a neat cement (sand and cement, no gravel) grout is often recommended, especially for surface sealing, shrinkage and cracking of the cement upon curing and weathering can create an improper seal. Shrink-resistant cement (such as Type K Expansive Cement) and mixtures of small amounts of bentonite with neat cement have been used successfully to help prevent cracking.

5.2.6 Development

Development is a facet of monitoring well construction that often is overlooked. During the drilling process, fine-grained materials smear on the sides of the borehole, forming a mud "cake" that reduces the hydraulic conductivity of the materials opposite the screened portion of the well. To facilitate entry of water into the monitoring well (a particularly important factor for low-yielding geologic materials), this mud cake must be broken down and the fine-grained materials removed from the well or well bore. Development also removes fluids, primarily water, which are introduced to the water-bearing formations during the drilling process.

Additionally, monitoring wells must be developed to provide water free of suspended solids for sampling. When sampling for metal ions and other inorganic constituents, water samples must be filtered and preserved at the well site at the time of sample collection. Improperly developed monitoring wells will produce samples containing suspended sediments that will both bias the chemical analysis of the collected samples and frequently cause clogging of the field filtering mechanisms.

The time and money spent for this important procedure will expedite sample filtration and result in samples more representative of water contained in the formation being monitored. The time saved in field filtration alone will more than offset the cost of development.

Successful development methods include bailing, surging, and flushing with air or water. The basic principle behind each method is to create reversals of flow in and out of the well (and/or bore hole) to break down the mud cake and draw the finer materials into the hole for removal. This process also helps remove the finer fraction of materials in proximity to the borehole, leaving behind a "natural" pack of coarser-grained materials.

Years ago, small-diameter well development was most commonly achieved through use of a bailer. The bailer was about the only "instrument" which had been developed for use in such wells. Rapidly dropping and retrieving the bailer in and out of the water caused a back-and-forth action of water in the well, moving some of the more loosely bound fine-grained materials into the well where they could be removed.

Depending on the depth of water in the well, the length of the well screen, and the volume of water the bailer could displace, this method was not always very efficient. "Surge blocks" which could fit inside 2-inch wells provided some improvement on bailing techniques. Such devices are simply plungers which, when given a vigorous up-and-down motion, transfer that energy to an in-and-out action on the water near the well screen. Surge blocks have the potential to move larger quantities of water with higher velocities but pose some risk to the well casing and screen if too tight a fit is made or if the up-and-down action becomes too vigorous. Improved surge block design has been the subject of some recent investigation (Schalla and Landick, 1985).

In more productive aquifers, "overpumping" was and is a popular method for well development. With this method, a pump is alternately turned on (usually at a slightly higher rate than the well can sustain) and off to simulate a surging action in the well. A problem with this method is that the outward movement of water normally created during surging efforts is not as pronounced with overpumping. This may tend to bridge the fine and coarse materials, limiting the movement of the fine materials into the well and thereby limiting the effectiveness of the method.

Pumping with air has also been used effectively (Figure 5-6). Better development has been accomplished by attaching differently shaped devices to the end of an airline to force the air out into the formation. An example of such a device is shown in Figure 5-7. Such a device causes a much more vigorous action on the movement of material in proximity to the well screen while also pushing water to the ground surface.

Air development techniques such as this may expose field crews to hazardous constituents when badly contaminated ground water is present. The technique may also cause chemical reactions with species present in the ground water, especially volatile organic compounds. Care must also be taken to filter the injected air to prevent contamination of the well environment with oil and other lubricants present in the compressor and airlines.

Development procedures for monitoring wells in relatively unproductive geologic materials is somewhat limited. Due to the low hydraulic conductivity of the materials, surging of water in and out of the well casing is extremely difficult. Also, when the well is pumped, the entry rate of the water is inadequate to effectively remove fines from the well bore and the gravel pack material outside the well screen.

In this type of geologic setting, where an open borehole can be sustained, *clean* water can be circulated down the well casing, out through the

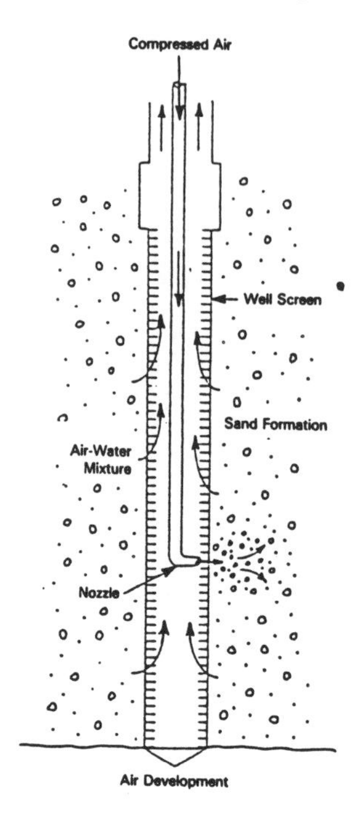

Figure 5-6 Well developments with compressed air.

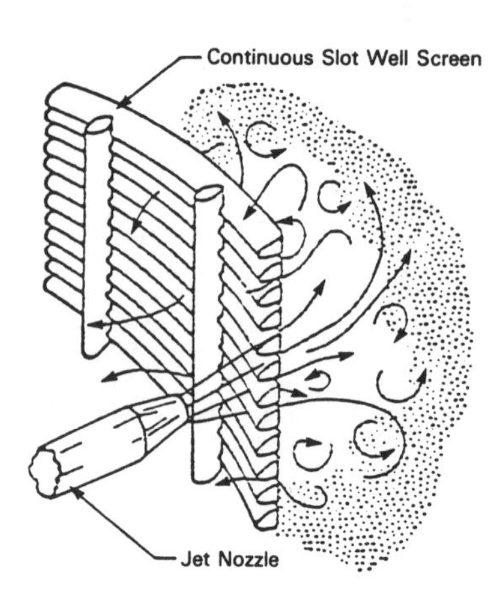

Figure 5-7 The effects of high-velocity jetting used for well development through openings in a continuous-slot well screen.

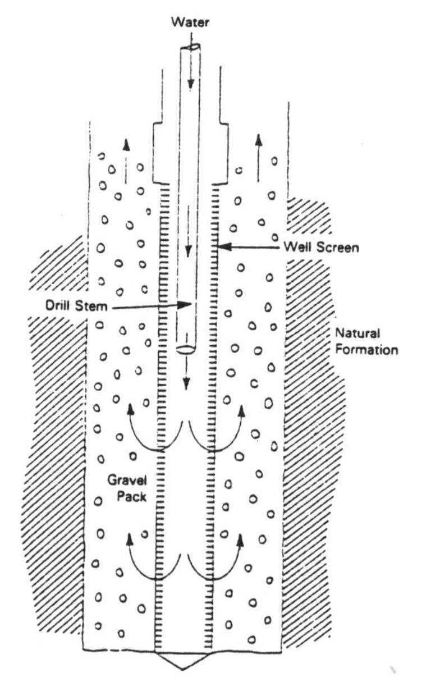

Figure 5-8 Well development by back-flushing with water.

screen, and back up the borehole (Figure 5-8). Relatively high water velocities can be maintained and the mud cake from the borehole wall can be broken down effectively and removed. Because of the low hydraulic conductivity of the geologic materials outside the well, only a small amount of water will penetrate the formation being monitored. This procedure can be done *before* and after placement of a gravel pack but must be conducted before a well seal has been placed. After the gravel pack has been placed, water should not be circulated too quickly or the gravel pack will be lifted out of the borehole as well. Immediately following development, the well should be sealed, backfilled, and pumped for a short period to stabilize the formation around the outside of the screen and to ensure that the well will produce fairly clear water.

5.2.7 Security

For most monitoring well installations, some precautions must be exercised to protect the surface portions of the well from damage. In many instances, inadvertent vehicular accidents do occur. Monitoring well installations seem particularly vulnerable to transgressions from grass mowers. Vandalism is often a major concern, from spontaneous "hunters" looking for a likely target to premeditated destruction of property associated with an unpopular operation. There are several simple solutions that can be employed to help minimize the damage due to accidental collisions. However, outwitting the determined vandal may be an impossible undertaking and certainly an expensive one.

The most basic problem to maintaining the physical condition of any monitoring well is being able to anticipate the hazards that might befall that particular installation. Some instances may call for making the well obvious to see whereas other instances may call for keeping the well inconspicuous.

Where the most likely problem is one of vehicular contact, be it mowers, construction traffic, or other types of two-, three-, or four-wheeled traffic, the first thing that can be done is to make the top of the well plain to see. Make sure it extends far enough above ground to be visible above grass, weeds, or small shrubs. If that is not practical, use a "flag" that extends above the well casing. This is also helpful for periods when leaves or snow have buried low-lying objects.

Paint the well casing a bright color (orange and yellow are the most visible). This not only makes the well more visible but also protects metal casing material from rusting. Care should be taken to make sure paint is not allowed inside the well casing or in threaded fittings that may contact sampling equipment.

Make sure the owners/operators of the site being monitored know where each installation is. Issue maps clearly and exactly indicating where the wells are located. Make certain their employees know the importance of those installations, the cost associated with them, and the difficulty involved in replacement.

The portions of the well that protrude from the ground can also be reinforced, particularly when the well is constructed of PVC or Teflon. The well could be constructed such that only the portion of the well above the water table is metal. In this manner, the integrity of the sample is maintained as ground water contacts only inert material and the physical condition of the well is maintained as the upper metal portion is better able to withstand impact.

There are two arguments to consider when constructing a well in this manner. The arguments are focused on the weak point in the well construction: at or near the juncture of the metal and nonmetal casings. One argument suggests that a longer section of metal casing is superior because the additional length of metal casing in the ground gives additional strength. This way a break is less likely to occur (although the casing is likely to be bent). The other argument suggests that should a break in the casing occur, a shorter length of metal casing is superior because a break nearer to ground surface is easier to repair. Each argument has its merits; only experience with site conditions is likely to produce the best solution.

The use of "well protectors" is another popular solution that involves the use of a larger diameter steel casing placed around the monitoring well at the ground surface and extending several feet below ground (Figure 5-9). The protectors are usually seated in the cement surface seal to a depth below the frost line.

Figure 5-9 Typical well protector installation.

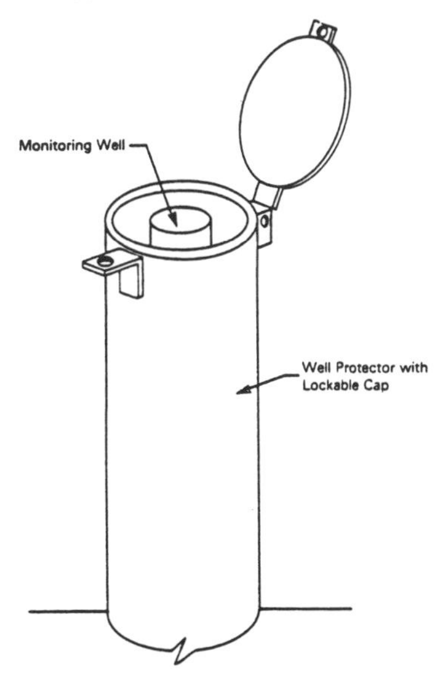

Well protectors are commonly equipped with a locking cap which insures against tampering with the inside of the well. Dropping objects down the well can create two potential problems: 1) impair the sampleability of the well by clogging the well screen or impeding the ability of the sampling device to reach water, and 2) altering the quality of the ground water, particularly where small quantities (perhaps drops) of an organic liquid may be sufficient to completely contaminate the well.

Problems associated with vandalism run from simple curiosity to outright wanton destruction. Obviously, sites within secured, fenced areas are less likely to be vandalized. However, there is probably no way to deter the determined vandal, short of posting a 24-hour guard. In such situations, well protectors are a must. The wells should be kept as inconspicuous as possible. However, the benefits of "hiding" monitoring wells must be weighed against the costs of delays in finding them for sampling and the potential costs for repairs or maintenance on untried security designs.

In some situations, it might be a good policy to notify the public of the need for the monitoring wells. If it is properly asserted that each well serves an environmental monitoring purpose and that the wells have been constructed to insure public well-being, it may create a civic conscience that would help minimize vandalism.

As with all the previously mentioned monitoring well components, no singular solution will best meet every different monitoring situation. Knowledge of the social, political, and economic conditions of the geographic area and circumstances surrounding the need for ground-water monitoring will dictate, to a large degree, the type of well protection needed.

5.3 Monitoring Well Drilling Methods

As might be expected, different drilling techniques can influence the quality of the ground-water sample produced from a particular formation in different ways. This applies to the drilling method employed (e.g., augered, driven, or rotary) as well as the driller. There is no substitute for a conscientious driller willing to take the extra time and care necessary to complete a good monitoring well installation.

Among the criteria used to select an appropriate drilling method are the following factors, listed in order of importance:

1) Hydrologic information
   a. type of formation
   b. depth of drilling
   c. depth of desired screen setting below top of zone of saturation
2) Types of pollutants expected
3) Location of drilling site, i.e., accessibility
4) Design of monitoring well desired
5) Availability of drilling equipment

Table 5-1 summarizes several different drilling methods, their advantages and their disadvantages when used for monitoring well construction. Several excellent publications are referenced for detailed discussions (Campbell and Lehr, 1973; Fenn *et al.*, 1977; Johnson, Inc., 1972; and Scalf *et al.*, 1981). The table also gives a concept of the advantages and disadvantages which need to be considered when choosing a drilling technique for different site and monitoring situations (see also, Lewis, 1982; Luhdorff and Scalmanini, 1982; Minning, 1982; and Voytek, 1983).

Hollow- and solid-stem augering is one of the most desirable drilling methods for constructing monitoring wells. No drilling fluids are used and disturbance to the geologic materials penetrated is minimal. Auger rigs are not typically used when consolidated rock must be penetrated and depths are usually limited to no more than 150 feet.

In formations where the borehole will not stand open, the monitoring well can be constructed inside the hollow-stem augers prior to removal from the hole. Generally, this limits the diameter of the well that can be built to 4 inches. The hollow-stem has an added advantage in offering the ability to collect continuous in situ geologic samples without removal of the auger sections.

The use of the solid-stem is most useful in fine-grained, unconsolidated materials that will not collapse when unsupported. The method is similar to the hollow-stem except that the augers must be removed from the hole to allow the insertion of the well casing and screen. Geologic cores cannot be collected when using a solid-stem. Therefore, geologic sampling must rely on cuttings which come to the surface, an undesirable method as the depth from which the cuttings come is not precisely known.

Cable-tool drilling is one of the oldest methods used in the water well industry. Even though the rate of penetration is rather slow, this method offers many advantages for monitoring well construction. With the cable-tool, excellent formation samples can be collected and the presence of thin permeable zones can be detected. As drilling progresses, a casing is normally driven and this provides an excellent temporary casing within which the monitoring well can be constructed.

In *air-rotary drilling*, air is forced down the drill stem and back up the borehole to remove the cuttings. This technique has been found to be particularly well suited to drilling in fractured rock formations. If the monitoring is intended for organic compounds, the air must be filtered to insure that oil from the air compressor is not introduced to the formation to be monitored. Air-rotary should not be used in highly contaminated environments because the water and cuttings blown out of the hole are difficult to control and can pose a hazard to the drill crew and observers. Where volatile compounds are of interest, air-rotary can volatilize those compounds and cause water samples withdrawn from the hole to be unrepresentative of in situ conditions. The use of foam additives to aid cuttings removal presents the opportunity for organic contamination of the monitoring well.

Air-rotary with percussion hammer increases the effectiveness of air-rotary for cavey or highly creviced formations. Addition of the percussion hammer gives air-rotary the ability to drive casing, cutting the loss of air circulation in fractured rock and maintaining an open hole in soft formations. The capability of constructing monitoring wells inside the driven casing prior to its being pulled adds to the appeal of air-percussion. However, the problems

Table 5-1 Advantages and Disadvantages of Selected Drilling Methods for Monitoring Well Construction.

Method	Drilling Principle	Advantages	Disadvantages
Drive Point	1.25 to 2 inch ID casing with pointed screen mechanically depth.	Inexpensive. Easy to install, by hand if necessary. Water samples can be collected as driving proceeds. Depending on overburden, a good seal between casing and formation can be achieved.	Difficult to sample from smaller diameter drive points if water level is below suction lift. Bailing possible. No formation samples can be collected. Limited to fairly soft materials. Hard to penetrate compact, gravelly materials. Hard to develop. Screen may become clogged if thick clays are penetrated. PVC and Teflon® casing and screen are not strong enough to be driven. Must use metal construction materials which may influence some water quality determinations.
Auger, Hollow- and Solid-stem	Successive 5-foot flights of spiral-shaped drill stem are rotated into the ground to create a hole. Cuttings are brought to the surface by the turning action of the auger.	Inexpensive. Fairly simple operation. Small rigs can get to difficult-to-reach areas. Quick set-up time. Can quickly construct shallow wells in firm, noncavey materials. No drilling fluid required. Use of hollow-stem augers greatly facilitates collection of split-spoon samples. Small-diameter wells can be built inside hollow-stem flights when geologic materials are cavey.	Depth of penetration limited, especially in cavey materials. Maximum depths 150 feet. Cannot be used in rock or well-cemented formations. Difficult to drill in cobbles/boulders. Log of well is difficult to interpret without collection of split spoons due to the lag time for cuttings to reach ground surface. Vertical leakage of water through borehole during drilling is likely to occur. Solid-stem limited to fine grained, unconsolidated materials that will not collapse when unsupported. With hollow-stem flights, heaving materials can present a problem. May need to add water down auger to control heaving or wash materials from auger before completing well.
Jetting	Washing action of water forced out of the bottom of the drill rod clears hole to allow penetration. Cuttings brought to surface by water flowing up the outside of the drill rod.	Inexpensive. Driller often not needed for shallow holes. In firm, noncavey deposits where hole will stand open, well construction fairly simple.	Somewhat slow, especially with increasing depth. Extremely difficult to use in very coarse materials, i.e., cobbles/boulders. A water supply is needed that is under enough pressure to penetrate the geologic materials present. Difficult to interpret sequence of geologic materials from cuttings. Maximum depth 150 feet, depending on geology and water pressure capabilities.
Cable-tool (Percussion)	Hole created by dropping a heavy "string" of drill tools into well bore, crusing materials at bottom. Cuttings are removed occasionally by bailer. Generally, casing is driven just ahead of the bottom of the hole; a hole greater than 6 inches in diameter is usually made.	Can be used in rock formations as well as unconsolidated formations. Fairly accurate logs can be prepared from cuttings if collected often enough. Driving a casing ahead of hole minimizes cross-contamination by vertical leakage of formation waters. Core samples can be obtained easily.	Requires an experienced driller. Heavy steel drive pipe used to keep hole open and drilling "tools" can limit accessibility. Cannot run some geophysical logs due to presence of drive pipe. Relatively slow drilling method.

Table 5-1 (continued)

Method	Drilling Principle	Advantages	Disadvantages
Hydraulic Rotary	Rotating bit breaks formation; cuttings are brought to the surface by a circulating fluid (mud). Mud is forced down the interior of the drill stem, out the bit, and up the annulus between the drill stem and hole wall. Cuttings are removed by settling in a "mud pit" at the ground surface and the mud is circulated back down the drill stem.	Drilling is fairly quick in all types of geologic materials. Borehole will stay open from formation of a mud wall on sides of borehole by the circulating drilling mud. Eases geophysical logging and well construction. Geologic cores can be collected. Virtually unlimited depths possible.	Expensive, requires experienced driller and fair amount of peripheral equipment. Completed well may be difficult to develop, especially small-diameter wells, because of mud wall on borehole. Geologic logging by visual inspection of cuttings is fair due to presence of drilling mud. Thin beds of sand, gravel, or clay may be missed. Presence of drilling mud can contaminate water samples, especially the organic, biodegradable muds. Circulation of drilling fluid through a contaminated zone can create a hazard at the ground surface with the mud pit and cross-contaminate clean zones during circulation.
Reverse Rotary	Similar to Hydraulic Rotary method except the drilling fluid is circulated down the borehole outside the drill stem and is pumped up the inside, just the reverse of the normal rotary method. Water is used as the drilling fluid, rather than a mud, and the hole is kept open by the hydrostatic pressure of the water standing in the borehole.	Creates a very "clean" hole, not dirtied with drilling mud. Can be used in all geologic formations. Very deep penetrations possible. Split-spoon sampling possible.	A large water supply is needed to maintain hydrostatic pressure in deep holes and when highly conductive formations are encountered. Expensive—experienced driller and much peripheral equipment required. Hole diameters are usually large, commonly 18 inches or greater. Cross-contamination from circulating water likely. Geologic samples brough to surface are generally poor, circulating water will "wash" finer materials from sample.
Air Rotary	Very similar to Hydraulic Rotary, the main difference being that air is used as the primary drilling fluid as opposed to mud or water.	Can be used in all geologic formations; most successful in highly fractured environments. Useful at any depth. Fairly quick. Drilling mud or water not required.	Relatively exoensive. Cross-contamination from vertical communication possible. Air will be mixed with water in the hole and that which is blown from the hole, potentially creating unwanted reactions with contaminants; may affect "representative" samples. Cuttings and water blown from the hole can pose a hazard to crew and surrounding environment if toxic compounds encountered. Organic foam additives to aid cuttings removal may contaminate samples.
Air-Percussion Rotary or Downhole-Hammer	Air Rotary with a reciprocating hammer connected to the bit to fracture rock.	Very fast penetrations. Useful in all geologic formations. Only small amounts of water needed for dust and bit temperature control. Cross-contamination potential can be reduced by driving casing.	Relatively expensive. As with most hydraulic rotary methods, the rig is fairly heavy, limiting accessibility. Vertical mixing of water and air creates cross-contamination potential. Hazard posed to surface environment if toxic compounds encountered. Organic foam additives for cuttings removal may contaminate samples.

with contamination and crew safety must still be considered.

Reverse-rotary drilling has limited application for monitoring well construction. Reverse-rotary requires that large quantities of water be circulated down the borehole and up the drill stem to remove cuttings. If permeable formations are encountered, significant quantities of water can move into the formation to be monitored, altering the quality of the water to be sampled.

Hydraulic rotary, or "mud" rotary, is probably the most popular method used in the water well industry. However, hydraulic rotary presents some disadvantages for monitoring well construction. With hydraulic rotary, a drilling mud (usually bentonite) is circulated down the drill stem and up the borehole to remove cuttings. The mud creates a wall on the side of the borehole which must be removed from the screened area by development procedures. With small diameter wells, complete removal of the drilling mud is not always achieved. The ion exchange potential of most drilling muds is high and may effectively reduce the concentration of trace metals in water entering the well. In addition, the use of biodegradable, organic drilling muds can introduce organic components to water sampled from the well.

Most ground-water monitoring wells will be completed in glaciated or unconsolidated materials and will be relatively shallow, perhaps less than 50 to 75 feet. In these applications, hollow-stem augering usually will be the method of choice. Solid-stem auger, cable-tool, and air-percussion also offer advantages depending on the geology and contaminant of interest.

5.3.1 Geologic Samples

Permit applications for disposal of waste materials often require that geologic samples be collected at the disposal site. Investigations of ground-water movement and contaminant transport should also include the collection of geologic samples for physical inspection and testing. Opportunity for stratigraphic sample collection is best afforded during monitoring well drilling.

Samples can be collected continuously, at each change in stratigraphic unit, or, in homogeneous materials, at regular intervals. These samples may later be classified, tested, and analyzed for physical properties such as particle size distribution, textural classification, and hydraulic conductivity, and for chemical analyses such as ion-exchange capacity, chemical composition, and specific parameter leachability.

Probably the most common method of material sampling is with a "split-spoon" sampler. This device is a 12- or 18-inch long hollow cylinder (2-inch diameter) which is split in half lengthwise. The halves are held together at each end with threaded couplings; the top end attaches to the drill rod, and the bottom end is a drive shoe (Figure 5-10a). The sampler is lowered to the bottom of the hole and driven ahead of the hole with a weighted "hammer" striking an anvil at the upper end of the drill rod to which the sampler is attached. Sample is forced up the inside of the tube and is held with a basket trap or flap valve that allow the sample to enter the sampler but not exit (though retention of noncohesive, sandy formations is often difficult). After the sampler is withdrawn from the hole, the sample is removed by unscrewing the ends and separating the sample collection tube.

Another common sampler is the thin wall tube or "Shelby" tube. These tubes are usually 2 to 5-1/2 inches in diameter and about 24 inches long. The cutting edge of the tube is sharpened and the upper end is attached to a coupling head by means of cap screws or a retaining pin (Figure 5-10b). A Shelby tube has a minimum ratio of wall area to sample area and creates the least disturbance to the sample of any drive-type sampler in current use (for hydraulic conductivity tests of low conductivity, $< 10^{-6}$ cm/sec materials, minimal disturbance is critical). After retraction, the tube is disconnected from the head and the sample is removed from the tube with a jack or press. If sample preservation is a major concern, the tube can be sealed and shipped to the laboratory.

Apart from permit requirements, material samples are very helpful for deciding at what depth to complete a monitoring well. Unexpected changes encountered during drilling can alter preconceived ideas concerning the local ground-water flow regime. In many instances, the driller will be able to detect a change in formation by a change in penetration rate, sound, or "feel" of the drilling rig. However, due to the lag time for cuttings to come to the surface and the amount of mixing the cuttings may undergo as they come up the borehole, the only way to truly know what the subsurface materials look like is to stop drilling and collect a sample.

5.3.2 Case History

Several different types of monitoring wells were constructed during the investigation of a volatile organic contaminant plume in northern Illinois (Wehrmann, 1984). A brief summary of the types of wells employed and the reasons for their use help illustrate how an actual ground-water quality monitoring problem was approached.

During the final weeks of a one-year study of ground-water nitrate quality in north central Illinois, the presence of a number of organic compounds was detected in the drinking water of all five homes sampled within a large rural residential subdivision. The principal compound found was trichloroethylene, TCE, at concentrations between 50 and 1,000

Figure 5-10 Cross-sectional views of (a) split spoon and (b) Shelby tube samplers (from Mobile Drilling Co., 1972).

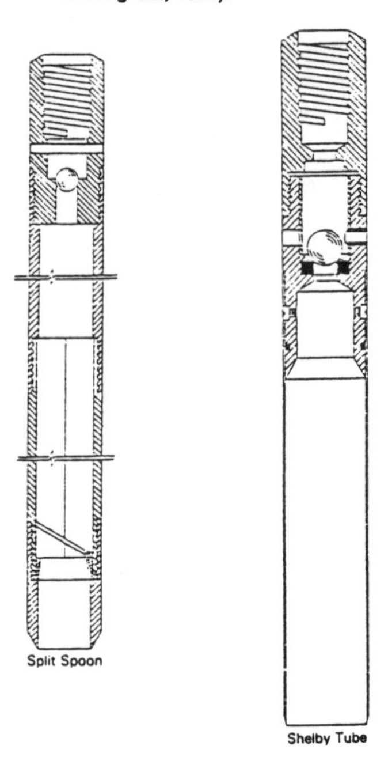

micrograms per liter (µg/l). All the homes in the subdivision utilized private wells tapping a surficial sand and gravel deposit at a depth of 65 to 75 feet. A geologic cross-section of the study area is shown in Figure 5-11.

Two immediate concerns needed to be addressed. First, how many other drinking wells were affected and, second, what was the contaminant source? Early thoughts connected the TCE contaminant to the contamination potential of the large number of septic systems in the subdivision. Previous work (Wehrmann, 1983) had established the ground-water flow direction beneath the affected subdivision was from north-northeast to south-southwest. Because the area upgradient of the subdivision was primarily farmland, several monitoring wells placed upgradient of the subdivision would help confirm or deny the possibility that the septic systems were the source of the VOC contamination.

Five "temporary" monitoring wells were constructed in the field upgradient of the affected subdivision. Original plans called for driving a 2-inch diameter sandpoint to depths from 40 to 70 feet. Samples would be collected at 10-foot intervals as the point was driven. Once 70 feet was reached, the sandpoint would be pulled, the hole properly abandoned, and the point driven at a new sampling location. The first hole was to be placed north (upgradient) of a domestic well found to be highly contaminated. Holes were to be placed successively in an upgradient direction proceeding across the field. In this manner, ground-water samples could be quickly collected at many depths and locations, the well materials recovered, and the field left relatively undisturbed.

Once drilling commenced, however, it became clear that it was not possible to drive 2-inch sandpoints into the coarse sand and gravel just below ground surface in this area. An air-percussion rig was brought on-site and a new approach was used. A 4-inch diameter screen (2 feet long) with a drive shoe was welded to a 4-inch diameter steel casing. This assembly was driven by air hammer to the desired sampling depth. The bottom of the drive shoe, being open, forced the geologic materials penetrated into the casing and screen. These materials were evacuated from the casing and screen by air rotary once the desired depth was reached. To avoid cross-contamination from using the same materials at several locations, all well materials were steam cleaned prior to use and between holes. Locations of the temporary well sites and the analytical results for TCE from samples bailed at depths of 40 and 50 feet are shown in Figure 5-12.

Results of the temporary sampling revealed the contaminant source was indeed outside of the subdivision. Due to the construction and sampling methods employed for these wells, emphasis was not placed on the quantitative aspects of the sampling results. However, important qualitative conclusions were made. The temporary wells confirmed the presence of VOCs directly upgradient of the subdivisions and provided information for the location and depth of nine permanent monitoring wells.

Due to the problems associated with organic compound leachability and adsorption from PVC casing and screen, flush-threaded stainless steel casing and screen, 2-inches in diameter, were used for the permanent sampling wells. The screens were 2 feet long with 0.01-inch wire-wound slot openings. All materials associated with the monitoring well construction, including the drill rig, were steam cleaned prior to the commencement of drilling to avoid organic contamination from cutting oils and grease. Prior to use, the casing and screen materials were kept off-site in a covered, protected area.

To insure that the sandy materials would not collapse the hole after drilling, casing lengths and the screen were screwed together above ground and placed down the inside of the augers before the auger flights were pulled out of the hole. The sand and gravel below the water table collapsed around the screen and casing as the augers were removed. To help prevent vertical movement of water down along the casing, a wet bentonite/cement mixture was placed in

Figure 5-11 East-west cross section across Rock River Valley at Roscoe (from Berg et al., 1981).

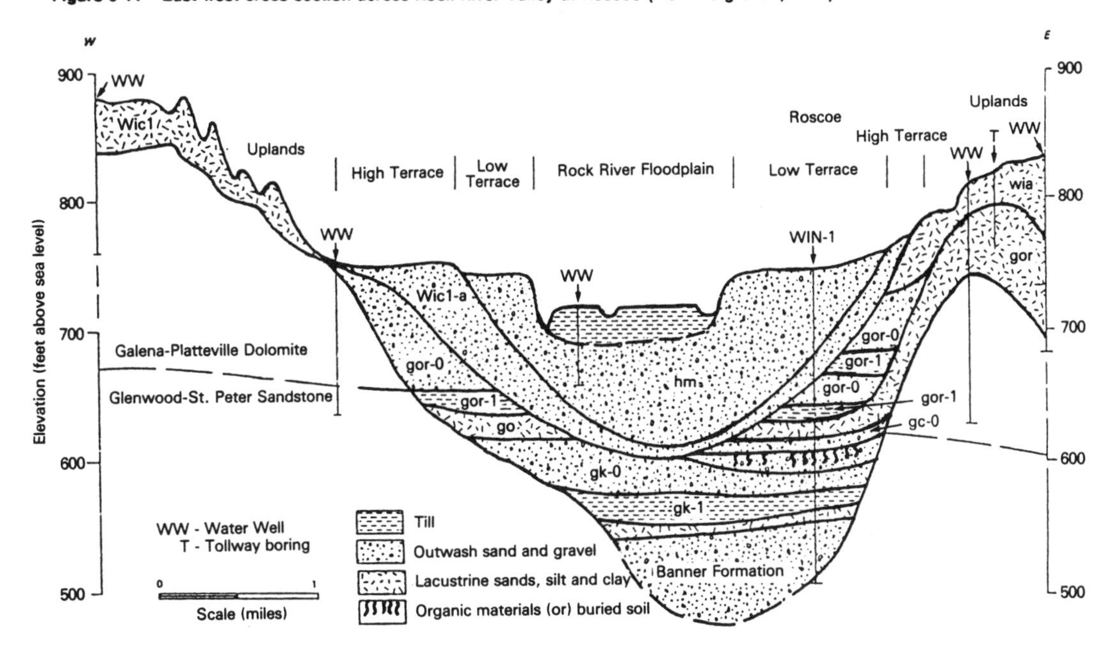

the annulus just above the water table to a thickness of 2 to 3 feet. Cuttings (principally clean, medium to fine sand) were backfilled above the bentonite/cement seal to within 4 feet of land surface. Another bentonite/cement mixture was placed to form a seal at ground surface, further preventing movement of water down along the well casing. A 4-inch diameter steel protective cover with locking cap was placed around the protruding casing and into the surface seal to protect against vandalism.

The nine wells were drilled at four locations with paired wells at three sites and a nest of three wells at one site (Figure 5-13). The locations were based on the analytical results of the samples taken from the temporary wells and basic knowledge of the ground-water flow direction. Locations were numbered as nests 1 through 4 in order of their construction. Nest 1, located immediately north of the affected subdivision, consists of three wells completed at approximately 60, 70 and 80 feet below ground surface. Nest 2 consists of two wells 50 and 60 feet deep. Nest 3 consists of two wells constructed to 40 and 55 feet and nest 4 consists of two wells completed at 50 and 60 feet.

Subsequent to the completion of these nine wells, it was felt an additional well constructed to 100 feet at the location of nest 1 was needed to further define the vertical extent of the contaminant plume. Because the hollow-stem auger rig was no longer available, arrangements were made to use a cable-tool rig to drill the hole. The well was constructed over a period of two days, somewhat slower than any of the other methods previously used (but typical of cable-tool speeds). With this method, a 6-inch casing was driven several feet, a bit was used to break up the materials inside the casing, then the materials were removed from the casing with a dart-valve bailer. This procedure was repeated until the desired depth of 100 feet was reached. Once this depth was reached, the well casing and screen were screwed together and lowered down the hole. The 6-inch casing was then pulled back allowing the hole to collapse about the well (the well was constructed of stainless steel exactly as the nine other monitoring wells). As before, all drilling equipment and well construction materials were steam cleaned prior to use.

Appraisal of the results of sampling these monitoring wells and the domestic wells in the area produced the pictorial representations shown in Figures 5-14 and 5-15. Figure 5-14 conceptually illustrates a cross section of the TCE plume looking in the general direction of ground-water flow in the vicinity of monitoring nests 2, 3, and 4. The likely extent of the VOC contaminant plume is shown in Figure 5-15. This map includes a limited amount of data from privately owned monitoring wells located on industrial property just upgradient of monitoring nests 2 and 4.

Figure 5-12 Locations and TCE concentrations for temporary monitoring wells at Roscoe, Illinois (from Wehrmann, 1984).

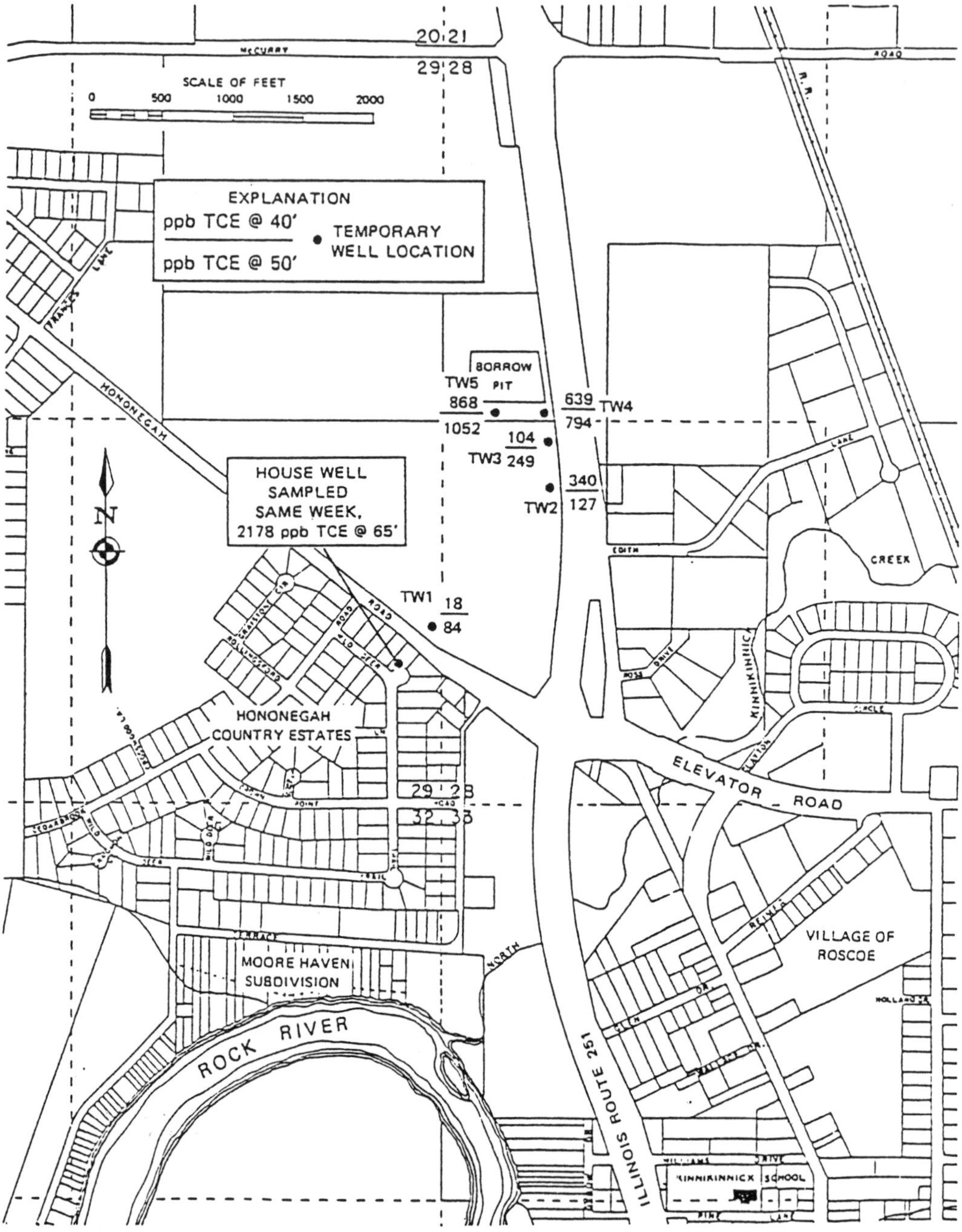

Figure 5-13 Location of monitoring well nests and cross-section A-A'at Roscoe, Illinois (from Wehrmann, 1984).

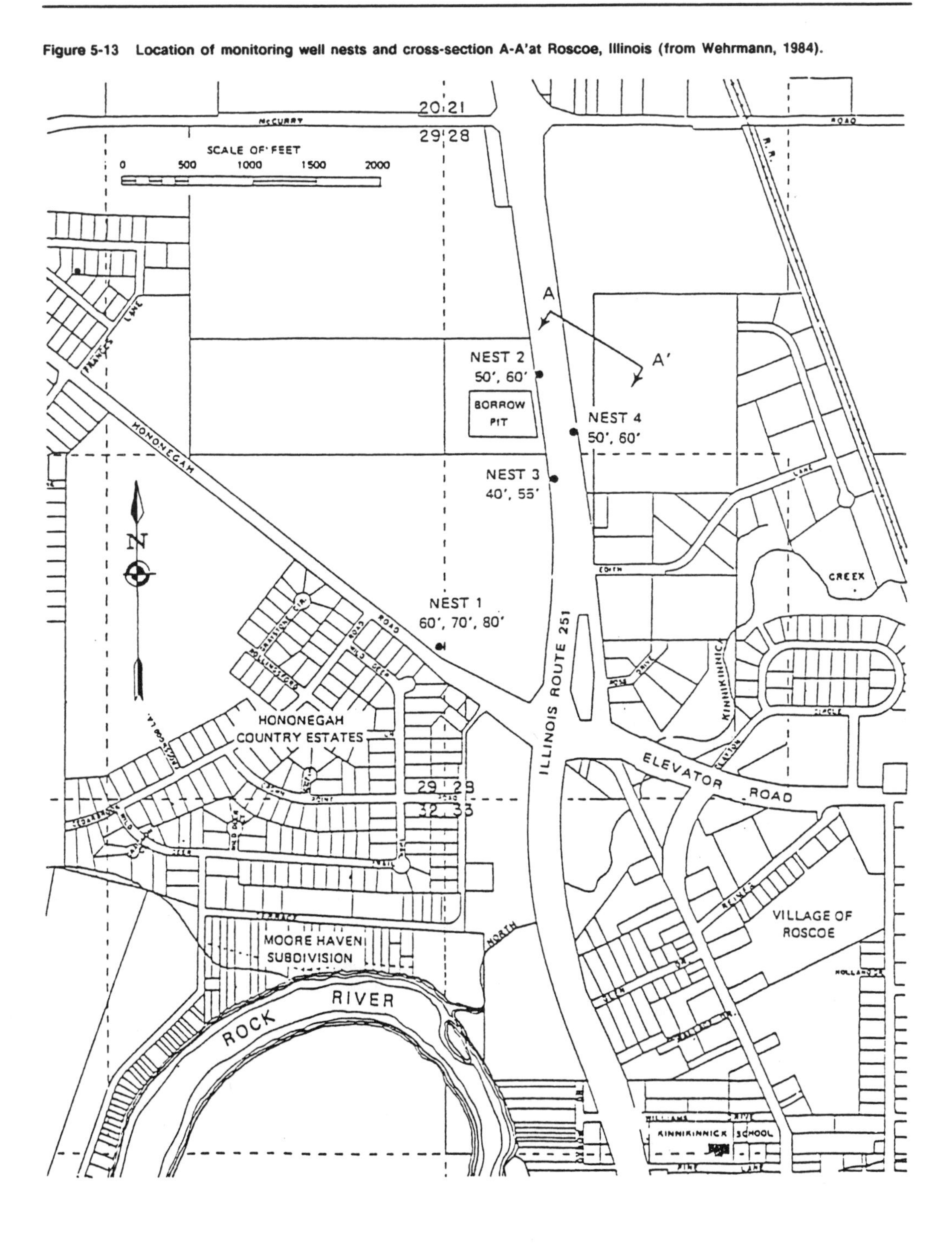

The dashed lines indicate the probable extent of the contaminant plume based on the dimensions of the plume as it passes beneath the developed area along the Rock River.

This monitoring situation clearly indicates the role different drilling and construction techniques can take in a ground-water sampling strategy. In each instance, much consideration was given to the effect the methods used for construction and sampling would have on the resultant chemical data. Where quantitative results for a fairly "quick" preliminary investigation were not necessary and, after determining that it was too difficult to drive sandpoints, it was felt using air-percussion rotary was acceptable. For the placement of the permanent monitoring wells, wells that may become crucial for contaminant source identification and possibly be involved in litigation, the hollow-stem auger was the technique of choice. Finally, when the hollow-stem auger was not available and it was decided another hole was needed, the cable-tool rig was chosen. Here, it was recognized that only one hole was to be drilled so the relative slowness of the method became less a factor. Also, the depth of completion (100 feet) in the cavey sand and gravel made cable-tool preferable over the hollow-stem. Note, too, that each method chosen was capable of maintaining an open hole without the use of drilling mud which could have affected the results of the organic compound analyses.

5.4 Summary

Critical considerations for the design of ground-water quality monitoring networks include alternatives for well design and drilling techniques. With a knowledge of the principal chemical constituents of interest, local hydrogeology, and an appreciation of subsurface geochemistry, appropriate selections of materials for well design and drilling techniques can be made. Whenever possible, physical disturbance and the amount of foreign material introduced into the subsurface should be minimized.

The choices of drilling methods and well construction materials are very important decisions to be made in every type of ground-water monitoring program. Details of network construction can introduce significant bias into monitoring data which frequently may be corrected only by repeating the process of well siting, installation, completion, and development. This can be quite costly in time, effort, money, and loss of information. Undue expense is avoidable if planning decisions are made cautiously with an eye to the future.

The expanding scientific literature on effective ground-water monitoring techniques should be read and evaluated on a continuing basis. This information will help supplement guidelines, such as this, for applications to specific monitoring efforts.

5.5 References

Barcelona, M.J., J.P. Gibb, J.A. Helfrich, and E.E. Garske. 1985. Practical Guide for Ground-Water Sampling. Illinois State Water Survey. U.S. Environmental Protection Agency, Robert S. Kerr Environmental Research Laboratory, Ada, OK and Environmental Monitoring and Support Laboratory, Las Vegas, NV.

Barcelona, M.J. 1984. TOC Determinations in Ground Water. Ground Water 22(1):18-24.

Barcelona, M.J., J.A. Helfrich, E.E. Garske, and J.P. Gibb. 1984. A Laboratory Evaluation of Ground Water Sampling Mechanisms. Ground Water Monitoring Review 4(2):32-41.

Barcelona, J.J., J.P. Gibb, and R.A. Miller. 1984. A Guide to the Selection of Materials for Monitoring Well Construction and Ground-Water Sampling. Illinois State Water Survey Contract Report 327. Illinois State Water Survey, Champaign, IL.

Berg, R.C., J.P. Kempton, and A.N. Stecyk. 1981. Geology for Planning in Boone and Winnebago Counties, Illinois. Illinois State Geological Survey Circular 531, Illinois State Geological Survey, Urbana, IL.

Brown, K.W., J. Green, and J.C. Thomas. 1983. The Influence of Selected Organic Liquids on the Permeability of Clay Liners. Proceedings of the Ninth Annual Research Symposium: Land Disposal, Incineration, and Treatment of Hazardous Wastes. U.S. Environmental Protection Agency SHWRD/EPCS, May 2-4, 1983, Ft. Mitchell, KY.

Burkland, P.W., and E. Raber. 1983. Method to Avoid Ground-Water Mixing Between Two Aquifers During Drilling and Well Completion Procedures. Ground Water Monitoring Review 3(4):48-55. Campbell, M.D. and J.H. Lehr. 1973. Water Well Technology. McGraw-Hill Book Company, New York, NY.

Fenn, D., E. Cocozza, J. Isbister, O. Braids, B. Yare, and P. Roux. 1977. Procedures Manual for Ground Water Monitoring at Solid Waste Disposal Facilities (SW-611). U.S. Environmental Protection Agency, Cincinnati, OH.

Gibb, J.P., R.M. Schuller, and R.A. Griffin. 1981. Procedures for the Collection of Representative Water Quality Data from Monitoring Wells. Cooperative Groundwater Report. Illinois State Water and Geological Surveys, Champaign, IL.

Gillham, R.W., M.J.L. Robin, J.F. Barker, and J.A. Cherry. 1983. Groundwater Monitoring and Sample Bias. American Petroleum Institute Publication 4367, Environmental Affairs Department.

Figure 5-14 Cross-Section A-A' through monitoring nests 2, 3, and 4, looking in the direction of ground-water flow (from Wehrmann, 1984).

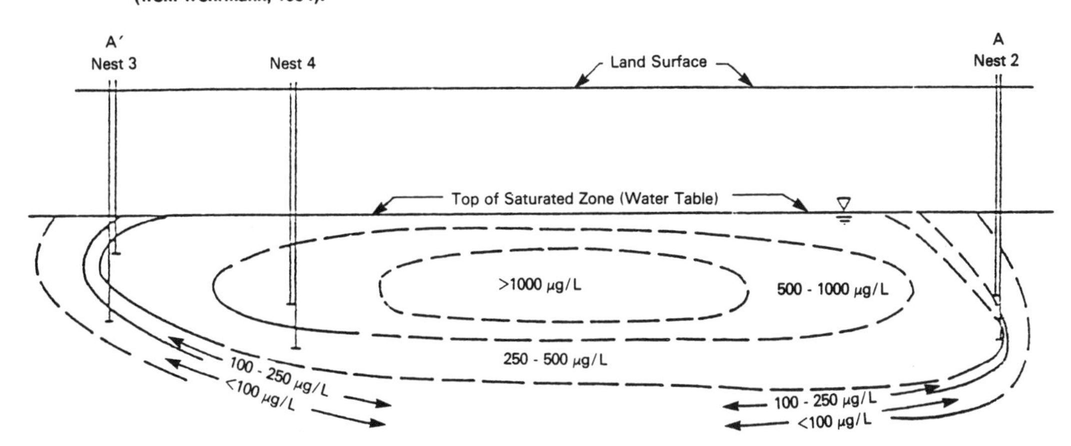

Illinois State Water Survey and Illinois State Geological Survey. 1984. Proceedings of the 1984 ISWS/ISGS Groundwater Monitoring Workshop. February 27-28, Champaign, IL.

Illinois State Water Survey and Illinois State Geological Survey. 1982. Proceedings of the 1982 ISWS/ISGS Groundwater Monitoring Workshop. Illinois Section of American Water Works Association. February 22-23, 1982, Champaign, IL.

Johnson, T.L. 1983. A Comparison of Well Nests vs. Single-Well Completions. Ground Water Monitoring Review 3(1):76-78.

Johnson, E.E., Inc. 1972. Ground Water and Wells. Johnson Division, Universal Oil Products Co., St. Paul, MN.

Keith, S.J., L.G. Wilson, H.R. Fitch, D.M. Esposito. 1982. Sources of Spatial-Temporal Variability in Ground-Water Quality Data and Methods of Control: Case Study of the Cortaro Monitoring Program, Arizona. Proceedings of the Second National Symposium on Aquifer Restoration and Ground Water Monitoring. National Water Well Association, May 26-28, 1982, Columbus, OH.

Lewis, R.W. 1982. Custom Designing of Monitoring Wells for Specific Pollutants and Hydrogeologic Conditions. Proceedings of the Second National Symposium on Aquifer Restoration and Ground Water Monitoring. National Water Well Association, May 26-28, 1982, Columbus, OH.

Luhdorff, E.E., Jr., and J.C. Scalmanini. 1982. Selection of Drilling Method, Well Design and Sampling Equipment for Wells for Monitor Organic Contamination. Proceedings of the Second National Symposium on Aquifer Restoration and Ground Water Monitoring, National Water Well Association, May 26-28, 1982, Columbus, OH.

Mackay, D.M., P.V. Roberts, and J.A. Cherry. 1985. Transport of Organic Contaminants in Groundwater. Environmental Science & Technology 19(5):384-392.

Miller, G.D. 1982. Uptake and Release of Lead, Chromium and Trace Level Volatile Organics Exposed to Synthetic Well Castings. Proceedings of Second National Symposium on Aquifer Restoration and Ground Water Monitoring. National Water Well Association, May 26-28, 1982, Columbus, OH.

Minning, R.C. 1982. Monitoring Well Design and Installation. Proceedings of the Second National Symposium on Aquifer Restoration and Ground Water Monitoring. National Water Well Association, May 26-28, 1982, Columbus, OH.

Mobile Drilling Company. 1972. Soil Sampling Equipment - Accessories. Catalog 650. Mobile Drilling Company, Indianapolis, IN.

Morrison, R.D. and P.E. Brewer. 1981. Air-Lift Samplers for Zone of Saturation Monitoring. Ground Water Monitoring Review 1(1):52-55.

Naymik, T.G. and M.E. Sievers. 1983. Groundwater Tracer Experiment (II) at Sand Ridge State Forest, Illinois. Illinois State Water Survey Contract Report 334. Illinois State Water Survey, Champaign, IL.

Naymik, T.G. and J.J. Barcelona. 1981. Characterization of a Contaminant Plume in Groundwater, Meredosia, Illinois. Ground Water 16(3):149-157.

Figure 5-15 General area of known TCE contamination (from Wehrmann, 1984).

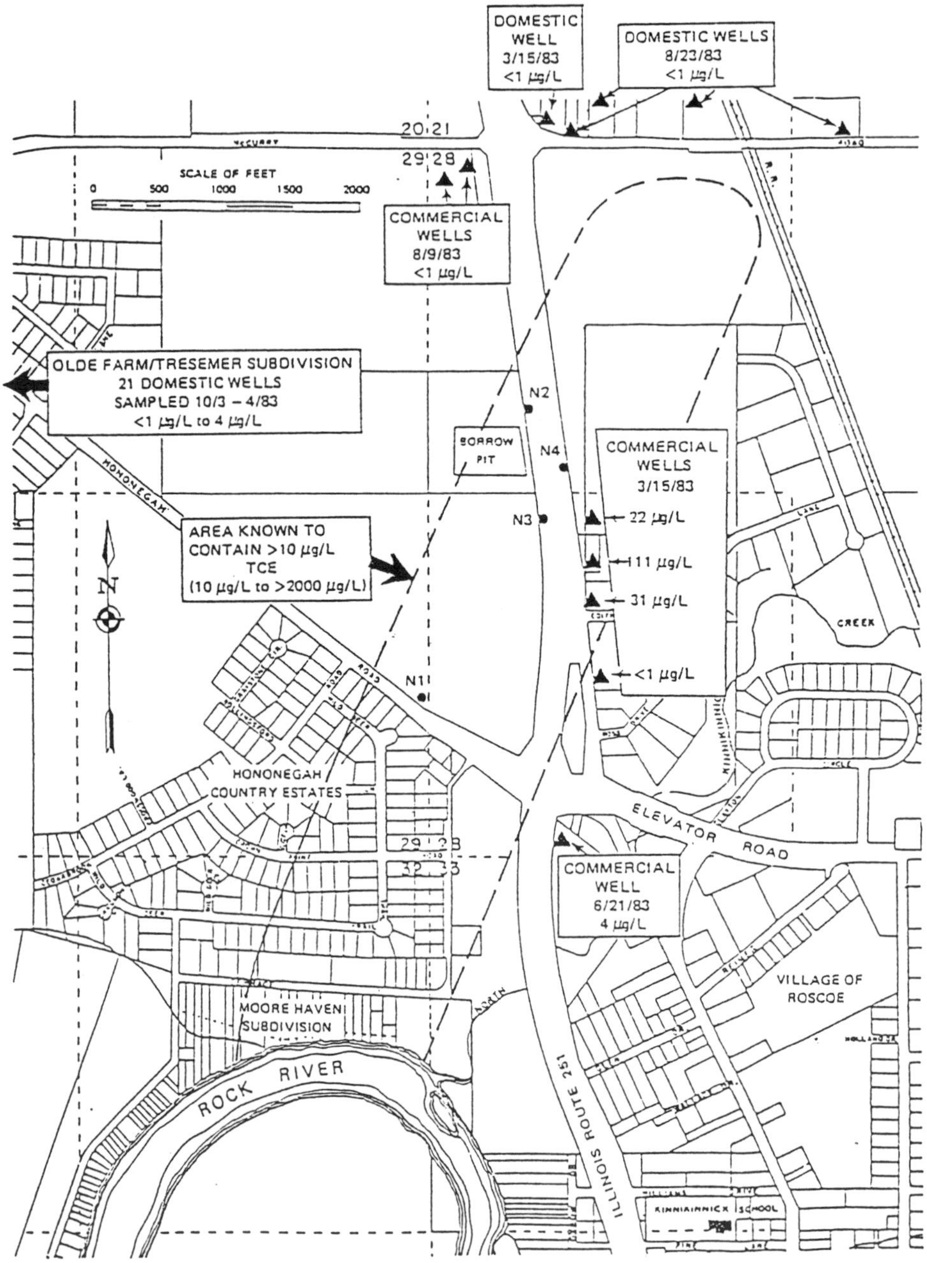

O'Hearn, M. 1982. Groundwater Monitoring at the Havana Power Station's Ash Disposal Ponds and Treatment Lagoon. Confidential Contract Report. Illinois State Water Survey, Champaign, IL.

Perry, C.A., and R.J. Hart. 1985. Installation of Observation Wells on Hazardous Waste Sites in Kansas Using a Hollow-Stem Auger. Ground Monitoring Review 5(4):70-73.

Pettyjohn, W.A., and A.W. Hounslow. 1982. Organic Compounds and Ground-Water Pollution. Proceedings of the Second National Symposium on Aquifer Restoration and Ground Water Monitoring. National Water Well Association, May 26-28, 1982, Columbus, OH.

Pettyjohn, W.A., W.J. Dunlap, R. Cosby, and J.W. Keeley. 1981. Sampling Ground Water for Organic Contaminants. Ground Water 19(2):180-189.

Pfannkuch, H.O. 1981. Problems of Monitoring Network Design to Detect Unanticipated Contamination. Proceedings of the First National Ground Water quality Monitoring Symposium and Exposition. National Water Well Association, May 29-30, 1981, Columbus, OH.

Pickens, J.F., J.A. Cherry, R.M. Coupland, G.E. Grisak, W.F. Merritt, and B.A. Risto. 1981. A Multi-Level Device for Ground-Water Sampling. Ground Water Monitoring Review 1(1):48-51.

Rinaldo-Lee, M.B. 1983. Small - vs. Large-Diameter Monitoring Wells. Ground Water Monitoring Review 3(1):72-75.

Scalf, M.R., J.F. McNabb, W.J. Dunlap, R.L. Cosby, and J. Fryberger. 1981. Manual of Ground-Water Sampling Procedures. NWWA/EPA Series, National Water Well Association, Worthington, OH.

Schalla, R., and R.W. Landick. 1985. A New Valved and Air-Vented Surge Plunger for Developing Small-Diameter Monitor Wells. Proceedings of the Third National Symposium and Exposition on Ground-Water Instrumentation. National Water Well Association, October 2-4, 1985, San Diego, CA.

Sosebee, J.B., Jr. *et al.* 1982. Contamination of Groundwater Samples with PVC Adhesives and PVC Primer from Monitor Wells. Environmental Science and Engineering, Inc., Gainesville, FL.

Torstensson, B.A. 1984. A New System for Ground Water Monitoring. Ground Water Monitoring Review 4(4):131-138.

Voytek, J.E., Jr. 1983. Considerations in the Design and Installation of Monitoring Wells. Ground Water Monitoring Review 3(1):70-71.

Walker, W.H. 1974. Tube Wells, Open Wells, and Optimum Ground-Water Resource Development. Ground Water 12(1):10-15.

Wehrmann, H.A. 1984. An Investigation of a Volatile Organic Chemical Plume in Northern Winnebago County, Illinois. Illinois State Water Survey Contract Report 346, Illinois State Water Survey, Champaign, IL.

Wehrmann, H.A. 1983. Monitoring Well Design and Construction. Ground Water Age 17(8):35-38.

Wehrmann, H.A. 1983. Potential Nitrate Contamination of Groundwater in the Roscoe Area, Winnebago County, Illinois. Illinois State Water Survey Contract Report 325, Illinois State Water Survey, Champaign, IL.

Wehrmann, H.A. 1982. Groundwater Monitoring for Fly Ash Leachate, Baldwin Power Station, Illinois Power Company. Confidential Contract Report. Illinois State Water Survey, Champaign, IL.

6. Ground-Water Sampling

6.1 Introduction

6.1.1 Background

Ground-water sampling is conducted to provide information on the condition of our subsurface water resources. Whether the goal of the monitoring effort is detection or assessment of contamination, the information gathered during sampling efforts must be of known quality and be well documented. The most efficient way to accomplish these goals for water quality information is the development of a sampling protocol which is tailored to the information needs of the program and the hydrogeology of the site or region under investigation. This sampling protocol incorporates detailed descriptions of sampling procedures and other techniques which of themselves are not sufficient to document data quality or reliability. Sampling protocols are central parts of networks or investigatory strategies.

The need for reliable ground-water sampling procedures has been recognized for years by a variety of professional, regulatory, public and private groups. The technical basis for the use of selected sampling procedures for environmental chemistry studies has been developed for surface-water applications over the last four decades. However, ground-water quality monitoring programs have unique needs and goals which are fundamentally different from previous investigative activities. The reliable detection and assessment of subsurface contamination require minimal disturbance of geochemical and hydrogeologic conditions during sampling.

At this time proven well construction, sampling and analytical protocols for ground-water sampling have been developed for many of the more problematic chemical constituents of interest. However, the acceptance of these procedures and protocols must await more careful documentation and firm regulatory guidelines for monitoring program execution. The time and expense of characterizing actual subsurface conditions place severe restraints on the methods which can be employed. Since the technical basis for documented, reliable drilling, sample collection and handling procedures is in the early stages of development, conscientious efforts to document method performance under real conditions should be a part of any ground-water investigation (Barcelona *et al.*, 1985; Scalf *et al.*, 1981).

6.1.2 Information Sources

Much of the literature on routine ground-water monitoring methodology has been published in the last 10 years. The bulk of this work has emphasized *ambient resource* or *contaminant resource monitoring* (detection and assessment) rather than case preparation or enforcement efforts. General references which are useful to the design and execution of sampling efforts are those of the U.S. Geological Survey (1977; Wood, 1976), the U.S. Environmental Protection Agency (Brass *et al.*, 1977; Dunlap *et al.*, 1977; Fenn *et al.*, 1977; Sisk, 1981) and others (National Council of the Paper Industry, 1982; Tinlin, 1976). In large part, these past works treat sampling in the context of overall monitoring programs, providing descriptions of available sampling mechanisms, sample collection and handling procedures. The impact of specific methodologies on the usefulness or reliability of the resulting data have received little discussion (Gibb *et al.*, 1981; Todd *et al.*, 1976).

High quality chemical data collection is essential in ground-water monitoring programs. The technical difficulties involved in "representative" sampling have been recognized only recently (Gibb *et al.*, 1981; Grisak *et al.*, 1978). It is clear that the long-term collection of high quality ground-water chemistry data is more involved than merely selecting a sampling mechanism and agreeing on sample handling procedures. Efforts to detect and assess contamination can be unrewarding without accurate (e.g., unbiased) and precise (e.g., comparable and complete) concentration data on ground-water chemical constituents. Also, the expense of data collection and management argue for documentation of data quality.

Gillham *et al.* (1983) have published a very useful reference on the principal sources of bias (i.e., inaccuracy) and imprecision (i.e., nonreproducibility) in ground-water monitoring results. Their treatment is extensive and stresses the minimization of random error which can enter into well construction, sample collection and sample handling operations.They further stress the importance of collecting precise data over time to maximize the effectiveness of trend analysis, particularly for regulatory purposes. Accuracy is also very important, since the ultimate reliability of statistical comparisons of results from different wells (e.g., upgradient versus downgradient samples) may depend on differences between mean values for selected constituents from relatively small replicate sample sets. Therefore, systematic error must be controlled by selecting proven methods for establishing sampling points and sample collection to insure known levels of accuracy.

6.1.3 The Subsurface Environment

The subsurface environment of ground water may be categorized broadly into two zones, the unsaturated (i.e., vadose) and saturated zones. The use of the term vadose zone is more accurate as isolated water-saturated regions may exist in the unsaturated zone above the table or most shallow confined aquifer.

Scientists and engineers have discovered recently that the subsurface is neither devoid of oxygen (Winograd and Robertson, 1982) nor sterile (Wilson and McNabb, 1983; Wilson *et al.*, 1983). These facts may have significant influence on the mobility and persistence of chemical species, as well as on the transformations of the original components of contaminant mixtures (Schwarzenbach *et al.*, 1985) which have been released to the subsurface.

The subsurface environment is also quite different from surface water systems in that vertical gradients in pressure and dissolved gas content have been observed within the usual depth ranges of monitoring interest (i.e., 1 to 150 m [3 to 500 ft]). These gradients can be linked to well-defined hydrologic or geochemical processes in some cases. However, reports of apparently anomalous geochemical processes have increased in recent years, particularly at contaminated sites (Barcelona and Garske, 1983; Heaton and Vogel, 1981; Schwarzenbach *et al.*, 1985; Winograd and Robertson, 1982; Wood and Petraitis, 1984).

The subsurface environment is not as readily accessible as surface water systems, and some disturbance is necessary to collect samples of earth materials or ground water. Therefore, "representative" (i.e., artifact or error free) sampling is really a function of the degree of detail needed to characterize subsurface hydrologic and geochemical conditions and the care taken to minimize disturbance of these conditions in the process (Claasen, 1982). Each well or boring represents a potential conduit for short-circuited contaminant migration or ground-water flow which must be considered a potential liability to investigative activities.

It is clear that the subsurface environment of ground water is dynamic over extended time frames and the processes of recharge and ground-water flow are very important to a thorough understanding of the system. Detailed descriptions of contaminant distribution, transport and transformation necessarily rely on the understanding of basic flow and fluid transport processes. It is important to keep in mind that short-term investigations may only provide a snapshot of contaminant levels or distributions. Since water quality monitoring data is normally collected on discrete dates, it is very important that reliable collection methods are used which assure high data quality over the course of the investigation. The reliability of the methods should be investigated thoroughly during the preliminary phase of monitoring network implementation.

Though the scope of this discussion is on sampling ground water for chemical analysis, it should be emphasized that the same data quality requirements apply to water level measurements and to hydraulic conductivity testing. These hydrologic determinations are the basis for the interpretation of chemical constituent data and may well limit the validity of fluid or solute transport model applications. Hydrologic measurements must be included in the development of the quality assurance/quality control program for ground-water quality monitoring networks.

6.1.4 The Sampling Problem and Parameter Selection

Cost-effective water quality sampling is difficult in ground-water systems, because proven field procedures have not been extensively documented. Regulations which call for "representative sampling" alone are not sufficient to insure high quality data collection. The most appropriate monitoring and sampling procedures for a ground-water quality network will depend on the specific purpose of the program. Resource evaluation, contaminant detection, remedial action assessments and litigation studies are purposes for which effective networks can be designed once the information needs have been identified. Due to the time, manpower and cost of most water-quality monitoring programs, the optimal network design should be phased so as to make the most of the available information as it is collected. This approach allows for the gradual refinement of program goals as the network is implemented.

There are two fundamental considerations which are common to most ground-water quality monitoring programs. These are establishing individual sampling points (i.e., in space and time) and the elements of the water sampling protocol which will be sufficient to meet the information needs of the overall program. The placement and number of sampling points can be phased to gradually increase the scale of the monitoring program. Similarly, the chemical constituents of initial interest should provide background ground-water quality data from which a list of likely contaminants may be prepared as the program progresses. Candidate chemical and hydrologic parameters for both detective and assessment monitoring activities are shown in Table 6-1. Special care should be taken to account for possible subsurface transformation of the principal pollutant species. Ground-water transport of contaminants can produce chemical distributions which vary substantially over time and space. Transformation of organic compounds in particular can change the identity of the original contaminant mixture substantially (Mackay *et al.*, 1985; Schwarzenbach *et al.*, 1985).

Table 6-1 Suggested Measurements for Ground-Water Monitoring Programs

Detective Monitoring
Chemical Parameters[a]
pH, Ω^{-1}, TOC, TOX, Alkalinity, TDS, Eh, Cl^-, NO_3^-, $SO_4^{=}$, $PO_4^{\equiv}$, SiO_2, Na^+, K^+, Ca^{++}, Mg^{++}, NH_4^+, Fe, Mn
Hydrologic Parameters
Water Level, Hydraulic Conductivity
Assessment Monitoring
Chemical Parameters[a]
pH, Ω^{-1}, TOC, TOX, Alkalinity, TDS, Eh, Cl^-, NO_3^-, $SO_4^{=}$, $PO_4^{\equiv}$, SiO_2, B, Na^+, K^+, Ca^{++}, Mg^{++}, NH_4^+, Fe, Mn, Zn, Cd, Cu, Pb, Cr, Ni, Ag, Hg, As, Sb, Se, Be
Hydrologic Parameters
Water Level, Hydraulic Conductivity

[a] Ω^{-1} = specific conductance, a measure of the charged species in solution.

Source: Barcelona et al., 1981.

Contaminant detection is generally the most important aspect of a water quality program which must be assured in network design. False negative contaminant detections due to the loss of chemical constituents or the introduction of interfering substances which mask the presence of the contaminants in water samples can be very serious. Such errors may delay needed remedial action and expose either the public or the environment to unreasonably high risk. False positive observations of contaminants may call for costly remedial actions or more intensive study which are not warranted by the actual situation. Reliable sample collection and data interpretation procedures are therefore central to an optimized network design.

In this respect, monitoring in the vadose zone is attractive because it should provide an element of "early" detection capability. The methodologies available for this type of monitoring have been under development for some time. However, there are distinct limitations to many of the available monitoring devices (Everett and McMillan, 1985; Everett *et al.*, 1982; Wilson, 1981; Wilson, 1982; Wilson, 1983) and it is frequently difficult to relate observed vadose-zone concentrations quantitatively to actual contaminant distributions in ground water (Everett *et al.*, 1984; Lindau and Spalding, 1984). Soil gas sampling techniques and underground storage tank monitors have been commercially developed, however, which can be extremely useful for source scouting. Given the complexity of vadose zone monitoring procedures and the need for additional investigation (Robbins and Gemmell, 1985), it may be difficult to implement these techniques in routine ground-water monitoring networks.

This chapter addresses water quality sampling in the saturated zone, reflecting the advanced state of monitoring technology appropriate for this compartment of the subsurface. There are a number of useful reference materials for the development of effective ground-water sampling protocols which include information on the types of drilling methods, well construction materials, sampling mechanisms and sample handling methods currently available (Barcelona *et al.*, 1985; Barcelona *et al.*, 1983; Gillham *et al.*, 1983; Scalf *et al.*, 1981; Todd *et al.*, 1976). In order to collect sensitive, high-quality contaminant concentration data, it is important to identify the type and magnitude of errors which may arise in ground-water sampling. A generalized diagram of the steps involved in sampling and principal sources of error is shown in Figure 6-1. It should be recognized that strict error control at each step is necessary for the collection of high quality data which is representative of the in situ condition.

Figure 6-1 Steps and sources of error in ground-water sampling.

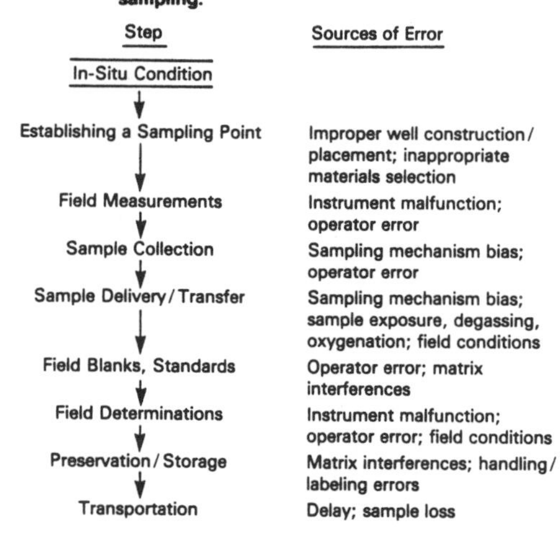

There are two major obstacles to achieving control over ground-water sampling errors. First, changes which may occur in the integrity of samples prior to sample delivery to the land surface cannot be accounted for by the use of field blanks, standards and split samples used in data quality assurance programs. Second, most of the sources of error which may affect sample integrity prior to sample delivery are not well documented in the literature for many of the contaminants of current interest. Among these sources of error are the contamination of the subsurface by drilling fluids, grouts or sealing materials, the sorptive or leaching effects on water samples due to well casing, pump or sampling tubing materials' exposures and the effects on the solution chemistry due to oxygenation, depressurization or gas exchange caused by the sampling mechanism. These sources of error have been investigated to some extent for volatile organic contaminants under laboratory conditions. However, to achieve confidence in field monitoring and sampling instrumentation for routine applications, common sense and a "research" approach to regulatory monitoring may be needed. Two of the most critical elements of a monitoring program are establishing both reliable sampling points and simple, efficient sampling protocols which will yield data of known quality.

6.2 Establishing a Sampling Point

If adequate care is taken in the selection of drilling methods, well construction materials and development techniques, it should be possible to at least approximate representative ground-water sampling from a monitoring well. The representative nature of the water samples can be maintained consistently with a trained sampling staff and good field-laboratory communication. Also, important hydrologic measurements (i.e., water level, hydraulic conductivity) can be made from the same sampling point. A representative water sample may then be defined as a minimally disturbed sample taken after proper well purging which will allow the determination of the chemical constituents of interest at predetermined levels of accuracy and precision. Sophisticated monitoring technology and sampling instrumentation are poor substitutes for an experienced sampling team which can follow a simple proven sampling protocol.

This section details some of the considerations which should be made in establishing a reliable sampling point. There are a number of alternative approaches to sampling point selection in monitoring network design. Arrays of either nested monitoring wells or multilevel devices (Barvenik and Cadwgan, 1983; Pickens *et al.*, 1978) deployed at various sites within the area of interest have their individual merits based on the ease of verifying sampling point isolation, durability, cost, ease of installation and site specific factors. Deciding which option is most effective for specific programs should be done with representative sampling criteria in mind. The sampling points must be durable, inert towards the chemical constituents of interest, allow for purging of stagnant water, provide sufficient water for analytical work with minimum disturbance, and permit the evaluation of the hydrologic characteristics of the formation of interest. Monitoring wells can be constructed to meet these criteria because a variety of drilling methods, materials, sampling mechanisms and pumping regimes for sampling and hydrologic measurements can be selected to meet the current needs of most monitoring programs.

The placement and number of wells will depend on the complexity of the hydrologic setting and the degree of spatial and temporal detail needed to meet the goals of the program. It is important to note that both the directions and approximate rates of ground-water movement must be known in order to interpret the chemical data satisfactorily. In this way it may also be possible to estimate the nature and location of pollutant sources (Gorelick *et al.*, 1983). Subsurface geophysical techniques can be very helpful in determining the optimum placement of monitoring wells under appropriate conditions and when sufficient hydrogeologic information is available (Evans and Schweitzer, 1984). Well placement should be viewed as an evolutionary activity which may expand or contract as the needs of the program dictate.

6.2.1 Well Design and Construction

Effective monitoring well design and construction requires considerable care and at least some understanding of the hydrogeology and subsurface geochemistry of the site. Preliminary borings, well drilling experience and the details of the operational history of a site can be very helpful. Monitoring well design criteria include depth, screen size, gravel pack specifications, and yield potential. These considerations differ substantially from those applied to production wells. The simplest, narrow diameter well completions which will permit development, accommodate the sampling gear and minimize the need to purge large volumes of potentially contaminated water are preferred for effective routine monitoring activities. Helpful references are in several publications (Barcelona *et al.*, 1983; Scalf *et al.*, 1981; Wehrmann, 1983).

6.2.2 Well Drilling

The selection of a particular drilling technique should depend on the geology of the site, the expected depths of the wells and the suitability of drilling equipment for the contaminants of interest. The various drilling and well completion methods have been reviewed with reference to these criteria in the previous chapter. Regardless of the technique used, every effort should be made to minimize subsurface disturbance. For critical applications, the drilling rig and tools should be steam cleaned to minimize the potential for cross-contamination between formations or successive borings. The use of drilling muds can be a liability for trace chemical constituent investigations because foreign organic matter will be introduced into the penetrated formations. Even "clay" muds without polymeric additives contain some organic matter which is added to stabilize the clay suspension and may interfere with some analytical determinations. Table 6-2 contains information on the total and soluble organic carbon contents of some common drilling and grouting materials (Barcelona, 1983). The effects of drilling muds on ground-water solution chemistry have not been investigated in detail. However, existing reports indicate that the organic carbon introduced during drilling can cause false water-quality observations for long periods of time (Barcelona, 1984; Brobst, 1984). The fact that the interferences are observable for gross indicators of levels of organic carbon compounds (i.e., TOC) and reduced substances (i.e., chemical oxygen demand) strongly suggests that drilling aids are a potential source of serious error. Innovative drilling techniques may be called for in special situations (Yare, 1975).

6.2.3 Well Development, Hydraulic Performance and Purging Strategy

Once a well is completed, the sampling point must be prepared for water sampling and measures must be taken to evaluate its hydraulic characteristics. These steps provide a basis for the maintenance of reliable sampling points over the duration of a ground-water monitoring program.

6.2.3.1 Well Development

The proper development of monitoring wells is essential to the collection of "representative" water samples. During the drilling process, fine particles are forced through the sides of the bore hole into the formation, forming a mud cake that reduces the hydraulic conductivity of the materials in the immediate area of the well bore. To allow water from the formation being monitored to freely enter the monitoring well, this mud cake must be broken down opposite the screened portion of the well and the fines removed from the well. This process also enhances the yield potential of the monitoring well, a critical factor when constructing monitoring wells in low-yielding geologic materials.

More importantly, monitoring wells must be developed to provide water free of suspended solids for sampling. When sampling for metal ions and other dissolved inorganic constituents, water samples must be filtered and preserved at the well site at the time of sample collection. Improperly developed monitoring wells will produce samples containing suspended sediments that may both bias the chemical analysis of the collected samples and cause frequent clogging of field filtering mechanisms. The additional time and money spent for well development will expedite sample filtration and result in samples that are more representative of water chemistry in the formation being monitored.

The development procedures used for monitoring wells are similar to those used for production wells. The first step in development involves the movement of water at alternately high and low velocity into and out of the well screen and gravel pack to break down the mud cake on the well bore and loosen fine

Table 6-2 Composition of Selected Sealing and Drilling Muds

	Ash (% by wt)	Organic Content (% by wt)	Soluble Carbon (% by wt)	Soluble Carbon in Total Organic Content (% by wt)
"Bentonite" muds/grouts				
Volclay[b] (~90% montmorillonite)	98.2	1.8	<0.001	94.4
Benseal[c]	88.5	11.5	<0.001	3.7
"Organic" muds/drilling aids				
Ez-Mud[c] (acrylamide-sodium acrylate copolymer dispersed in food-grade oil [normally used in 0.25% dilution])	11.5	21.5	17.9	2.1
Revert[d] (guar bean starch-based mixture)	1.6	98.4	33.8	85.6

[a]All percentages determined on a moisture-free basis.
[b]Trademark of American Colloid Co.
[c]Trademark of NL Baroid/NL Industries Inc.
[d]Trademark of Johnson Division, UOP Inc.
Source: Wood, 1976.

particles in the borehole. This step is followed by pumping to remove these materials from the well and the immediate area outside the well screen. This procedure should be continued until the water pumped from the well is visually free of suspended materials or sediments.

6.2.3.2 Hydraulic Performance of Monitoring Wells

The importance of understanding the hydraulics of the geologic materials at a site cannot be overemphasized. Collection of accurate water level data from properly located and constructed wells provides information on the direction of ground-water flow (Chapter 4). The success of a monitoring program also depends on knowledge of the rates of travel of both the ground water and solutes. The response of a monitoring well to pumping also must be known to determine the proper rate and length of time of pumping prior to collecting a water sample.

Hydraulic conductivity measurements provide a basis for judging the hydraulic connection of the monitoring well and adjacent screened formation to the hydrogeologic setting. These measurements also allow an experienced hydrologist to estimate an optimal sampling frequency for the monitoring program (Barcelona *et al.*, 1985).

Traditionally, hydraulic conductivity testing has been conducted by collecting drill samples which were then taken to the laboratory for testing. Several techniques using laboratory permeameters are routinely used. Falling head or constant head permeameter tests on recompacted samples in fixed wall or triaxial test cells are among the most common. The relative applicability of these techniques is dependent on both operator skill and methodology since calibration standards are not available. The major problem with laboratory test procedures is that one collects data on recompacted geologic samples rather than geologic materials under field conditions. Only limited work has been done to date on performing laboratory tests on "undisturbed" samples to improve the field applicability of laboratory hydraulic conductivity results.

Hydraulic conductivity is most effectively determined under field conditions by pumping or slug testing. The water level drawdown can be measured during pumping. Alternatively, water levels are measured after the static water level is depressed by application of gas pressure or elevated by the introduction of a slug of water. These procedures are rather straightforward for wells which have been properly developed. The utility of these measurements is obvious when one considers the extent of pumping necessary to remove stagnant water from the monitoring well and how much water must be removed to establish a representative volume of formation water for sampling.

Figure 6-2a Example of well purging requirement estimating procedure (Barcelona et al., 1985).

Given:
- 48-foot deep, 2-inch diameter well
- 2-foot long screen
- 3-foot thick aquifer
- static water level about 15 feet below land surface
- hydraulic conductivity = 10^{-2} cm/sec

Assumptions:
- A desired purge rate of 500 mL/min and sampling rate of 100 mL/min will be used.

Calculations:

One well volume = (48 ft - 15 ft) x 613 mL/ft (2-inch diameter well)
= 20.2 liters

Aquifer Transmissivity = hydraulic conductivity x aquifer thickness
= 10^{-4} m/sec x 1 meter
= 10^{-4} m²/sec or 8.64 m²/day

From Figure 6-2b:

At 5 minutes: 95% aquifer water and
(5 min x 0.5 L/min)/20.2 L
= 0.12 well volumes

At 10 minutes: 100% aquifer water and
(10 min x 0.5 L/min)/20.2 L
= 0.24 well volumes

It appears that a high percentage of aquifer water can be obtained within a relatively short time of pumping at 500 mL•min^{-1}. This pumping rate is below that used during well development to prevent well damange or further development.

Figure 6-2b Percentage of aquifer water versus time for different transmissivities.

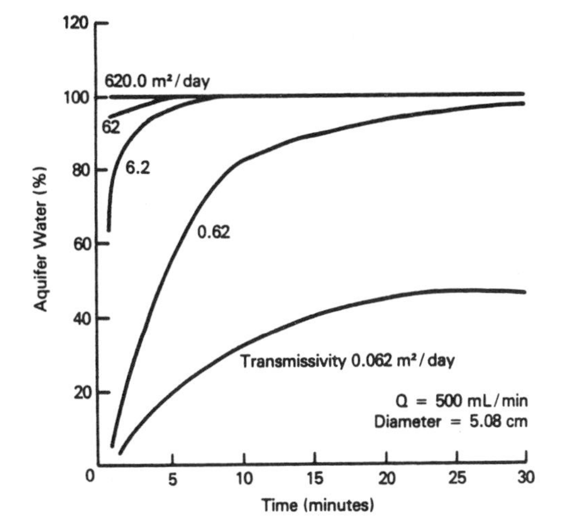

6.2.3.3 Well Purging Strategies

The number of well volumes to be pumped from a monitoring well prior to the collection of a water sample must be tailored to the hydraulic properties of the geologic materials being monitored, the well construction parameters, the desired pumping rate, and the sampling methodology to be employed. There is no one single number of well volumes to be pumped that is best or fits all situations. The goal in establishing a well purging strategy is to obtain water from the geologic materials being monitored while minimizing the disturbance of the regional flow system and the collected sample. To accomplish this goal a basic understanding of well hydraulics and the effects of pumping on the quality of water samples is essential. Water that has remained in the well casing for extended periods of time (i.e., more than about two hours) has had the opportunity to exchange gases with the atmosphere and to interact with the well casing material. The chemistry of water stored in the well casing is unrepresentative of that in the aquifer and it should not be collected for analysis. Purge volumes and pumping rates should be evaluated on a case by case basis.

Gibb (1981) has shown how the measurements of hydraulic conductivity can be used to estimate the well purging requirement. An example of this procedure is shown in Figures 6-2a and 6-2b. In practice, it may be necessary to test the hydraulic conductivity of several wells within a network. The calculated purging requirement should then be verified by measurements of pH and specific conductance during pumping to signal equilibration of the water being collected.

The selection of purging rates and volumes of water to be pumped prior to sample collection can also be influenced by the anticipated water quality. In hazardous environments where purged water must be contained and disposed of in a permitted facility, it is desirable to minimize the amount of purged water. This can be accomplished by pumping the wells at very low pumping rates (100 ml/min) to minimize the drawdown in the well and maximize the percent aquifer water delivered to the surface in the shortest period of time. Pumping at low rates, in effect, isolates the column of stagnant water in the well bore and negates the need for its removal. This approach is only valid in cases where the pump intake is placed at the top of, or in, the well screen.

In summary, well purging strategies should be established by (1) determining the hydraulic performance of the well; (2) calculating reasonable purging requirements, pumping rates, and volumes based on hydraulic conductivity data, well construction data, site hydrologic conditions, and anticipated water quality; (3) measuring the well purging parameters to verify chemical "equilibrated" conditions; and (4) documenting the entire effort (actual pumping rate, volumes pumped, and purging parameter measurements before and after sample collection).

6.2.3.4 Sampling Materials and Mechanisms

In many monitoring situations it is not possible to predict the requirements which either materials for well casings, pumps and tubing or pumping mechanisms that the head and lift conditions must meet in order to provide error-free samples of ground water. Ideally these components of the system should be durable and inert towards the chemical properties of samples or the subsurface so as to neither contaminate nor remove chemical constituents from the water samples. Due to the long duration of regulatory programs' requirements, well casing materials in particular must be sufficiently durable and nonreactive to last several decades. It is generally much easier to substitute more appropriate sampling pumps or pump/tubing materials as knowledge of subsurface conditions improves than to drill additional wells to replace inadequate well casing or screen materials. Also there is no simple way to account for errors which occur prior to handling a sample at the land surface. Therefore, it is good practice to carefully choose these components of the sampling system which make up the rigid materials in well casing/screens or pumps and the flexible materials used in sample delivery tubing.

Rigid Materials. An experienced hydrologist can plan well construction details based mainly on hydrogeologic criteria, even for challenging situations where a separate contaminant phase may be present (Villaume, 1985). However, the question of the best material for a specific monitoring application must be addressed by considering subsurface geochemistry and the likely contaminants of interest. Therefore, strength, durability and inertness should be balanced with cost considerations in the choice of rigid materials for well casing, screens, pumps, etc. Common well casing materials include TFE (Teflon), PVC (polyvinyl chloride), stainless steel, and other ferrous materials. The strength, durability and potential for sorptive or leaching interferences on chemical constituent determinations have been reviewed in detail for these materials (Barcelona *et al.*, 1985; Barcelona *et al.*, 1983). Unfortunately, there is very little documentation of the severity or magnitude of well casing interferences from actual field investigations. This is the point at which optimized monitoring network design takes on an element of "research" or the need for systematic evaluation of the components of the monitoring installation.

Polymeric materials have the potential to absorb dissolved chemical constituents and leach either previously sorbed substances or components of the polymer formulations. Similarly, ferrous materials may adsorb dissolved chemical constituents and leach metal ions or corrosion products which may introduce errors into the results of chemical analysis. This potential in both cases is real, yet not completely understood. The recommendations in the references noted above can be summarized as follows:

- Teflon[R] is the well casing material least likely to cause significant error in ground-water monitoring programs focused on either organic or inorganic chemical constituents. It has sufficient strength for most applications at shallow depth (i.e., <100 m) and is among the most inert materials ever made. For deeper installations, it can be linked to another material above the highest seasonal, stagnant water level.
- Stainless steel (either 316 or 304 type) well casing can be expected to, under noncorrosive conditions, be the second least likely material to cause significant error for organic chemical constituent monitoring investigations. The release of Fe, Mn or Cr may occur under corrosive conditions. Organic constituent sorption effects may also be significant sources of error after corrosion processes have altered the virgin surface.
- Rigid PVC well casing material with National Sanitation Foundation approval should be used in monitoring well applications when noncemented or threaded joints are used and organic chemical constituents are not expected to be of present or future interest. Significant losses of strength, durability and inertness (i.e., sorption or leaching) may be expected under conditions where organic contaminants are present in high concentration. It should perform adequately in inorganic chemical constituent studies when organic constituents are not present in high concentrations and tin or antimony species are not target chemical constituents.

Monitoring wells made of appropriate materials and screened over discrete sections of the saturated thickness of geologic formations can yield a wealth of chemical and hydrologic information. Whether or not this level of performance is achieved may depend frequently on the care taken in evaluating the hydraulic performance of the sampling point.

Flexible Materials. Pump components and sample delivery tubing may contact a water sample more intimately than other components of a sampling system, including storage vessels and well casing. Similar considerations of inertness and noncontaminating properties apply to tubing, bladder, gasket and seal materials. Experimental evidence (Barcelona *et al.*, 1985) has supported earlier recommendations drawn from manufacturers' specifications (Barcelona *et al.*, 1983). A summary is provided in Table 6-3. Again, the care taken in materials selection for the specific needs of the sampling program can pay real dividends and provides greater assurance of error-free sampling.

Sample Mechanisms. It is important to remember that sampling mechanisms themselves are not protocols. The sampling protocol for a particular monitoring

Table 6-3 Recommendations for Flexible Materials in Sampling Applications

Materials	Recommendations
Polytetrafluoroethylene (Teflon*)	Recommended for most monitoring work, particularly for detailed organic analytical schemes. The material least likely to introduce significant sampling bias or imprecision. The easiest material to clean in order to prevent cross-contamination.
Polypropylene Polyethylene (linear)	Strongly recommended for corrosive high dissolved solids solutions. Less likely to introduce significant bias into analytical results than polymer formulations (PVC) or other flexible materials with the exception of Teflon*.
PVC (flexible)	Not recommended for detailed organic analytical schemes. Plasticizers and stabilizers make up a sizable percentage of the material by weight as long as it remains flexible. Documented interferences are likely with several priority pollutant classes.
Viton* Silicone (medical grade only) Neoprene	Flexible elastomeric materials for gaskets, O-rings, bladder, and tubing applications. Performance expected to be a function of exposure type and the order of chemical resistance as shown. Recommended only when a more suitable material is not available for the specific use. Actual controlled exposure trials may be useful in assessing the potential for analytical bias.

*Trademark of DuPont, Inc.

Source: Barcelona et al., 1981.

network is basically a step-by-step written description of the procedures used for well purging, sample delivery to the surface and the handling of the samples in the field. Once the protocol has been developed and used in a particular investigation, it provides a basis for modifying the program if the extent or type of contamination requires more intensive work. An appropriate sampling mechanism is an important part of any protocol. Ideally, the pumping mechanism should be capable of purging the well of stagnant water at rates of liters or gallons per minute and also delivering ground water to the surface so that sample bottles may be filled at low flow rates (i.e., ~100 ml/min^{-1} to minimize turbulence and degassing of the sample. In this way the criteria for representative sampling can be met while keeping the purging and sample collection steps simple. Nielsen & Yeates (1985) have reviewed the types of sample collection mechanisms commercially available (Anonymous, 1985) in line with the results of research studies of their performance (Barcelona *et al.*, 1984; Stoltzenburg and Nichols, 1985). Examples of types of pumps or other samplers are shown in Figure 6-3 and they are described fully in a number of references (Barcelona *et al.*, 1985; Gillham *et al.*, 1983; Scalf *et al.*, 1981). Given all of the varied hydrogeologic settings and potential chemical constituents of interest several types of pumps or sampling mechanisms may be suitable for specific applications. Figure 6-4 contains some recommendations for reliable sampling mechanisms given the sensitivity of the sample to error. The main criteria for sampling pumps are the capabilities to purge stagnant water from the well and deliver the water samples to the surface with minimal loss of sample integrity. Clearly, a mechanism that is proven to provide accurate and precise samples for volatile organic compound determinations should be suitable for most chemical constituents of interest.

Now that a sampling point and the means to collect a sample have been established, the next step is the development of the detailed sampling protocol.

6.3 Elements of the Sampling Protocol

There are few aspects of this subject which generate more controversy than the sampling steps which make up the sampling protocol. Efforts to develop reliable protocols and optimize sampling procedures require particular attention to sampling mechanism effects on the integrity of ground-water samples (Barcelona *et al.*, 1984; Stolzenburg and Nichols, 1985), as well as the potential errors involved in well purging, delivery tubing exposures (Barcelona *et al.*, 1985; Ho, 1983), sample handling and the impact of sampling frequency on both the sensitivity and reliability of chemical constituent monitoring results. Quality assurance measures including field blanks, standards, and split control samples cannot account for errors in these steps of the sampling protocol (Barcelona *et al.*, 1985). Actually, the sampling protocol is the focus of the overall study network design (Nacht, 1983) and should be prepared flexibly to be refined as the information on site conditions improves.

Each step within the protocol has a bearing on the quality and completeness of the information being collected. This is perhaps best shown by the progression of steps depicted in Figure 6-5. Corresponding to each step is a goal and recommendation for achieving the goal. The principal utility of this description is that it provides an outlined agenda for high quality chemical and water quality data.

To insure maximum utility of the sampling effort and resulting data, documentation of the sampling protocol as performed in the field is essential. In addition to noting the obvious information (i.e., persons conducting the sampling, equipment used, weather conditions, and documentation of adherence to the protocol and unusual observations) three basic elements of the sampling protocol should be recorded: (1) water level measurements made prior to sampling, (2) the volume and rate at which water is removed from the well prior to sample collection (well purging), and (3) the actual sample collection including measurement of well-purging parameters, sample preservation, sample handling and chain of custody.

6.3.1 Water-Level Measurement

Prior to the purging of a well or sample collection, it is extremely important to measure and record the water level in the well to be sampled. These measurements are needed to estimate the amount of water to be pumped from the well prior to sample collection. In addition, this information can be useful when interpreting monitoring results. Low water levels may reflect the influence of a nearby production well. High water levels compared to measurements made at other times of the year are indicative of recent recharge events. In relatively shallow monitoring settings high water levels from recent natural recharge events may result in dilution of the total dissolved solids in the collected sample. Conversely, if contaminants are temporarily held in an unsaturated zone above the geologic zone being monitored, recharge events may "flush" these contaminants into the shallow ground water system and result in higher levels of some constituents.

Documenting the nonpumping water levels for all wells at a site will provide historical information on the hydraulic conditions at the site. Analysis of this information will reveal changes in flow paths and serve as a check on the effectiveness of the wells to monitor changing hydrologic conditions. It is very useful to develop an understanding of the seasonal

Figure 6-3 **Schematic diagrams of common ground-water sampling devices (Neilsen and Yeates, 1985).**

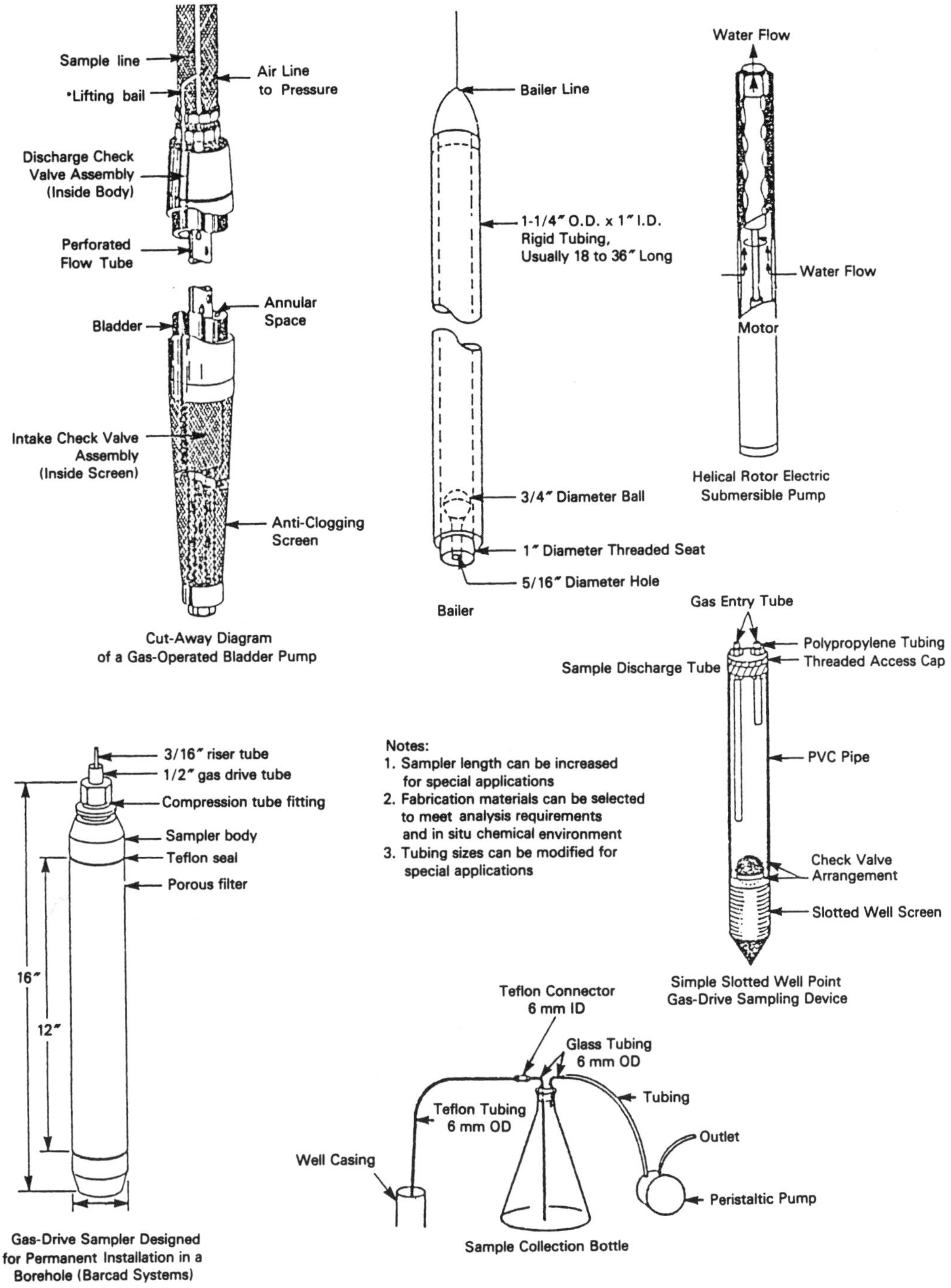
Sample line
•Lifting bail
Air Line to Pressure
Discharge Check Valve Assembly (Inside Body)
Perforated Flow Tube
Bladder
Annular Space
Intake Check Valve Assembly (Inside Screen)
Anti-Clogging Screen
Cut-Away Diagram of a Gas-Operated Bladder Pump
Bailer Line
1-1/4" O.D. x 1" I.D. Rigid Tubing, Usually 18 to 36" Long
3/4" Diameter Ball
1" Diameter Threaded Seat
5/16" Diameter Hole
Bailer
Water Flow
Water Flow
Motor
Helical Rotor Electric Submersible Pump
Gas Entry Tube
Polypropylene Tubing
Threaded Access Cap
Sample Discharge Tube
PVC Pipe
Check Valve Arrangement
Slotted Well Screen
Simple Slotted Well Point Gas-Drive Sampling Device
3/16" riser tube
1/2" gas drive tube
Compression tube fitting
Sampler body
Teflon seal
Porous filter
16"
12"
Gas-Drive Sampler Designed for Permanent Installation in a Borehole (Barcad Systems)
Notes:
1. Sampler length can be increased for special applications
2. Fabrication materials can be selected to meet analysis requirements and in situ chemical environment
3. Tubing sizes can be modified for special applications
Teflon Connector 6 mm ID
Glass Tubing 6 mm OD
Tubing
Outlet
Teflon Tubing 6 mm OD
Well Casing
Peristaltic Pump
Sample Collection Bottle

Figure 6-4 Matrix of sensitive chemical constituents and various sampling mechanisms.

INCREASING RELIABILITY OF SAMPLING MECHANISMS ← ; INCREASING SAMPLE SENSITIVITY ↑

Type of constituent	Example of constituent	Positive displacement bladder pumps	Thief, in situ or dual check valve bailers	Mechanical positive displacement pumps	Gas-drive devices	Suction mechanisms
Volatile Organic Compounds; Organometallics	Chloroform TOX CH_3Hg	Superior performance for most applications	May be adequate if well purging is assured	May be adequate if design and operation are controlled	Not recommended	Not recommended
Dissolved Gases; Well-purging Parameters	O_2, CO_2; pH, Ω^{-1} Eh	Superior performance for most applications	May be adequate if well purging is assured	May be adequate if design and operation are controlled	Not recommended	Not recommended
Trace Inorganic Metal Species; Reduced Species	Fe, Cu; NO_2^-, S^-	Superior performance for most applications	May be adequate if well purging is assured	Adequate	May be adequate	May be adequate if materials are appropriate
Major Cations & Anions	Na^+, K^+, Ca^{++} Mg^{++}; Cl^-, SO_4^-	Superior performance for most applications	Adequate; May be adequate if well purging is assured	Adequate	Adequate	Adequate

Figure 6-5 Generalized ground-water sampling protocol.

Step	Goal	Recommendations
Hydrologic Measurements	Establish nonpumping water level.	Measure the water level to ± 0.3 cm (± 0.01 ft).
Well Purging	Remove or isolate stagnant H_2O which would otherwise bias representative sample.	Pump water until well purging parameters (e.g., pH, T, Ω^{-1}, Eh) stabilize to $\pm 10\%$ over at least two successive well volumes pumped.
Sample Collection	Collect samples at land surface or in well-bore with minimal disturbance of sample chemistry.	Pumping rates should be limited to ~100 mL/min for volatile organics and gas-sensitive parameters.
Filtration/Preservation	Filtration permits determination of soluble constituents and is a form of preservation. It should be done in the field as soon as possible after collection.	Filter: Trace metals, inorganic anions/cations, alkalinity. Do not filter: TOC, TOX, volatile organic compound samples; other organic compound samples only when required.
Field Determinations	Field analyses of samples will effectively avoid bias in determining parameters/constituents which do not store well; e.g., gases, alkalinity, pH.	Samples for determining gases, alkalinity and pH should be analyzed in the field if at all possible.
Field Blanks/Standards	These blanks and standards will permit the correction of analytical results for changes which may occur after sample collection: preservation, storage, and transport.	At least one blank and one standard for each sensitive parameter should be made up in the field on each day of sampling. Spiked samples are also recommended for good QA/QC.
Sample Storage/Transport	Refrigerate and protect samples to minimize their chemical alteration prior to analysis.	Observe maximum sample holding or storage periods recommended by the Agency. Documentation of actual holding periods should be carefully performed.

changes in water levels and associated chemical concentration variability at the monitored site.

6.3.2 Purging

The volume of stagnant water which should be removed from the monitoring well should be calculated from the analysis of field hydraulic conductivity measurements. Rule-of-thumb guidelines for the volume of water which should be removed from a monitoring well prior to sample collection can cause time delays and unnecessary pumping of excess contaminated water. These rules (i.e. three-, five- or 10-well volume) largely ignore the hydraulic characteristics of individual wells and geologic settings. One advantage of using the same pump to both purge stagnant water and collect samples is the ability to measure pH and specific conductance Ω^{-1} in an in-line flow cell. These parameters aid in the verification of the purging efficiency and also provide a consistent basis for comparisons of samples from a single well or wells at a particular site. Since pH is a standard variable for aqueous solutions that is affected by degassing and depressurization (i.e., loss of CO_2), in-line measurements provide more accurate and precise determinations than discrete samples collected by grab sampling mechanisms.

The following example illustrates some of the other advantages of verifying the purge requirement for monitoring wells.

For example, the calculated well purging requirement (e.g., >90 percent aquifer water) calls for the removal of five well volumes prior to sample collection for a particular well. Field measurements of the well purging parameters have historically confirmed this recommended procedure. During a subsequent sampling effort, 12 well volumes were pumped before stabilized well purging parameter readings were obtained. Several possible causes could be explored: (1) A limited plume of contaminants may have been present at the well at the beginning of sampling and inadvertently discarded while pumping in an attempt to obtain stabilized indicator parameter readings; (2) The hydraulic properties of the well have changed due to silting or encrustation of the screen, indicating the need for well rehabilitation or maintenance; (3) The flow-through device used for measuring the indicator parameters was malfunctioning; or (4) The well may have been tampered with by the introduction of a contaminant or relatively clean water source in an attempt to bias the sample results.

Documentation of the actual well purging process employed should be a part of a standard field sampling protocol.

6.3.3 Sample Collection and Handling

Water samples should be collected when the solution chemistry of the ground water being pumped has stabilized as indicated by pH, Eh, Ω^{-1} and T readings.

In practice, stable sample chemistry is indicated when the purging parameter measurements have stabilized over two successive well volumes. First, samples for volatile constituents, TOC, TOX and those constituents which require field filtration or field determination should be collected. Then large volume samples for extractible organic compounds, total metals or nutrient anion determinations should be collected.

All samples should be collected as close as possible to the well head. A "tee" fitting placed ahead of the in-line device for measuring the well purging parameters makes this more convenient. Regardless of the sample mechanism in use or the components of the sampling train, wells that are located upgradient of a site, and therefore are expected to be representative of background quality, should be sampled first to minimize the potential for cross-contamination followed by the wells that are located downgradient of a site and may in fact contain contaminants from the site. Laboratory detergent solutions and distilled water should be used to clean the sampling train between samples. An acid rinse (0.1 N HCl) or solvent rinse (i.e., hexane or methanol) may be used to supplement these cleaning steps if necessary. Cleaning procedures should be followed by distilled water rinses which may be saved to check cleaning efficiency.

The order in which samples are taken for specific types of chemical analyses should be decided by the sensitivity of the samples to handling (i.e., most sensitive first) and the need for specific information. For example, the flow chart shown in Figure 6-6 depicts a priority order for a generalized sample collection effort. The samples for organic chemical constituent determinations are taken in decreasing order in relation to sensitivity to handling errors, while the inorganic chemical constituents, which may require filtration, are taken afterwards.

There are instances which arise, even with properly developed monitoring wells, that call for the filtration of water samples. It should be evident, however, that well development procedures which require two to three hours of bailing, swabbing, pumping or air purging at each well will save many hours of time in sample filtration. Well development may have to be repeated at periodic intervals to minimize the collection of turbid samples. In this respect, it is important to minimize the disturbance of fines which accumulate in the well bore. This can be achieved by careful placement of the sampling pump intake at the top of the screened interval, low pumping rates, and by avoiding the use of bailing techniques which disturb sediment accumulations at the bottom of the well.

Figure 6-6 Generalized flow diagram of ground-water sampling steps (Barcelona et al., 1985).

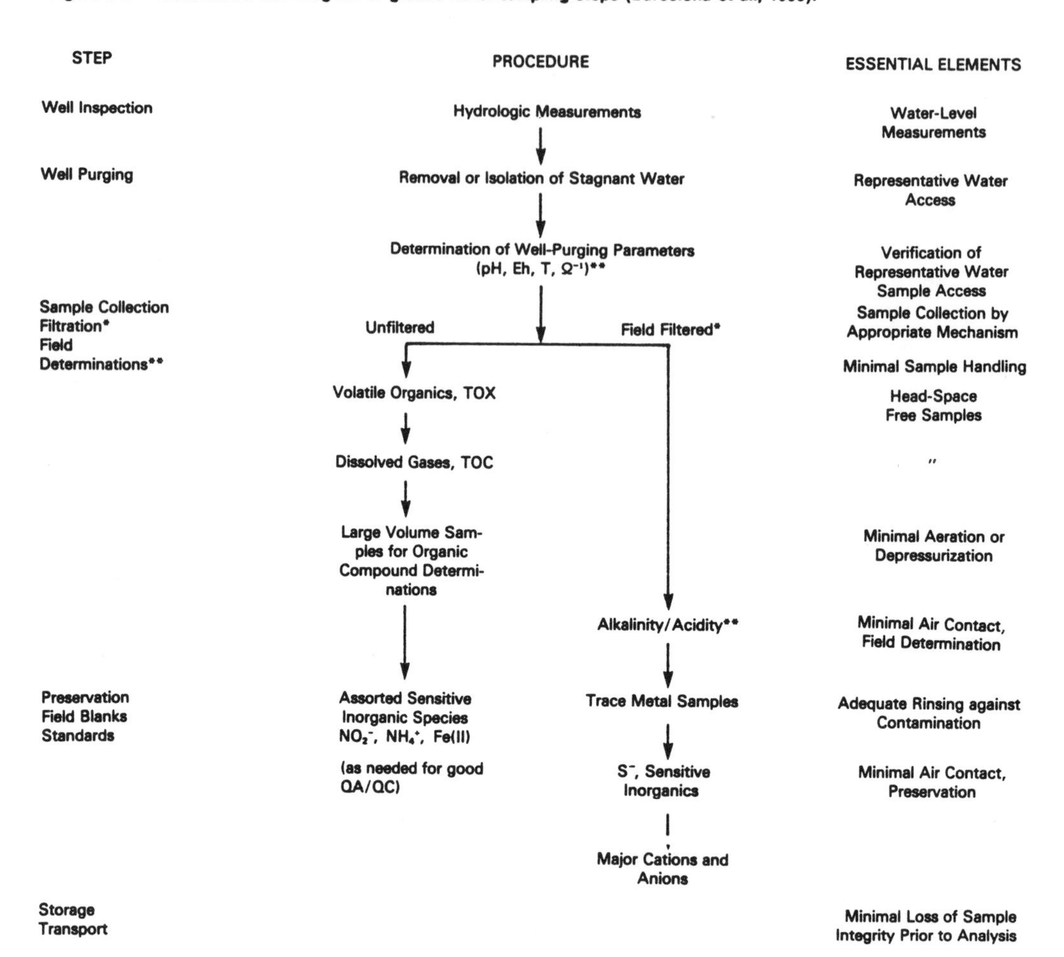

* Denotes samples which should be filtered in order to determine dissolved constituents. Filtration should be accomplished preferably with in-line filters and pump pressure or by N_2 pressure methods. Samples for dissolved gases or volatile organics should not be filtered. In instances where well development procedures do not allow for turbidity-free samples and may bias analytical results, split samples should be spiked with standards before filtration. Both spiked samples and regular samples should be analyzed to determine recoveries from both types of handling.

** Denotes analytical determinations which should be made in the field.

It is advisable to refrain from filtering TOC, TOX or other organic compound samples as the increased handling required may result in the loss of chemical constituents of interest. Allowing the samples to settle prior to analysis followed by decanting the sample is preferable to filtration in these instances. If filtration is necessary for the determination of extractable organic compounds, the filtration should be performed in the laboratory by the application of nitrogen pressure. It may be necessary to run parallel sets of filtered and unfiltered samples with standards to establish the recovery of hydrophobic compounds when samples must be filtered. All of the materials' precautions used in the construction of the sampling train should be observed for filtration apparatus. Vacuum filtration of ground-water samples is not recommended.

Water samples for dissolved inorganic chemical constituents (e.g., metals, alkalinity and anionic species) should be filtered in the field. The preferred arrangement is an in-line filtration module which utilizes sampling pump pressure for its operation. These modules have tubing connectors on the inlet and outlet parts and range in diameter from 2.5 to 15 cm. Large diameter filter holders, which can be rapidly disassembled for filter pad replacement, are the most convenient and efficient designs (Kennedy *et al.*, 1976; Skougstad and Scarbo, 1968).

Representative sampling is the result of the execution of a carefully planned sampling protocol which establishes necessary hydrologic and chemical data for each sample collection effort. An important consideration for maintaining sample integrity after collection is to minimize sample handling which may bias subsequent determinations of chemical constituents. Since opportunities to collect high quality data for the characterization of site conditions may be limited by time, it is prudent to conduct sample collection as carefully as possible from the beginning of the sampling period. It is preferable to emphasize the need to risk error on the conservative side when the doubt exists as to the sensitivity of specific chemical constituents to sampling or handling errors. Repeat sampling or analysis cannot make up for lost data collection opportunities.

Samples collected for specific chemical constituents may require modifications of recommended sample handling and analysis procedures. Samples that contain several chemicals and extended storage periods can cause significant problems in this regard. It is frequently more effective to perform a rapid field determination of specific inorganic constituents (e.g., alkalinity, pH, ferrous iron, sulfide, nitrite or ammonium) than to attempt sample preservation followed by laboratory analysis of these samples.

Many samples can be held for the U.S. EPA recommended maximum holding times after proper preservation. These are shown in Table 6-4 which has been modified slightly from Scalf *et al.* (1981).

6.3.4 Quality Assurance/Quality Control

Planning for valid water quality data collection depends upon both the knowledge of the system and continued refinement of all sample handling or collection procedures. As discussed in Section 6.2 of this chapter, the need to begin QA/QC planning with the installation of the sampling point cannot be over-emphasized.

The use of field blanks, standards and spiked samples for field QA/QC performance is analogous to the use of laboratory blanks, standards and procedural or validation standards. The fundamental goal of field QC is to insure that the sample protocol is being executed faithfully and that situations leading to error are recognized before they seriously impact the data. The use of field blanks and standards and spiked samples can account for changes in samples which occur during sample collection.

Field blanks and standards enable quantitative correction for bias (i.e., systematic errors), which arise due to handling, storage, transport and laboratory procedures. Spiked samples and blind controls provide the means to correct combined sampling and analytical accuracy or recoveries for the actual conditions to which the samples have been exposed.

All QC measures should be performed for at least the most sensitive chemical constituents for each sampling date. Examples of sensitive constituents would be benzene or trichloroethylene as volatile organic compounds and lead or iron as metals. It is difficult to use laboratory blanks alone for the determination of the limits of detection or quantitation. Laboratory distilled water may contain apparently higher levels of volatile organic compounds (e.g., methylene chloride) than those of uncontaminated ground-water samples. The field blanks and spiked samples should be used for this purpose, conserving the results of lab blanks as checks on elevated laboratory background levels.

Whether the ground water is contaminated with interfering compounds or not, spiked samples provide a basis for both the identification of the constituents of interest and the correction of their recovery (or accuracy) based on the recovery of the spiked standard compounds. For example, if trichloroethylene in a spiked sample is recovered at a mean level of 80 percent (-20 percent bias), the concentrations of trichloroethylene determined in the samples for this sampling date may be corrected by a factor of 1.2 for low recovery. Similarly, if 50 percent recovery (-50 percent bias) is reported for the spiked standard, it is likely that sample handling or analytical procedures are out of control and corrective measures should be taken at once. It is important to

Table 6-4 Recommended Sample Handling and Preservation Procedures for a Detective Monitoring Program

Parameters (Type)	Volume Required (mL) 1 Sample*	Container (Material)	Preservation Method	Maximum Holding Period
Well Purging				
pH (grab)	50	T,S,P,G	None; field det.	<1 hr**
Ω^{-1} (grab)	100	T,S,P,G	None; field det.	<1 hr**
T (grab)	1000	T,S,P,G	None; field det.	None
Eh (grab)	1000	T,S,P,G	None; field det.	None
Contamination Indicators				
pH, Ω^{-1} (grab)	As above	As above	As above	As above
TOC	40	G,T	Dark, 4°C	24 hr
TOX	500	G,T	Dark, 4°C	5 days
Water Quality				
Dissolved gases (O_2, CH_4, CO_2)	10 mL minimum	G,S	Dark, 4°C	<24 hr
Alkalinity/Acidity	100	T,G,P	4°C/None	<6 hr**/ <24 hr
	Filtered under pressure with appropriate media			
(Fe, Mn, Na^+, K^+, Ca^{++}, Mg^{++})	All filtered 1000 mL	T,P	Field acidified to pH <2 with HNO_3	6 months***
(PO_4^-, Cl^-, Silicate)	@50	(T,P,G glass only)	4°C	24 hr/ 7 days; 7 days
NO_3^-	100	T,P,G	4°C	24 hr
SO_4^-	50	T,P,G	4°C	7 days
NH_4^+	400	T,P,G	4°C/H_2SO_4 to pH <2	24 hr/ 7 days
Phenols	500	T,G	4°C/H_3PO_4 to pH <4	24 hr
Drinking Water Suitability				
As, Ba, Cd, Cr, Pb, Hg, Se, Ag	Same as above for water quality cations (Fe, Mn, etc.)	Same as above	Same as above	6 months
F^-	Same as chloride above	Same as above	Same as above	7 days
Remaining Organic Parameters	As for TOX/TOC, except where analytical method calls for acidification of sample			24 hr

*It is assumed that at each site, for each sampling date, replicates, a field blank and standards must be taken at equal volume to those of the samples.
**Temperature correction must be made for reliable reporting. Variations greater than ±10% may result from longer holding period.
***In the event that HNO_3 cannot be used because of shipping restrictions, the sample should be refrigerated to 4°C, shipped immediately, and acidified on receipt at the laboratory. Container should be rinsed with 1:1 HNO_3 and included with sample.

Note: T = Teflon; S = stainless steel; P = PVC, polypropylene, polyethylene; G = borosilicate glass.

From Scalf et al., 1981.

know if the laboratory has performed these corrections or taken corrective action when they report the results of analyses. It should be further noted that many regulatory agencies require evidence of QC and analytical performance but do not generally accept data which has been corrected.

Field blanks, standards and blind control samples provide independent checks on handling and storage as well as the performance of the analytical laboratory. It should be noted that ground-water analytical data is incomplete unless the analytical performance data (e.g., accuracy, precision, detection, and quantitation limits) are reported along with each set of results. Discussions of whether significant changes in ground-water quality have occurred must be tempered by the accuracy and precision performance for specific chemical constituents.

Table 6-5 is a useful guide to the preparation of field standards, and spiking solutions for split samples. It is important that the field blanks and standards be made on the day of sampling and are subjected to all conditions to which the samples are exposed. Field spiked samples or blind controls should be prepared by spiking with concentrated stock standards in an appropriate background solution prior to the collection of any actual samples. Additional precautions should be taken against the depressurization of samples during air transport and the effects of undue exposure to light during sample handling and storage. All of the QC measures noted above will provide both a basis for high quality data reporting and a known degree of confidence in data interpretation. Well planned overall quality control programs will also minimize the uncertainty in long-term trends when different personnel have been involved in sample collection and analysis.

6.3.5 *Sample Storage and Transport*

The storage and transport of ground-water samples are often the most neglected elements of the sampling protocol. Due care must be taken in sample collection, field determinations and handling. Transport should be planned so as not to exceed sample holding time before laboratory analysis. Every effort should be made to inform the laboratory staff of the approximate time of arrival so that the most critical analytical determinations can be made within recommended storage periods. This may require that sampling schedules be adjusted so that the samples arrive at the laboratory during working hours.

The documentation of actual sample storage and treatment may be handled by the use of chain of custody procedures. An example of a chain of

Table 6-5 Field Standard and Sample Spiking Solutions

				Stock Solution for Field Spike of Split Samples		
Sample Type	Volume	Composition	Field Standard (Concentration)	Solvent	Concentration of Components	Field Spike Volume
Alkalinity	50 mL	Na^+, HCO_3^-	10.0; 25 (ppm)	H_2O	10,000; 25,000 (ppm)	(50 μL)
Anions	1 L	K^+, Na^+, Cl^-, SO_4^- F^-, NO_3^-, $PO_4^{\equiv}$, SI	25, 50 (ppm)	H_2O	25,000; 50,000 (ppm)	(1 mL)
Cations	1 L	Na^+, K^+ Ca^{++}, Mg^{++}, Cl^-, NO_3^-	5.0; 10.0 (ppm)	H_2O, H^+ (acid)	5,000; 10,000 (ppm)	(1 mL)
Trace Metals	1 L	Cd^{++}, Cu^{++}, Pb^{++} Cr^{+++}, Ni^{2+}, Ag^+ Fe^{+++}, Mn^{++}	10.0; 25.0 (ppm)	H_2O, H^+ (acid)	10,000; 25,000 (ppm)	(1 mL)
TOC	40 mL	Acetone KHP	0.2; 0.5 (ppm-C) 1.8; 4.5 (ppm-C)	H_2O	200; 500 (ppm-C) 1,800; 4,500 (ppm-C)	(40 μL)
TOX	50 mL	Chloroform 2,4,6 Trichlorophenol	12.5; 25 (ppb) 12.5; 25 (ppb)	H_2O/poly* (ethylene glycol)	12,500; 25 (ppm) 12,500; 25 (ppm)	(500 μL)
Volatiles	40 mL	Dichlorobutane, Toluene Dibromopropane, Xylene	25; 50 (ppb)	H_2O/poly* (ethylene glycol)	25; 50 (ppm)	(40 μL)
Extractables A	1 L	Phenol Standards	25; 50 (ppb)	Methanol**	25; 50 (ppm)	(1 mL)
Extractables B	1 L	Polynuclear Aromatic Standards	25; 50 (ppb)	Methanol	25; 50 (ppm)	(1 mL)
Extractables C	1 L	Standards as Required	25; 50 (ppb)	Methanol	25; 50 (ppm)	(1 mL)

*75:25 water/polyethylene glycol (400 amu) mixture.

**Glass distilled methanol.

Source: Barcelona et al., 1981.

custody form is shown in Figure 6-7. Briefly, the chain of custody record should contain the dates and times of collection, receipt and completion of all the analyses on a particular set of samples. It frequently is the only record that exists of the actual storage period prior to the reporting of analytical results. The sampling staff members who initiate the chain of custody should require that a copy of the form be returned to them with the analytical report. Otherwise, verification of sample storage and handling will be incomplete.

Sample shipment arrangements should be planned to insure that samples are neither lost nor damaged enroute to the laboratory. There are several commercial suppliers of sampling kits which permit refrigeration by freezer packs and include proper packing. It may be useful to include special labels or distinctive storage vessels for acid-preserved samples to accommodate shipping restrictions.

6.4 Summary

Ground-water sampling is conducted for a variety of reasons ranging from detection or assessment of the extent of a contaminant release to evaluations of trends in regional water quality. Reliable sampling of the subsurface is inherently more difficult than either air or surface water sampling because of the inevitable disturbances which well-drilling or pumping can cause and the inaccessibility of the sampling zone. Therefore, "representative" sampling generally requires minimal disturbance of the subsurface environment and the properties of a representative sample are therefore scale dependent. For any particular case, the applicable criteria should be set at the beginning of the effort to judge representativeness.

Reliable sampling protocols are based on the hydrogeologic setting of the study site and the degree of analytical detail required by the information needs of the monitoring program. Quality control over water quality data begins with the evaluation of the hydraulic performance of the sampling point or well and the proper selection of mechanisms and materials for well purging and sample collection. All other elements of the program and variables which effect data validity which follow sample collection may be accounted for by field blanks, standards and control samples.

Although research is needed on a host of topics involved in ground-water sampling, defensible sampling protocols can be developed to insure the collection of data of known quality for many types of programs. If properly planned and developed, long-term sampling efforts can benefit from the refinements which research progress will bring. Careful documentation will provide the key to this opportunity.

6.5 References

Anonymous. 1985. Monitoring Products: A Buyers Guide. Ground Water Monitoring Review 5(3):33-45.

Barcelona, M.J., J.P. Gibb, J.A. Helfrich, and E.E. Garske. 1985. Practical Guide for Ground-Water Sampling. State Water Survey Contract Report 374, U.S. Environmental Protection Agency, Robert S. Kerr Environmental Research Laboratory, Ada, OK and U.S. Environmental Protection Agency, Environmental Monitoring and Support Laboratory, Las Vegas, NV.

Barcelona, M.J., J.A. Helfrich, E. E. Garske, and J.P. Gibb. 1984. A Laboratory Evaluation of Ground Water Sampling Mechanisms. Ground Water Monitoring Review 4(2):32-41.

Barcelona, M.J. 1984. TOC Determinations in Ground Water. Ground Water 22(1):18-24.

Barcelona, M.J., and E.E. Garske. 1983. Nitric Oxide Interference in the Determination of Dissolved Oxygen by the Azide-Modified Winkler Method. Analytical Chemistry 55:965-967.

Barcelona, M.J., J.P. Gibb, and R.A. Miller. 1983. A Guide to the Selection of Materials for Monitoring Well Construction and Ground-Water Sampling. Illinois State Water Survey Contract Report, USEPA-RSKERL, EPA-600/S2-84-024. 78 pp.

Barcelona, M.J. 1983. Chemical Problems in Ground-Water Monitoring. Proceedings of the Third National Symposium on Aquifer Rehabilitation and Ground Water Monitoring, May 24-27, 1983, Columbus, OH.

Barcelona, M.J., J.A. Helfrich, and E.E. Garske. 1985. Sampling Tubing Effects on Ground Water Samples. Analytical Chemistry 47(2):460-464.

Barvenik, M.J., and R.M. Cadwgan. 1983. Multi-Level Gas-Drive Sampling of Deep Fractured Rock Aquifers in Virginia. Ground Water Monitoring Review 3(4):34-40.

Brass, H.J., M.A. Feige, T. Halloran, J.W. Mellow, D. Munch, and R.F. Thomas. 1977. The National Organic Monitoring Survey: Samplings and Analyses for Purgeable Organic Compounds. In: Drinking Water Quality Enhancement through Source Protection, edited by R.B. Pojasek. Ann Arbor Science Publishers, Ann Arbor, MI.

Brobst, R.B. 1984. Effects of Two Selected Drilling Fluids on Ground Water Sample Chemistry. Monitoring Wells, Their Place in the Water Well Industry Educational Session, NWWA National Meeting and Exposition, Las Vegas, NV.

Claasen, H.C. 1982. Guidelines and Techniques for Obtaining Water Samples That Accurately Represent

Figure 6-7 Sample chain of custody form.

CHAIN OF CUSTODY RECORD

Sampling Date ____________ Site Name ______________________________

Well or Sampling Points: ____

____

Sample Sets for Each: Inorganic, Organic, Both

Inclusive Sample Numbers:

Company's Name __________________________ Telephone (__) ________

Address ____
number street city state zip

Collector's Name ________________________ Telephone (__) ________

Date Sampled __________ Time Started ______ Time Completed ________

Field Information (Precautions, Number of Samples, Number of Sample Boxes, Etc.):

1. ____
name organization location

2. ____
name organization location

Chain of Possession (After samples are transported off-site or to laboratory):

1. ____________________ ______________________ ____ ____ (IN)
signature title
____________________ ______________________ ____ ____(OUT)
name (printed) date/time of receipt

2. ____________________ ______________________ ____ ____ (IN)
signature title
____________________ ______________________ ____ ____(OUT)
name (printed) date/time of receipt

Analysis Information:

	Aliquot	Analysis Begun (date/time)	Initials	Analysis Completed (date/time)	Initials
1.					
2.					
3.					
4.					
5.					

the Water Chemistry of an Aquifer. U.S. Geological Survey Open File Report, Lakeland, CO.

Dunlap, W.J., J.F. McNabb, M.R. Scalf, and R.L. Cosby. 1977. Sampling for Organic Chemicals and Microorganisms in the Subsurface. U.S. Environmental Protection Agency, Robert S. Kerr Environmental Research Laboratory, Ada, OK.

Evans, R.B., and G.E. Schweitzer. 1984. Assessing Hazardous Waste Problems. Environmental Science and Technology 18(11):330A-339A.

Everett, L.G., and L.G. McMillion. 1985. Operational Ranges for Suction Lysimeters. Ground Water Monitoring Review 5(3):51-60.

Everett, L.G., L.G. Wilson, E.W. Haylman, and L.G. McMillion. 1984. Constraints and Categories of Vadose Zone Monitoring Devices. Ground Water Monitoring Review 4(4).

Everett, L.G., L.G. Wilson, and L.G. McMillion. 1982. Vadose Zone Monitoring Concepts for Hazardous Waste Sites. Ground Water 20(3):312-324.

Fenn, D., E. Cocozza, J. Isbister, O. Braids, B. Yare, and P. Roux. 1977. Procedures Manual for Ground Water Monitoring at Solid Waste Disposal Facilities. EPA-530/SW611, U.S. Environmental Protection Agency, Cincinnati, OH.

Gibb, J.P., R.M. Schuller, and R.A. Griffin. 1981. Procedures for the Collection of Representative Water Quality Data from Monitoring Wells. Illinois State Water Survey Cooperative Report 7, Illinois State Water Survey and Illinois State Geological Survey, Champaign, IL.

Gillham, R.W., M.J.L. Robin, J.F. Barker, and J.A. Cherry. 1983. Ground Water Monitoring and Sample Bias. API Pub. 4367, American Petroleum Institute.

Grisak, G.E., R.E. Jackson, and J.F. Pickens. 1978. Monitoring Gro undwater Quality: The Technical Difficulties. Water Resources Bulletin 6:210-232.

Gorelick, S.M., B. Evans, and I. Remsan. 1983. Identifying Sources of Ground Water Pollution: An Optimization Approach. Water Resources Research 19(3):779-790.

Heaton, T.H.E., and J.C. Vogel. 1981. "Excess Air" in Ground Water. Journal Hydrology 50:201-216.

Ho, J.S-Y. 1983. Effect of Sampling Variables on Recovery of Volatile Organics in Water. Journal American Water Works Association 12:583-586.

Kennedy, V.C., E.A. Jenne, and J.M. Burchard. 1976. Backflushing Filters for Field Processing of Water Samples Prior to Trace-Element Analysis. Open-File Report 76-126. U.S.Geological Survey Water Resources Investigations.

Lindau, C.W., and R. F. Spalding. 1984. Major Procedural Discrepancies in Soil Extracted Nitrate Levels and Nitrogen Isotopic Values. Ground Water 22(3):273-278.

Mackay, D.M., P.V. Roberts, and J.A. Cherry. 1985. Transport of Organic Contaminants in Ground Water. Environmental Science and Technology 19(5):384-392.

Nacht, S.J. 1983. Monitoring Sampling Protocol Considerations. Ground Water Monitoring Review Summer:23-29.

National Council of the Paper Industry for Air and Stream Improvement. 1982. A Guide to Groundwater Sampling. Technical Bulletin 362, NCASI, New York, NY.

Nielsen, D.M., and G.L. Yeates. 1985. A Comparison of Sampling Mechanisms Available for Small-Diameter Ground Water Monitoring Wells. Ground Water Monitoring Review 5(2):83-99.

Pickens, J.F., J.A. Cherry, G.E. Grisak, W.F. Merritt, and B.A. Risto. 1978. A Multilevel Device for Ground-Water Sampling and Piezometric Monitoring. Ground Water 16(5):322-327.

Robbins, G.A., and M.M. Gemmell. 1985. Factors Requiring Resolution in Installing Vadose Zone Monitoring Systems. Ground Water Monitoring Review 5(3):75-80.

Scalf, M.R., J.F. McNabb, W.J. Dunlap, R.L. Cosby, and J. Fryberger. 1981. Manual of Ground Water Quality Sampling Procedures. National Water Well Association, OH.

Schwarzenbach, R.P. *et al.* 1985. Ground-Water Contamination by Volatile Halogenated Alkanes: Abiotic Formation of Volatile Sulfur Compounds Under Anaerobic Conditions. Environmental Science and Technology 19:322-327.

Sisk, S.W. 1981. NEIC Manual for Groundwater/Subsurface Investigations at Hazardous Waste Sites. U.S. Environmental Protection Agency, Office of Enforcement, National Enforcement Investigations Center, Denver, CO.

Skougstad, M.W., and G.F. Scarbo, Jr. 1968. Water Sample Filtration Unit. Environmental Science and Technology 2(4):298-301.

Stolzenburg, T.R., and D.G. Nichols. 1985. Preliminary Results on Chemical Changes in Ground Water Samples Due to Sampling Devices. Report to Electric Power Research Institute, Palo Alto, California, EA-4118. Residuals Management Technology, Inc., Madison, WI.

Tinlin, R.M., ed. 1976. Monitoring Groundwater Quality: Illustrative Examples. EPA-600/4-76-036,

U.S. Environmental Protection Agency, Environmental Monitoring and Support Laboratory, Las Vegas, NV.

Todd, D.K., R.M. Tinlin, K.D. Schmidt, and L.G. Everett. 1976. Monitoring Ground-Water Quality: Monitoring Methodology. EPA-600/4-76-026, U.S. Environmental Protection Agency, Las Vegas, NV.

U.S. Geological Survey. 1977. National Handbook of Recommended Methods for Water-Data Acquisition. U.S. Geological Survey, Office of Water Data Coordination, Reston, VA.

Villaume, J.R. 1985. Investigations at Sites Contaminated with Dense, Non-Aqueous Phase Liquids (NAPLS). Ground Water Monitoring Review 5(2):60-74.

Wehrmann, H.A. 1983. Monitoring Well Design and Construction. Ground Water Age 4:35-38.

Wilson, J.T., and J.F. McNabb. 1983. Biological Transformation of Organic Pollutants in Ground Water. EOS 64(33):505-506.

Wilson, J.T. *et al.* 1983. Biotransformation of Selected Organic Pollutants in Ground Water. In: Volume 24 Developments in Industrial Microbiology, Society for Industrial Microbiology.

Wilson, L.G. 1983. Monitoring in the Vadose Zone: Part III. Ground Water Monitoring Review 3(2):155-166.

Wilson, L.G. 1982. Monitoring in the Vadose Zone: Part II. Ground Water Monitoring Review 2(1):31-42.

Wilson, L.G. 1984. Monitoring in the Vadose Zone: Part I. Ground Water Monitoring Review 1(3):32-41.

Winograd, I.J., and F.N. Robertson. 1982. Deep Oxygenated Ground Water: Anomaly or Common Occurrence? Science 216:1227-1230.

Wood, W., and M.J. Petraitis. 1984. Origin and Distribution of Carbon Dioxide in the Unsaturated Zone of the Southern High Plains of Texas. Water Resources Research 20(9):1193-1208.

Wood, W.W. 1976. Guidelines for Collection and Field Analysis of Groundwater Samples for Selected Unstable Constituents. In: Techniques for Water Resources Investigations, of the U.S. Geological Survey.

Yare, B.S. 1975. The Use of a Specialized Drilling and Ground-Water Sampling Technique for Delineation of Hexavalent Chromium Contamination in an Unconfined Aquifer, Southern New Jersey Coastal Plain. Ground Water 13(2):151-154.

7. Ground-Water Tracers

The material presented in this chapter has been condensed from the report *Ground-Water Tracers* (Davis *et al.*, 1985).

7.1 General Characteristics of Tracers

As used in hydrogeology, a tracer is matter or energy carried by ground water which will give information concerning the direction of movement and/or velocity of the water and potential contaminants which might be transported by the water. If enough information is collected, the study of tracers can also help with the determination of hydraulic conductivity, porosity, dispersivity, chemical distribution coefficients, and other hydrogeologic parameters. A tracer can be entirely natural, such as the heat carried by hot-spring waters; it can be accidentally introduced, such as fuel oil from a ruptured storage tank; or it can be introduced intentionally, such as dyes placed in water flowing within limestone caverns.

Understanding the potential chemical and physical behavior of the tracer in ground water is the most important criterion in selecting a tracer. A tracer should travel with the same velocity and direction as the water and not interact with solid material. For most uses, a tracer should be nontoxic. It should be relatively inexpensive to use and should be, for most practical problems, easily detected with widely available and simple technology. The tracer should be present in concentrations well above background concentrations of the same constituent in the natural system which is being studied. Finally, the tracer itself should not modify the hydraulic conductivity or other properties of the medium being studied.

No one ideal tracer has been found. Because the natural systems to be studied are so complex and the requirements for the tracers themselves so numerous, the selection and use of tracers is almost as much an art as it is a science.

7.2 Public Health Considerations

Artificial introduction of tracers must be done with a careful consideration of possible health implications. Usually, investigations using artificially introduced tracers must have the approval of local or State health authorities, local citizens must be informed of the tracer injections, and the results should be made available to the public. Under some circumstances, analytical work associated with tracer studies must be done in appropriately certified laboratories.

7.3 Direction of Water Movement

To complete a tracer test using more than one well, the general direction of ground-water movement should be known. This is particularly true if the travel of tracers is to be studied using two wells with ground water flowing under a natural gradient.

Unfortunately, local flow directions may diverge widely from directions predicted on the basis of widely spaced water wells (Figure 7-1). It is not at all uncommon to inject a tracer in a well and not be able to intercept that tracer in another well just a few meters away, particularly if the tracer flows under the natural hydraulic gradient which is not disturbed by pumping.

7.4 Travel Time

Travel time of a water particle can be estimated using the equation:

$$t = \frac{n_e (\Delta L)^2}{K \Delta h}$$

where:

- t = time taken by the average water particle to move through distance ΔL
- n_e = effective porosity
- K = hydraulic conductivity
- Δh = hydraulic head drop.

If a tracer travels with the water, t is also the travel time of the tracer. The use of this equation is illustrated in Figure 7-2.

As can be seen in Equation 7-1, the expected travel time for a given head drop is a function of the distance squared $(\Delta L)^2$ and therefore increases very rapidly with the distance, ΔL. Thus, a tracer test in one region using a specific hydraulic head drop of Δh over a distance of 1,000 m would take 10,000 times

Figure 7-1 Divergence from predicted direction of ground water.

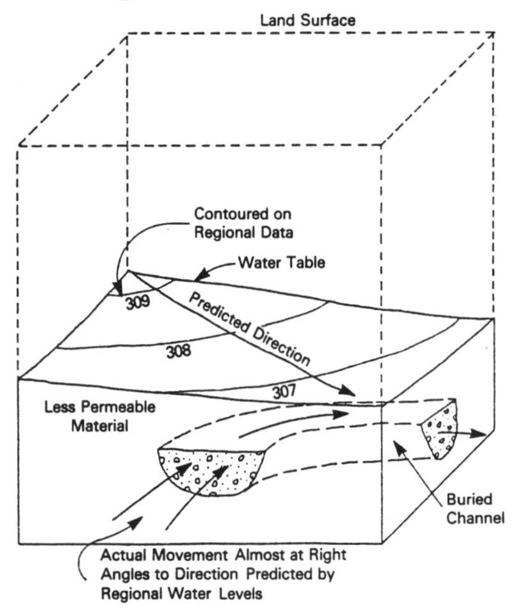

Figure 7-2 Example of water particle (and tracer) travel time calculation.

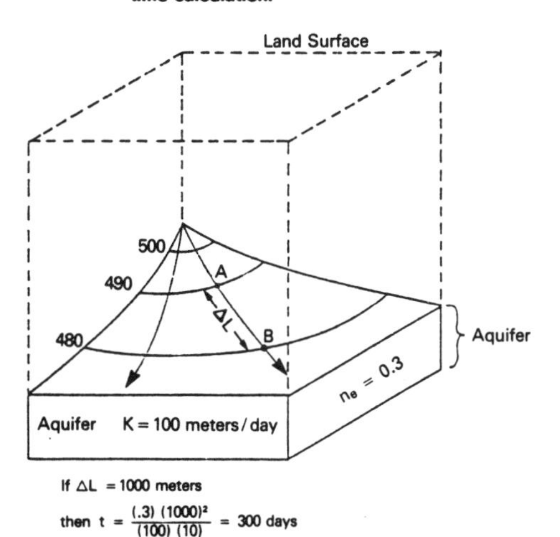

as long as a test in another region over a distance of 10 m which has the same head drop, provided the effective porosities and hydraulic conductivities are identical.

7.5 Sorption of Tracers and Related Phenomena

Sorption occurs when a dissolved ion or molecule becomes attached to the surface of a solid or dissolves in the solid. The term "sorption," as used here, includes the sum of the physical-chemical phenomena of ion exchange, induced dipole moments, hydrogen bonding, ligand exchange, and chemical bonding. Two processes of sorption are adsorption, a strictly surficial phenomenon, and absorption, a phenomenon which involves movement of material from solution to sites within the structure of the solid phase. Most sorption processes discussed here are relatively fast, reversible reactions; that is, the dissolved constituent which is sorbed from the water can be released to the water again under favorable circumstances. Cation exchange is probably the most familiar type of adsorption, and is a good example of reversible sorption.

Molecules of some tracers have a tendency to be sorbed on the surfaces of solids for brief periods, after which they move off the solid and into the water again. If the water is moving, the tracer molecules move at a slower rate than the water molecules, because tracer molecules spend part of their time sorbed on solids. Thus, the sorptive characteristics of a tracer must be known in order to design meaningful tracer experiments.

Certain tracers will be virtually unaffected by sorptive processes. Those tracers are commonly called "conservative" tracers because their concentrations, and hence their direct relation to the moving ground water, will be conserved if hydrodynamic dispersion is not considered.

Although unlikely in most artificially introduced tracer experiments, the possibility of mineral dissolution or precipitation should always be kept in mind. As a simple example, if the sulfate ion is used as a tracer in water which moves through a natural bed of gypsum, dissolution of the gypsum will undoubtedly add sulfate to the ground water and may confuse the interpretation of the experiment.

7.6 Hydrodynamic Dispersion and Molecular Diffusion

Two natural phenomena, hydrodynamic dispersion and molecular diffusion, always work together to dilute the concentrations of artificially injected tracers. These phenomena are complex and their effects are difficult to separate in field experiments. The two phenomena are, however, theoretically quite distinct.

Hydrodynamic dispersion is produced by natural differences in the local ground-water velocities related to the local differences in permeabilities (Figure 7-3). Molecular diffusion is produced by differences in chemical concentrations which tend to be erased in time by the random motion of molecules (Figure 7-4). Generally, short-term tracer experiments in permeable material are affected almost exclusively by hydrodynamic dispersion. In contrast, the concentrations of natural tracers moving very slowly in highly heterogeneous materials are affected profoundly by molecular diffusion.

7.7 Practical Aspects

7.7.1 Planning a Test

The purpose and practical constraints of a potential tracer test must be understood clearly prior to actual planning of tracer tests. Is only the direction of water flow to be determined? Are other parameters such as travel time, porosity, and hydraulic conductivity of interest? How much time is available for the test? If answers must be obtained within a few weeks, then tracer tests using only the natural hydraulic gradient between two wells which are more than about 20 meters apart would normally be out of the question because of the long time period needed for the tracer to flow between the wells. Another primary consideration is the budget. If several deep holes are to be drilled, if packers are to be set to control sampling or injection, and if hundreds of samples must be analyzed in an EPA-certified laboratory, then total costs could easily exceed $1 million. In contrast, some short-term tracer tests may be possible at costs of less than $1,000.

The initial step in determining the physical feasibility of a tracer test is to collect as much hydrogeologic information as possible concerning the field area. The logs of the wells at the site to be tested, or logs of the wells closest to the proposed site, should be obtained. Logs will give some idea of the homogeneity of the aquifer, layers present, fracture patterns, porosity, and boundaries of the flow system. Local or regional piezometric maps, or any published reports on the hydrology of the area (including results of aquifer tests) are valuable, as they may give an indication of the hydraulic gradient and hydraulic conductivity.

The hydrogeologic information is used to estimate the direction and magnitude of the ground-water velocity in the vicinity of the study area (Fetter, 1981). One method to arrive at a local velocity estimate is the use of water-level maps together with Darcy's Law if transmissivity, aquifer thickness, and head values are available. The second method involves using a central well with satellite boreholes, and running a preliminary tracer test. The classical method for determining the regional flow direction is to drill three boreholes at extremities of a triangle, with the sides

Figure 7-3 Variations in ground-water flow and distribution of tracer due to hydrodynamic dispersion.

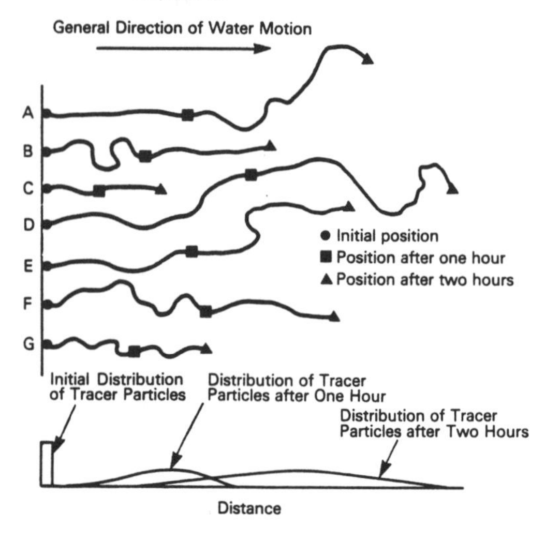

Figure 7-4 Movement by molecular diffusion.

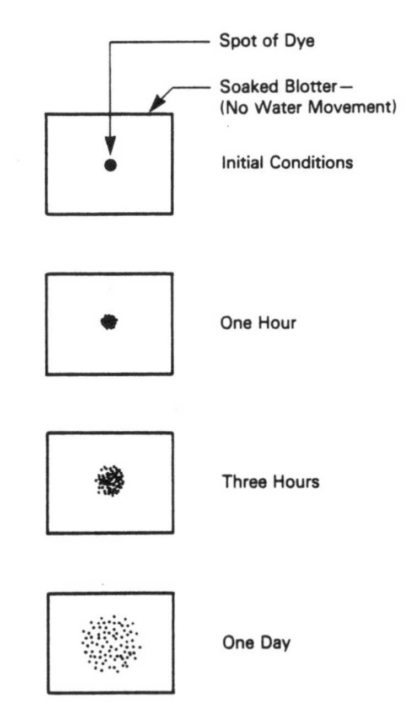

100 to 200 m apart (Figure 7-5). The water levels are measured and the line of highest slope gives the direction of flow. However, regional flow is generally not as important as local flow in most tracer tests, and the importance of having an accurate flow direction cannot be overemphasized. Gaspar and Oncescu (1972) described a method to determine local flow direction by drilling five to six satellite wells in the general direction of flow. The advantage of knowing the general flow direction is that fewer observation wells will eventually be drilled. If a preliminary value of the magnitude of the natural velocity of the aquifer is available, then the injection or pumping rate necessary to obtain radial flow can be determined. Also, when a velocity magnitude is obtained from the preliminary test or available data, a decision as to the distance from the injection well to observation well(s) can be made. This decision depends on whether the test is a natural flow or induced flow (injection or pumping) type test. Natural flow tests are less common due to the greater amount of time involved.

Figure 7-5 Determining the direction of ground-water flow.

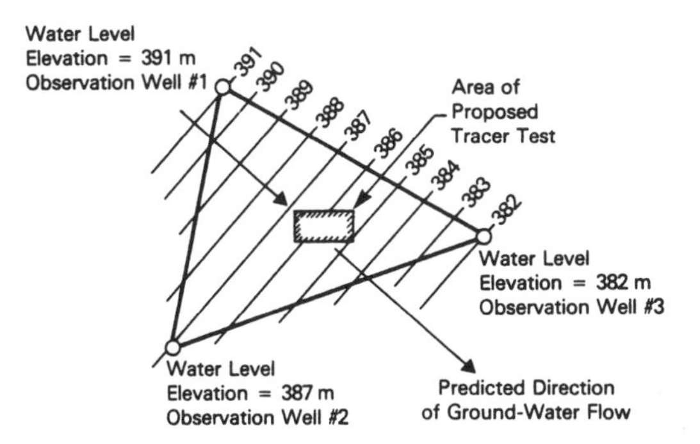

A second major consideration when planning a test is which tracers are the best for the conditions at the site and the objectives of the test. Samples should be analyzed for background values of relevant parameters, such as temperature, major ions, natural fluorescence, fluorocarbons, etc. Choice of a tracer will depend partially on which analytical techniques are easily available and which background constituents might interfere with these analyses. Various analytical techniques incorporate different interferences, and consultation with the chemist or technician who will analyze the samples is necessary. Determination of the amount of tracer to inject is based on the natural background concentrations detection limit for the tracer and the dilution expected. If a value for porosity can be estimated, the volume of voids in the medium can be calculated as the volume of a cylinder with one well at the center and the other a distance away. Adsorption, ion exchange, and dispersion will decrease the amount of tracer arriving at the observation well, but recovery is usually not less than 20 percent (of the injected mass) for two-hole tests using a forced recirculation system and conservative tracers. The concentration should not be increased so much that density effects become a problem. Lenda and Zuber (1970) gave graphs which can be used to estimate the approximate quantity of tracer needed. The values are based on estimates of the porosity and dispersion coefficient of the aquifer.

7.7.2 Types of Tracer Tests

The variety of tracer tests is almost infinite when one considers the various combinations of tracer types, local hydrologic conditions, injection methods, sampling methods, and the geological setting of the site.

Some of these varieties are shown in Figure 7-6. The following sections discuss a few of the more common types of tracer tests. Differences in the tests are due to the parameters (such as velocity, dispersion coefficient, and porosity) which are to be determined, the scale of the test, and the number of wells to be used.

7.7.2.1 Single-Well Techniques

Two techniques, injection/withdrawal and point dilution, give values of parameters which are valid at a local scale. Advantages of single-well techniques are:

- Less tracer is required than for two-well tests
- The assumption of radial flow is generally valid, so natural aquifer velocity can be ignored, making solutions easier
- Knowledge of the exact direction of flow is not necessary.

Injection/Withdrawal. The single-well injection/withdrawal (or pulse) technique results in a value of pore velocity and the longitudinal dispersion coefficient. The method assumes that porosity is known or can be estimated with reasonable accuracy. A given quantity of tracer is instantaneously added to the borehole, the tracer is mixed, and then two to three borehole volumes of fresh water are pumped in to force the tracer to penetrate the aquifer. Only a small quantity is injected so that natural flow is not disturbed.

After a certain time, the borehole is pumped out at a constant rate which is large enough to overcome the natural ground-water flow. Tracer concentration is measured with time or pumped volume. If concentration is measured at various depths with point samplers, relative permeability of layers can be determined. The dispersion coefficient is obtained by matching experimental breakthrough curves with

Figure 7-6 Common configurations and uses for groundwater tracing.

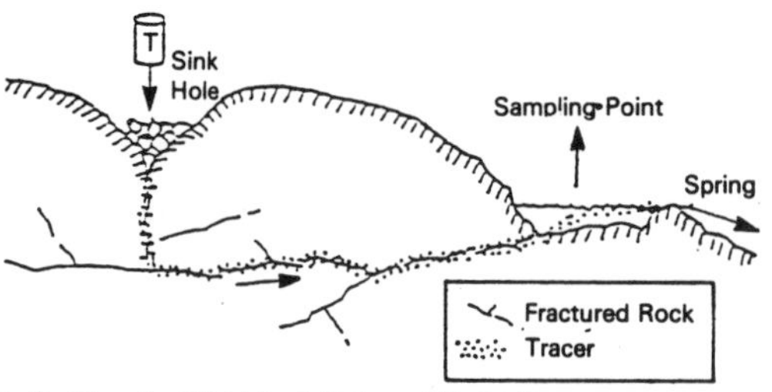

A. To determine if trash in sinkhole contributes to contamination of spring.

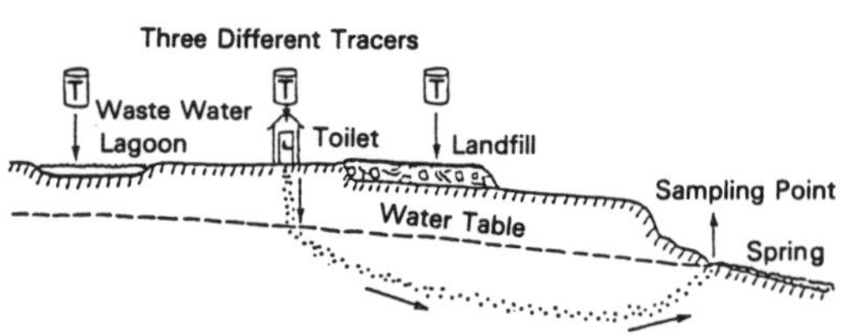

E. To determine source of pollution from three possibilities.

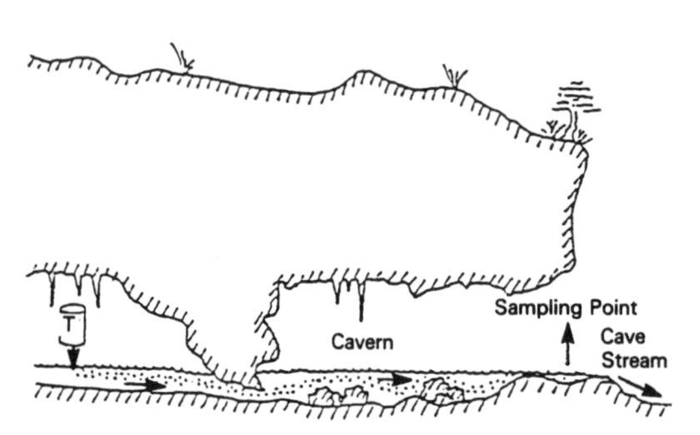

B. To measure velocity of water in cave stream.

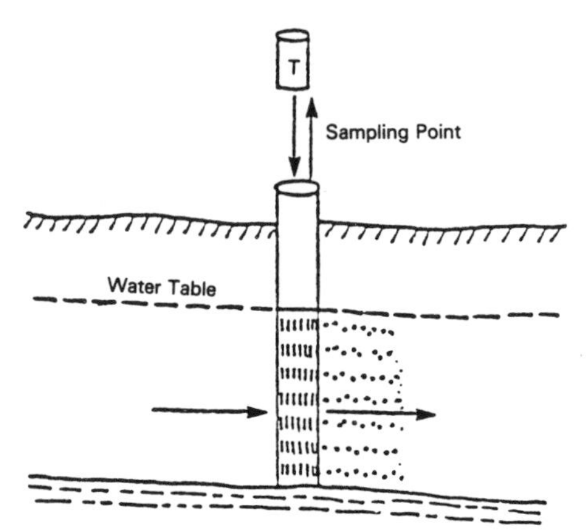

F. To determine velocity and direction of ground-water flow under natural conditions. Injection followed by sampling from same well.

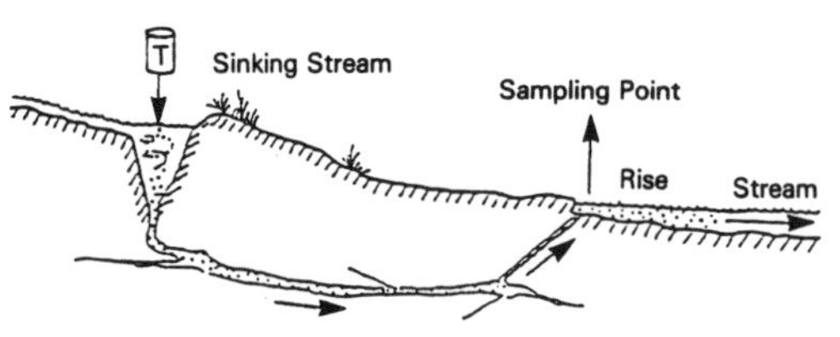

C. To check source of water at rise in stream bed.

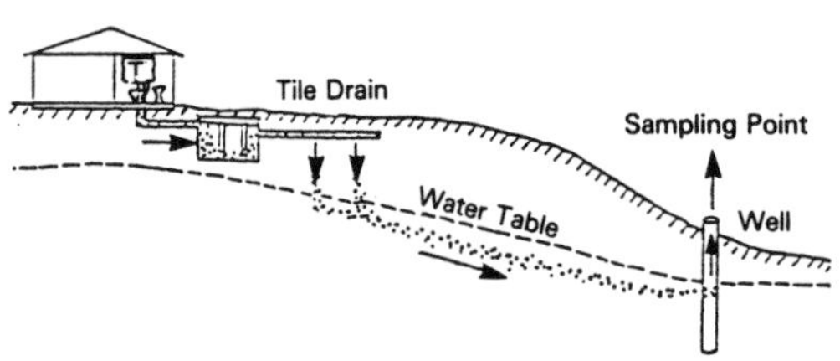

D. To determine if tile drain from septic tank contributes to contamination of well.

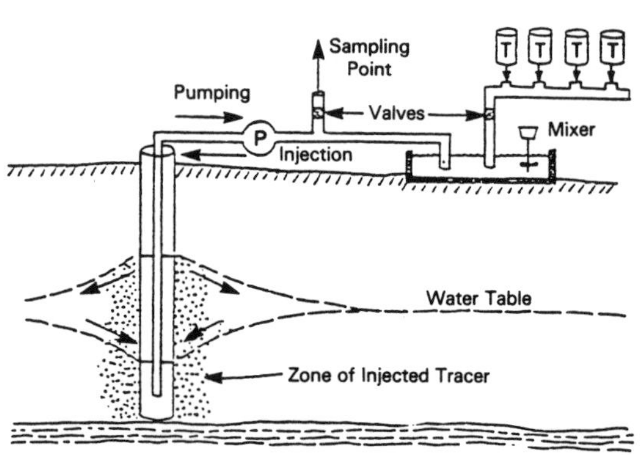

G. To test precipitation of selected constituents on the aquifer material by injecting multiple tracers into aquifer then pumping back the injected water.

Figure 7-6 (continued)

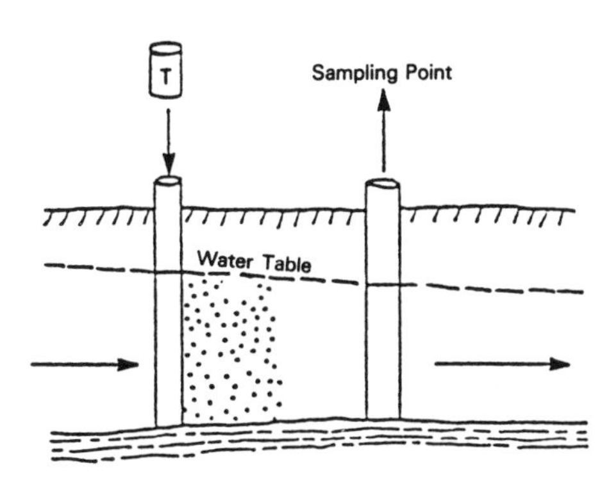

H. To test velocity of movement of dissolved material under natural ground-water gradients.

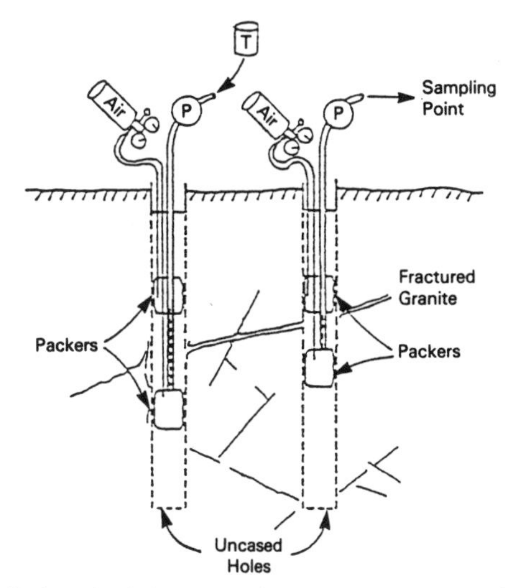

K. To determine the interconnect fractures between two uncased holes. Packers are inflated with air and can be positioned as desired in the holes.

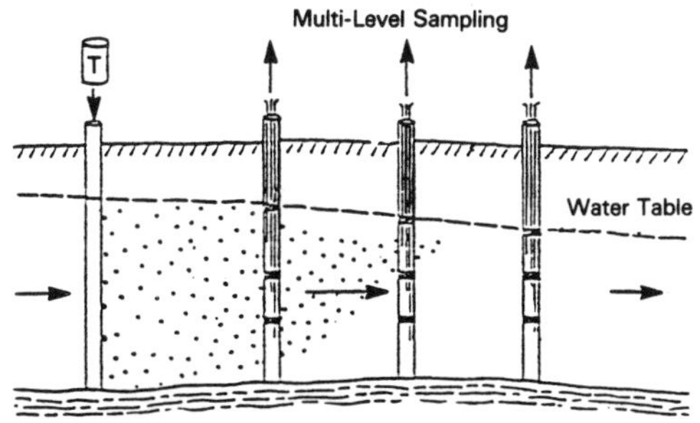

I. To test hydrodynamic dispersion in aquifer under natural ground-water gradients.

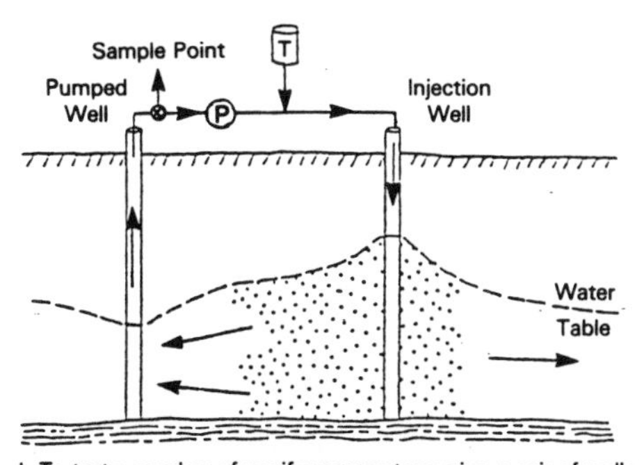

J. To test a number of aquifer parameters using a pair of wells with forced circulation between wells.

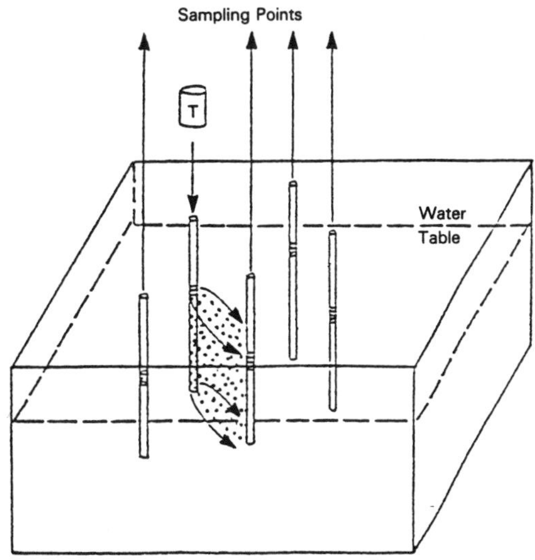

L. To determine the direction and velocity of natural ground-water flow by drilling an array of sampling wells around a tracer injection well.

Figure 7-6 (continued)

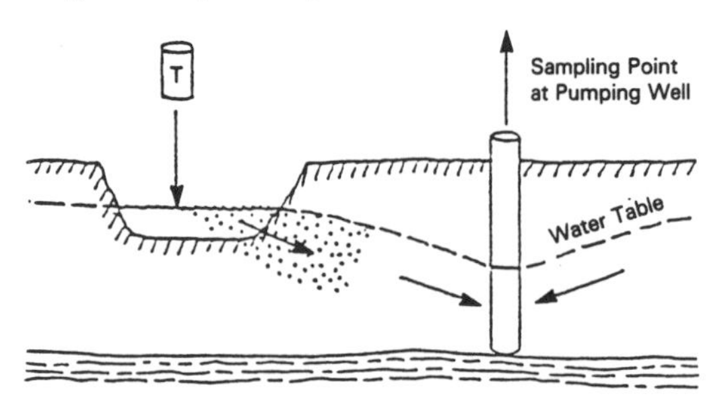

M. To verify connection between surface water and well.

theoretical curves based on the general dispersion equation. A finite difference method is used to simulate the theoretical curves (Fried, 1975).

Fried concluded that the method is useful for local information (2 to 4 m) and for detecting the most permeable strata. An advantage of this test is that nearly all of the tracer is removed from the aquifer at the end of the test.

Borehole Dilution. This technique can be used to measure the magnitude and direction of horizontal tracer velocity and vertical flow.

The procedure is to introduce a known quantity of tracer instantaneously into the borehole, mix it well, and then measure the concentration decrease with time. The tracer is generally introduced into an isolated volume of the borehole using packers. Radioactive tracers have been used for borehole dilution tests, but other tracers can be used.

Some factors to keep in mind when conducting a point dilution test are the homogeneity of the aquifer, effects of drilling (mudcake, etc.), homogeneity of the mixture of the tracer and the well water, degree of tracer diffusion, and density effects.

The ideal condition for conducting the test is to use a borehole with no screen or gravel pack. If a screen is used, it should be next to the borehole as dead space alters the results. Samples should be very small in volume so that flow is not disturbed by its removal.

The direction of ground-water flow can be measured in a single borehole by a method similar to point dilution. A tracer (often radioactive) is introduced slowly and without mixing. A section of the borehole is usually isolated by packers. After some time, a compartmental sampler (four to eight compartments) within the borehole is opened. The direction of minimum concentration corresponds to the flow direction. A similar method is to introduce a radioactive tracer and subsequently measure its adsorption on the borehole or well screen walls by means of a counting device in the hole. The method is described in more detail in Gaspar and Oncescu (1972).

7.7.2.2 Two-Well Techniques

There are two methods, one testing for uniform (natural) flow and the other for radial flow. The parameters measured (dispersion coefficient and porosity) are assumed to be the same for both types of flow.

Uniform Flow. A tracer is placed in one well without disturbing the flow field and a signal is measured at observation wells. This test can be used at a local (2 to 5 m) or intermediate (5 to 100 m) scale, but it requires much more time than radial tests. The direction and magnitude of the velocity must be known quite precisely, or a large number of observation wells are needed. The quantity of tracer needed to cover a large distance can be expensive. On a regional scale environmental tracers are generally used, including seawater intrusion, radionuclides, or stable isotopes of hydrogen and oxygen. Manmade pollution has also been used. For regional problems, a mathematical model is calibrated with concentration versus time curves from field data, and the same model is used to predict future concentration distributions.

Analysis of local or intermediate scale uniform flow problems can be done analytically, semianalytically, or by curve-matching. Layers of different permeability can cause distorted breakthrough curves, which can usually be analyzed (Gaspar and Oncescu, 1972). One- or two-dimensional models are available. Analytical solutions can be found in Fried (1975) and Lenda and Zuber (1970).

Radial Flow. These techniques are based on imposing a velocity on the aquifer, and generally solutions are easier if radial flow is much greater than uniform flow. A value for natural ground-water velocity is not obtained, but porosity and the dispersion coefficient are obtained.

A diverging test involves constant injection of water into an aquifer with a slug or continuous flow of tracer introduced instantaneously into the injected water. The tracer is detected at an observation well which is not pumping. Very small samples are taken at the observation well so that flow is not disturbed. Packers can be used in the injection well to isolate an interval. Sampling can be done with point samplers or an integrated sample can be taken.

Converging tests involve introduction of the tracer at an observation well, and another well is pumped. Concentrations are monitored at the pumped well. The tracer is often injected between two packers or below one packer, and then two to three well bore volumes are injected to push the tracer out into the aquifer. At the pumping well, intervals of interest are isolated (particularly in fractured rock), or an integrated sample is obtained.

A recirculating test is similar to a converging test, but the pumped water is injected back into the injection well. This tests a significantly greater part of the formation because the wells inject to and pump from 360 degrees. The flow lines are longer, partially canceling out the advantage of a higher gradient. Theoretical curves are available for recirculating tests (see Sauty, 1980).

7.7.3 Design and Construction of Test Wells

In many tracer tests the construction of test wells is the single most expensive part of the work. It also can be the source of major difficulty if the construction is not done properly.

Five common types of problems are encountered with tracer tests. The first problem relates to site selection. If heavy equipment is to be moved into an area, lack of overhead clearance, narrow roads, poor bearing capacities of bridges, and the lack of flat ground at the site can all be major problems. Also, overhead electrical power lines at the site should be avoided. One of the most common hazards is accidental grounding of power lines by drill rigs and auger stems with subsequent electrocution of workers.

The second problem relates to the improper choice of drilling equipment and the use of drilling fluids which will affect the tracer tests. Certain drilling muds and mud additives have a very high capacity for the sorption of most types of tracers. The muds could also clog small pores and alter the permeability of the aquifer near the drill hole. The use of compressed air for drilling may avoid some of these problems.

A third problem is the choice of casing diameter. Ideally, packers should be used to isolate the zones being sampled from the rest of the water in the well. For a number of reasons which include economics, insufficient time, and lack of technical training, packers are often not used in tracer tests. In this case, the diameter of the sampling well should be as small as possible in order to minimize the amount of "dead" water in the well during sampling. The diameter, however, cannot be too small because the well must be adequately cleaned after installation and the well must accommodate bailers, pumps, and other sampling equipment. Common casing diameters used range from about 1 in to 4 in for relatively shallow test holes to as much as 6 in to 8 in for very deep tests.

The type of casing to be used is a fourth concern, primarily if low-level concentrations of tracers are to be used and particularly, if these tracers are organic compounds or metallic cations. For plastic casings, Teflon absorbs and releases less organics than does PVC. Adhesives used to connect sections of plastic pipes may be also a troublesome source of interfering organic compounds. Metal casing could release trace metals but it is generally superior to plastic casing in terms of strength and sorptive characteristics. Inexpensive metal casing, however, will have a short life if ground waters are corrosive.

A fifth problem is the choice of filter construction for the wells, which depends on the aquifer and the type of test to be completed. If the aquifer being tested is a very permeable coarse gravel and if the casing diameter is small, then numerous holes drilled in the solid casing may be adequate. In contrast, for a single-well test with an alternating cycle of injection and pumping of large volumes of water into and out of loose, fine-grained sand, an expensive well screen with a carefully placed gravel pack may be required. Regardless of the type of filter used, it is absolutely essential that the casing perforations, gravel pack, or screen, as well as the aquifer at the well, be cleaned of silt, clay, drilling mud, and other material which would prevent the free movement of water in and out of the well. This process of cleaning or development is so critical that it should be specified in clear terms in any contract related to well construction.

7.7.4 Injection and Sample Collection

Injection equipment depends on the depth of the borehole and the funds available. In very shallow holes, the tracer can be lowered through a tube, placed in an ampule which is lowered into the hole and broken, or it can be just poured in. Mixing is desirable and important for most types of tests and is simple for very shallow holes. For example, a plunger can be surged up and down in the hole or the release of the tracer can be through a pipe with many perforations. Flanges on the outer part of the pipe will allow the tracer to be mixed by raising and lowering the pipe. For deeper holes, tracers must be injected under pressure and equipment can be quite sophisticated. The equipment used in work conducted in fractured rock by the Department of Hydrology at the University of Arizona is described in detail in Simpson *et al.* (1983).

Sample collection can also be simple or sophisticated. For tracing thermal pulses, only a thermistor needs to be lowered into the ground water. For chemical tracers at shallow depths, a hand pump may be sufficient. Bailers can also be used, but they mix the tracer in the borehole which, for some purposes, should be avoided. A Teflon bottom-loading bailer is described in Buss and Bandt (1981). It may be desirable to clear the borehole before taking a sample, in which case a gas-drive pump can be used to evacuate the well. For a nonpumping system, deciding how much water must be withdrawn from a borehole in order to obtain a representative sample of the water adjacent to the borehole is not a trivial

Figure 7-7 Results of tracer tests at the Sand Ridge State Forest, Illinois.

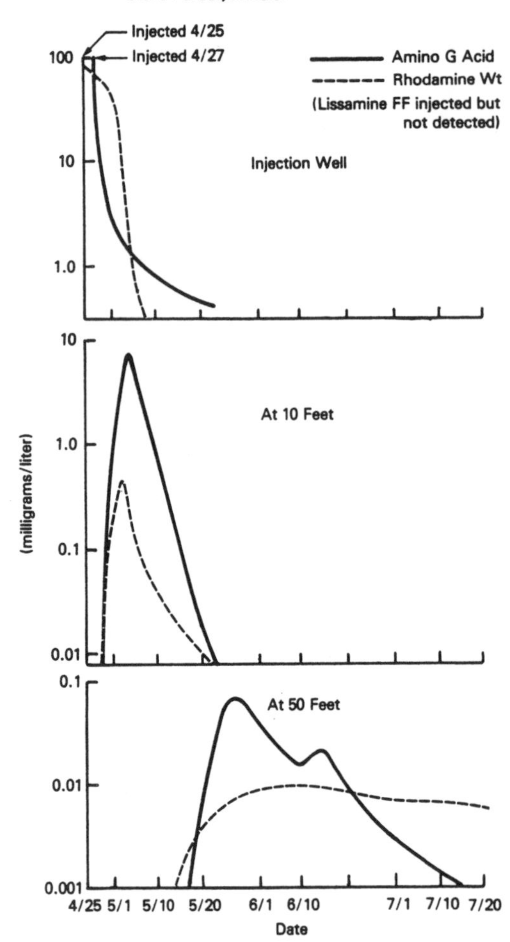

problem. If not enough water is withdrawn, the sample composition will be influenced by semistatic water, which will normally fill much of the well. If too much water is drawn, a gradient towards the well will be created and the natural movement of the tracer will be distorted. A common rule of thumb is to pump out four times the volume of water in the well before the sample is taken.

If existing wells which have been drilled for water-supply purposes are used for tracer tests, extreme care is required because of the complex relationship among such variables as pumping rates, patterns of water circulation within the well, and the yields of different parts of the aquifers which are penetrated. This complexity is usually reflected in the variability of water chemistry as a well is being pumped (Keith *et al.*, 1982; Schmidt, 1977). Stated simply, for wells drawing water from complex aquifers or a series of aquifers, an analysis of a single water sample taken at a given point in time cannot yield definitive information about the water chemistry of any individual zone.

The preservation and analysis of samples is covered in Chapter 6 of this publication. Keith *et al.* (1982) also cover some of the practical problems involved with sample collection, analyses, and quality control.

7.7.5 *Interpretation of Results*

The following remarks and figures are intended only as a brief qualitative introduction to the interpretation of the results of tracer tests. More extensive and quantitative treatments are found in the works of such authors as Halevy and Nir (1962), Theis (1963), Fried (1975), Custodio (1976), Sauty (1978), Grisak and Pickens (1980), and Gelhar (1982).

The basic plot of the concentration of a tracer as a function of time or water volume passed through the system is called a breakthrough curve. The concentration is either plotted as the actual concentration (Figure 7-7) or, quite commonly, as the ratio of the measured tracer concentration at the sampling point, C, to the input tracer concentration, C_0 (Figure 7-8).

The measured quantity which is fundamental for most tracer tests is the first arrival time of the tracer as it goes from an injection point to a sampling point. The first arrival time conveys at least two bits of information. First, it indicates that a connection for ground-water flow actually exists between the two points. For many tracer tests, particularly in karst regions, this is all the information which is desired. Second, an approximation of the maximum velocity of ground-water flow between the two points may be obtained if the tracer used is conservative.

Interpretations more elaborate than the two simple ones mentioned depend very much on the type of aquifer being tested, the velocity of ground-water flow, the configuration of the tracer injection and sampling systems, and the type of tracer or mixture of tracers used in the test.

After the first arrival time, interest is most commonly centered on the arrival time of the peak concentration for a slug injection or, for a continuous feed of tracers, the time since injection when the concentration of the tracer changes most rapidly as a function of time (Figure 7-9). In general, if conservative tracers are used, this time is close to the theoretical transit time of an average molecule of ground water traveling between the two points.

If a tracer is being introduced continuously into a ditch penetrating an aquifer, as shown in Figure 7-8,

Figure 7-8 Tracer concentration at sampling well, C, measured against tracer concentration at input, C_o.

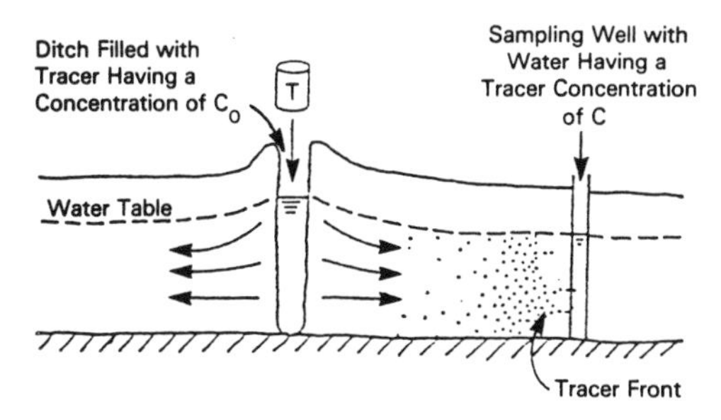

A. Tracer movement from injection ditch to sampling well.

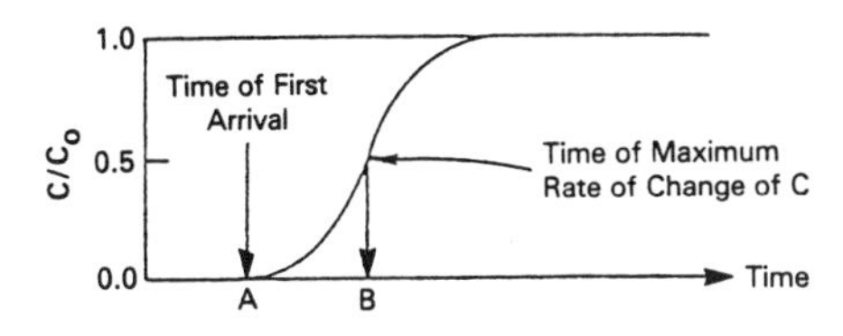

B. Breakthrough Curve.

Figure 7-9 Incomplete saturation of acquifer with tracer.

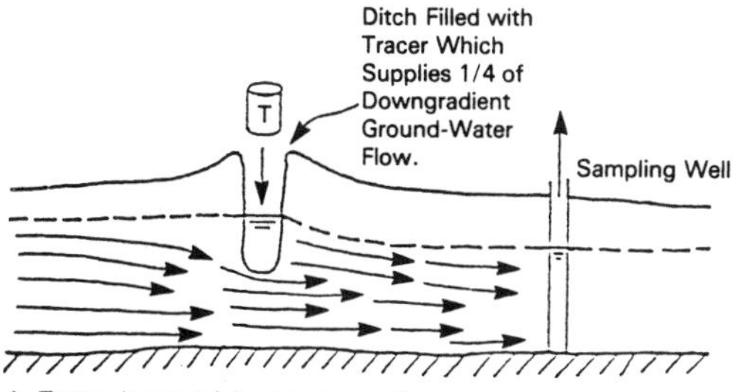

A. Tracer does not fully saturate aquifer.

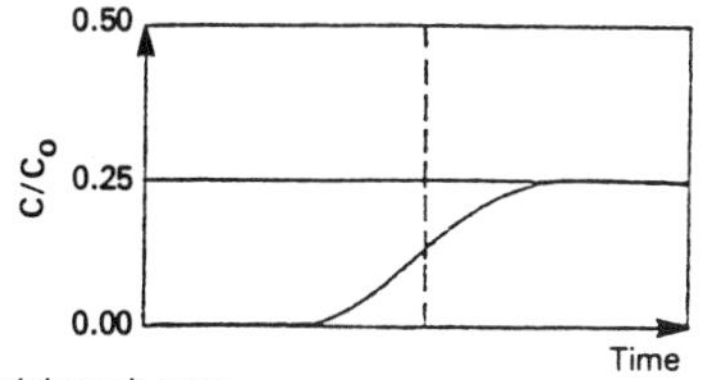

B. Breakthrough curve.

then the ratio C/C_o will approach 1.0 after the tracer starts to pass the sampling point. The ratio of 1.0 is rarely approached in most tracer tests in the field, however, because waters are mixed by dispersion and diffusion in the aquifer and because wells used for sampling will commonly intercept far more ground water than has been tagged by tracers (Figure 7-9). Ratios of C/C_o in the range of between 10^{-5} and 2×10^{-1} are often reported from field tests.

If a tracer is introduced passively into an aquifer but is recovered by pumping a separate sampling well, then various mixtures of the tracer and the native ground water will be recovered depending on the amount of water pumped, the transmissivity of the aquifer, the slope of the water table, and the shape of the tracer plume. Keely (1984) has presented this problem graphically with regards to the removal of contaminated water from an aquifer.

With an introduction of a mixture of tracers, possible interactions between the tracers and the solid part of the aquifer may be studied. If interactions take place, they can be detected by comparing breakthrough curves of a conservative tracer with the curves of the other tracers being tested (Figure 7-10). A common strategy for these types of tracer tests is to inject and subsequently remove the water containing mixed tracers from a single well. If injection is rapid and pumping to remove the tracer follows immediately, then a recovery of almost all the injected conservative tracer is possible. If the pumping is delayed, the injected tracer will drift downgradient with the general flow of the ground water and the percentage of the recovery of the conservative tracer will be less as time progresses. Successive tests using longer delay times between injection and pumping can then be used to estimate ground-water velocities in permeable aquifers with moderately large hydraulic gradients.

The methods of quantitative analyses of tracer breakthrough curves are generally by curve-matching of computer-generated type curves, or by analytical methods.

7.8 Types of Tracers

7.8.1 Water Temperature

The temperature of water changes slowly as it migrates through the subsurface because water has a high specific heat capacity compared to most natural materials. For example, temperature anomalies associated with the spreading of warm wastewater in the Hanford Reservation in south central Washington have been detected more than 8 km (5 mi) from the source (U.S. Research and Development Administration, 1975).

Water temperature is a potentially useful tracer, although it has not been used frequently. The method should be applicable in granular media, fractured

Figure 7-10 Breakthrough curves for conservative and nonconservative tracers.

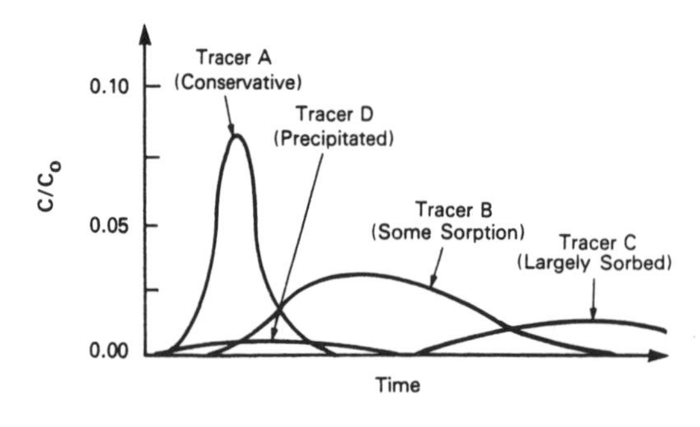

Figure 7-11 Results of field test using a hot water tracer.

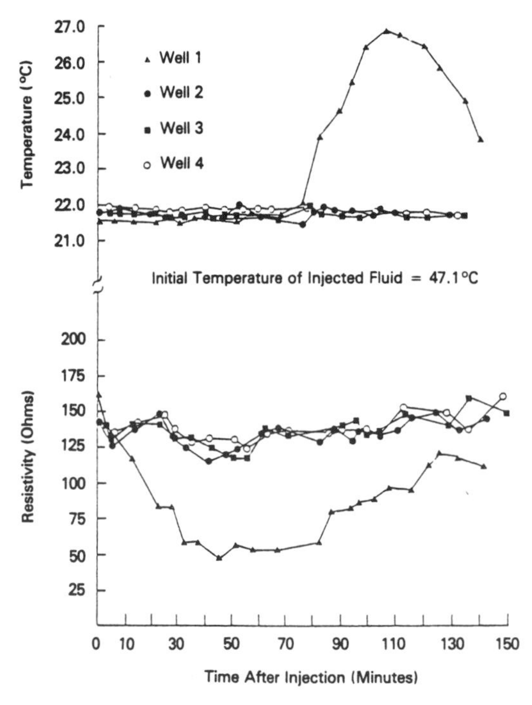

rock, or karst regions. Keys and Brown (1978) traced thermal pulses resulting from the artificial recharge of playa lake water into the Ogallala Formation in Texas. They described the use of temperature logs (temperature measurements at intervals in cased holes) as a means of detecting hydraulic conductivity differences in an aquifer. Temperature logs have also been used to determine vertical movement of water in a borehole (Keys and MacCary, 1971; Sorey, 1971).

Changes in water temperature are accompanied by changes in density and viscosity of the water. This in turn alters the velocity and direction of flow of the water. For example, injected ground water with a temperature of 40°C will travel more than twice as fast in the same aquifer under the same hydraulic gradient as water at 5°C. Because the warm water has a slightly lower density than cold water, buoyant forces give rise to flow which "floats" on top of the cold water. In order to minimize problems of temperature-induced convection, small temperature differences with very accurate temperature measurements should be used if hot or cold water is in the introduced tracer.

Temperature was used as a tracer for small-scale field tests, using shallow drive-point wells two feet apart in an alluvial aquifer. The transit time of the peak temperature was about 107 min, while the resistivity data indicated a travel time of about 120 min (Figure 7-11). The injected water had a temperature of 38°C, while the ground-water temperature was 20°C. The peak temperature obtained in the observation well was 27°C.

In these tests, temperature served as an indicator of breakthrough of the chemical tracers, aiding in the timing of sampling. It was also useful as a simple, inexpensive tracer for determining the correct placement of sampling wells.

Another application of water-temperature tracing is the detection of river recharge in an aquifer. Most rivers have large seasonal water temperature fluctuations. If the river is recharging an aquifer, the seasonal fluctuations can be detected in the ground water adjacent to the river (Rorabaugh, 1956).

7.8.2 Solid Particles

Solid material in suspension can be a useful tracer in areas where water flows in large conduits such as some basalt, limestone, or dolomite aquifers. Aley (1976) reported that geese, bales of hay, and wheat chaff have been used in Missouri in karst regions. In the past decade, small particulate tracers such as bacteria have been used successfully in porous media.

Paper and Simple Floats. Some examples of these tracers are small bits of paper (as punched out from computer cards, for example), or multicolored polypropylene floats. Due to the large size of these tracers, they are useful only when flow is through large passages. The particles must be of such a size and density as to pass through shallow sections of flow without settling out. Because these particulates generally float on the surface, they travel faster than the water's mean velocity. These tracers are most useful for approximating the flow velocity and establishing the flow path.

Dunn (1963) described the use of polypropylene floats of approximately 3/32-in diameter and 1-in length.

Signal-Emitting Floats. These are delayed time bombs which float through a cave system. When the bomb explodes, the location of the explosion is determined by seismic methods at the surface (Arandjelovic, 1969 and 1977). Problems with this method include noise interference from wind, traffic, and surface streams. Because these methods are relatively expensive, they have seldom been used.

Yeast. The use of baker's yeast (*Saccharomyces cerevisiae*) as a ground-water tracer in a sand and gravel aquifer has been reported by Wood and Ehrlich (1978). Yeast is a single-celled fungus which is ovoid in shape. The diameter of a yeast cell is 2 to 3 µm, which closely approximates the size of pathogenic bacterial cells. This tracer is probably most applicable in providing information concerning the potential movement of bacteria.

Wood and Ehrlich (1976) found that the yeast penetrated more than 7 m into a sand and gravel aquifer in less than 48 hours after injection. The tracer is very inexpensive, as is analysis. The lack of environmental concerns related to this tracer is another of its advantages.

Bacteria. Bacteria are the most commonly used microbial tracers, due to their ease of growth and simple detection. Keswick *et al*. (1982) reviewed case studies of bacteria used as tracers. Some of the bacteria which have been used successfully are *Escherichia coliform* (E. coli), *Streptococcus faecalis*, *Bacillus stearothermophilus*, *Serratia marcescens*, and *Serratia indica*. They range in size from 1 to 10 m and have been used in a variety of applications.

A fecal coliform, E. coli, has been used to indicate fecal pollution at pit latrines, septic fields, and sewage disposal sites. A "marker" such as antibiotic resistance or H_2S production is necessary to distinguish the tracer from background organisms.

The greatest health concern in using these tracers is that the bacteria must be nonpathogenic to man. Even E. coli has strains which can be pathogenic, and Davis *et al*. (1970) reported that Serratia marcescens may be pathogenic. Antibiotic-resistant strains are another concern. The antibiotic resistance can be transferred to potential human pathogens. This can be avoided by using bacteria which cannot transfer this genetic information. As is true with most other injected tracers, permission to use bacterial tracers should be obtained from the proper Federal, State, and local health authorities.

Viruses. Animal, plant, and bacterial viruses have been used as ground-water tracers. Viruses are generally much smaller than bacteria, ranging from 0.2 to 1.0 µm (see Table 7-1). In general, human enteric viruses cannot be used due to disease potential. Certain vaccine strains, such as a type of polio virus, have been used but are considered risky. Most animal enteric viruses are considered safer as they are not known to infect man (Keswick *et al*., 1982). However, neither human nor most animal viruses are generally considered to be suitable tracers for field work because of their potential to infect man.

Table 7-1 Comparison of Microbial Tracers

Tracer	Size (µm)	Time Required for Assay (days)	Essential Equipment Required
Bacteria	1-10	1-2	Incubator*
Spores	25-33	1/2	Microscope Plankton nets
Yeast	2-3	1-2	Incubator*
Viruses:			
Animal (enteric)	0.2-0.8	3-5	Incubator Tissue culture Laboratory
Bacterial	0.2-1.0	1/2-1	Incubator*

*Many may be assayed at room temperature.

Spores. Lycopodium spores have been used as a water tracer since the early 1950s, and the techniques are well developed. Spore tracing was initiated by Mayr (1953) and Maurin and Zotl (1959) and modified by Drew (1968). Lycopodium is a clubmoss which has spores that are nearly spherical in shape, with a mean diameter of 33 µm. It is composed of cellulose and is slightly denser than water, requiring some turbulence to keep the material in suspension. Some advantages of lycopodium are:

- The spores are relatively small
- They are not affected by water chemistry or adsorbed by clay or silt
- They travel at approximately the velocity of the surrounding water
- The injection concentration can be very high (e.g., 8×10^6 spores per cm^3)
- They pose no health threat
- The spores are easily detectable under the microscope
- At least five dye colors may be used, allowing five tracings to be conducted simultaneously in a karst system.

Some disadvantages associated with its use include the large amount of time required for preparation and analysis of the spores, and the problem of spores being filtered by sand or gravel if flow is not sufficiently turbulent.

The basic procedure involves the addition of a few kilograms of dyed spores to a cave or sinking stream. The movement of the tracer is monitored by sampling downstream in the cave or at a spring, with plankton nets installed in the stream bed. The sediment caught in the net is concentrated and treated to remove organic matter. The spores are then examined under the microscope.

Tracing by lycopodium spores is most useful in open joints or solution channels (karst terrain). It is not useful in wells or boreholes unless the water is pumped continuously to the surface and filtered. A velocity of a few miles per hour has been found sufficient to keep the spores in suspension. According to Smart and Smith (1976), lycopodium is preferable to dyes for use in large-scale water resource reconnaissance studies in karst areas. This holds if skilled personnel are available to sample and analyze the spores and a relatively small number of sampling sites are used.

The spores survive well in polluted water, but do not perform well in slow flow or in water with a high sediment concentration. Lycopodium spores have been used extensively in the United States, Great Britain, and other countries to determine flow paths and to estimate time of travel in karst systems.

7.8.3 Ions

Ionic compounds such as common salts have been used extensively as ground-water tracers. This category of tracers includes those compounds which undergo ionization in water resulting in separation into charged species possessing a positive charge (cations) or a negative charge (anions). The charge on an ion affects its movement through aquifers by numerous mechanisms.

Ionic tracers have been used as tools for a wide range of hydrologic problems dealing with the determination of flow paths and residence time and the measurement of aquifer properties.

Specific characteristics of individual ions or ionic groups may approach those of an ideal tracer, particularly in the case of dilute concentrations of certain anions.

In most situations, anions (negatively charged ions) are not affected by the aquifer medium. Mattson (1929), however, has shown that the capacity of clay minerals for holding anions increases with decreasing pH. Under conditions of low pH, anions in the presence of clay, other minerals, or organic detritus may undergo anion exchange. Other effects which may occur include anion exclusion and precipitation/dissolution reactions. Cations (positively charged ions) react much more frequently with clay minerals through the process of cation exchange which in turn displaces other cations such as sodium and calcium into solution. For this reason, little work has been done with cations due to the interaction with the aquifer media. Kaufman (1956) has shown that when permeabilities and flow rates are low, often indicative of a large clay fraction, the solid phase may have a considerable adsorption of an ionic component. This is significant for cationic tracers and may have some significance for certain anionic tracers.

One advantage of the simple ionic tracers is that they do not decompose and thus are not lost from the system. However, a large number of ions (including Cl^- and NO_3^-) have high natural background concentrations, thus requiring the injection of a tracer of high concentration. This may result in density separation and gravity segregation during the tracer test (Grisak, 1979). Density differences will alter flow patterns, the degree of ion exchange, and secondary chemical precipitation, which may change the aquifer permeability.

Various applications of ionic tracers have been described in the literature. Methods similar to those used for Cl^- were also postulated for ions such as nitrate (NO_3^-), dichromate (CR_2O_7), and ammonium (NH_4^+) (Haas, 1959). Murray (1981) used lithium bromide (LiBr) in carbonate terraine to establish hydraulic connection between a landfill and a fresh-water spring where use of rhodamine WT dye tracer proved inappropriate. Sodium chloride (NaCl) was used by Mather (1969) to investigate the influence of mining subsidence on the pattern of ground-water flow. Tennyson (1980) used bromide (Br^-) to evaluate pathways and transit time of recharge through soil at a proposed sewage effluent irrigation site. Chloride (Cl^-) and calcium (Ca^+) were used by Grisak (1979) to study solute transport mechanisms in fractures. Potassium (K^+) was used to determine leachate migration and extent of dilution by receiving waters located by a waste disposal site (Ellis, 1980).

7.8.4 Dyes

Dyes are relatively inexpensive, simple to use, and effective. Nonfluorescent dyes include congo red and malachite green, which have been used in conjunction with cotton strip detectors (Drew, 1968) or with visual detection, often in soil studies. The extensive use of fluorescent dyes for ground-water tracing began around 1960. Fluorescent dyes are preferable to nonfluorescent varieties due to much better detectability.

Although fluorescent dyes exhibit many of the properties of an ideal tracer, a number of factors interfere with concentration measurement. Fluorescence is used to measure dye concentration, but it may vary with suspended sediment load, temperature, pH, $CaCO_3$ content, salinity, etc. Other variables which affect tracer test results are "quenching" (some emitted fluorescent light is reabsorbed by other molecules), adsorption, and photochemical and biological decay. Another disadvantage of fluorescent dyes is their poor performance in tropical climates due to chemical reactions with dissolved carbon dioxide.

The advantages of using these dyes include their very high detectability, rapid field analysis, and relatively low cost and low toxicity.

Fluorescence intensity is inversely proportional to temperature. Smart and Laidlaw (1977) described the numerical relationship and provided temperature correction curves. The effect of pH on rhodamine WT fluorescence is shown in Figure 7-12. An increase in the suspended sediment concentration generally causes a decrease in fluorescence. Adsorption on kaolinite caused a decrease in the measured fluorescence of several dyes, as measured by Smart and Laidlaw.

The detected fluorescence may decrease or actually increase due to adsorption. If dye is adsorbed onto suspended solids, and the fluorescence measurements are taken without separating the water samples from the sediment, the dye concentration is a measurement of sediment content and not of water flow. Adsorption can occur on organic matter, clays (bentonite, kaolinite, etc.), sandstone, limestone, plants, plankton, and even glass sample bottles. These adsorption effects are a strong incentive to choose a nonsorptive dye for the type of medium tested. The sorption of orange dyes on bentonite clay is shown in Table 7-2.

Dyes travel slower than water due to adsorption, and are generally not as conservative as the ionic or radioactive tracers. Drew (1968) compared lycopodium, temperature, and fluorescein as karst tracers and found fluorescein breakthrough to be the slowest (Figure 7-13).

Although only one test is generally run due to economic considerations, it may be advisable to run several tests to check reproducibility if accuracy is important. Brown and Ford (1971) obtained some very interesting results by running three identical dye tracer tests in the same karst system. These yielded three different flow-through times. One of the values differed by 50 percent from the original test value.

Fluorescein, also known as uranin, sodium fluorescein, and pthalien, has been one of the most widely used green dyes. Like all green dyes, its use is commonly complicated by high natural background fluroescence, which lowers sensitivity of analyses and makes interpretation of results more difficult. Feuerstein and Selleck (1963) recommended that

Table 7-2 Measured Sorption of Dyes on Bentonite Clay

Dye	Losses Due to Adsorption on Clay (%)
Rhodamine WT	28
Rhodamine B	96
Sulfo Rhodamine B	65

Source: Repogle et al., 166.

Figure 7-12 The effect of pH on rhodamine WT (adapted from Smart and Laidlaw, 1977).

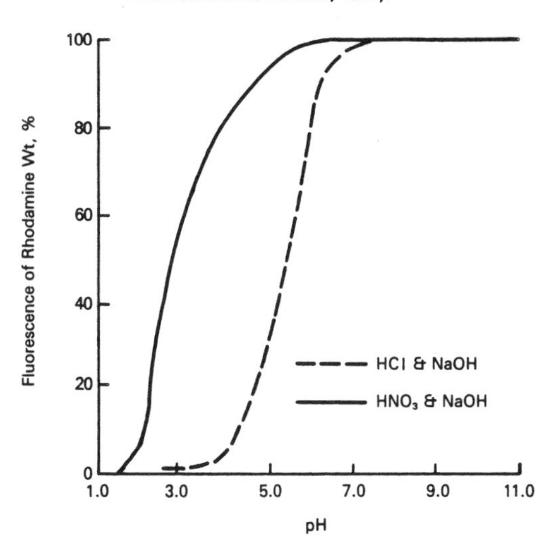

Figure 7-13 A comparison of the results of three simultaneous tracer tests in a karst system (data from Drew, 1968).

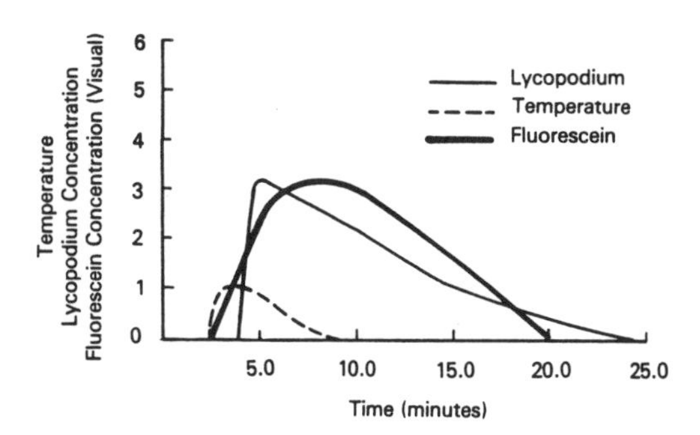

Table 7-3 Sensitivity and Minimum Detectable Concentrations for the Tracer Dyes

Dye	Sensitivity* µg/L Per Scale Unit	Background Reading** Scale Units 0-100	Minimum Detectability*** µg/L
Amino G Acid	0.27	19.0	0.51
Photine CU	0.19	19.0	0.36
Fluorescein	0.11	26.5	0.29
Lissamine FF	0.11	26.5	0.29
Pyranine	0.333	26.5	0.087
Rhodamine B	0.010	1.5	0.010
Rhodamine WT	0.013	1.5	0.013
Sulfo Rhodamine B	0.061	1.5	0.061

For a Turner 111 filter fluorometer with high-sensitivity door and recommended filters and lamp at 21°C.

* At a pH of 7.5.

** For distilled water.

*** For a 10 percent increase over background reading or one scale unit, whichever is larger.

Adapted from Smart and Laidlaw, 1977.

fluorescein be restricted to short-term studies of only the highest quality water.

Lewis *et al.* (1966) used fluorescein in a fractured rock study. Another example is a mining subsidence investigation in South Wales, where more than one ton of fluorescein was used in a sandstone tracer test (Mather *et al.*, 1969). Tester *et al.* (1982) used fluorescein to determine fracture volumes and diagnose flow behavior in a fractured granitic geothermal reservoir. He found no measurable adsorption or decomposition of the dye during the 24 hr exposures to rocks at 392°F. Omoti and Wild (1979) stated that fluorescein is one of the best tracers for soil studies, but Rahe *et al.* (1978) did not recover any injected dye in their hillslope studies, even at a distance of 2.5 m downslope from the injection point. The same experiment used bacterial tracers successfully. Figure 7-13 compares fluorescein, lycopodium, and temperature as karst tracers.

The approximate sensitivity and minimum detection limit for a number of dyes are given in Table 7-3.

Another green fluorescent dye, pyranine, has been used in several soil studies, and Reynolds (1966) found it to be the most stable dye used in an acidic, sandy soil. Omoti and Wild (1979) recommended pyranine and fluorescein as the best tracers for soil tests, although pyranine is relatively unstable if the organic matter content of the soil is high. Drew and Smith (1969) stated that pyranine is not as easily detectable as fluorescein, but is more resistant to decoloration and adsorption. Pyranine has a very high photochemical decay rate, and is strongly affected by pH in the range found in most natural waters (McLaughlin, 1982).

The orange dye rhodamine WT, is thought to be slightly less toxic than rhodamine B and sulfo rhodamine B (Smart and Laidlaw, 1977). This source notes that rhodamine WT and fluorescein are of comparable toxicity, but Aley and Fletcher (1976) stated that rhodamine WT is not as "biologically safe" as fluorescein.

This dye has been considered one of the most useful tracers for quantitative studies, based on minimum detectability, photochemical and biological decay rates, and adsorption (Smart and Laidlaw, 1977; Wilson, 1968; and Knuttson, 1968). Hubbard *et al.* (1982) stated that it is the most conservative of dyes available for stream or karst tracing.

Some recent uses of rhodamine WT include projects by Burden (1981), Aulenbach *et al.* (1978), Brown and Ford (1971), Gann (1975), and Aulenbach and Clesceri (1980). Burden successfully used the dye in a water contamination study in New Zealand in an alluvial aquifer. Aulenbach and Clesceri also found rhodamine WT very successful in a sandy medium. Gann (1975) used rhodamine WT for karst tracing in a limestone and dolomite system in Missouri. Three fluorescent dyes (rhodamine B, rhodamine WT, and fluorescein) were used by Brown and Ford (1971) in a karst test in the Maligne Basin in Canada. The highest recovery of dye (98%) was obtained for rhodamine WT. The fluorescein was not recovered at all. Aulenbach *et al.* (1978) compared rhodamine B, rhodamine WT, and tritium as tracers in a delta sand. The project involved tracing effluent from a sewage treatment plant. The rhodamine B was highly adsorbed, while the rhodamine WT and tritium yielded similar breakthrough curves. Rhodamine WT seems to be adsorbed less than rhodamine B or sulfo rhodamine B (Table 7-3). Wilson (1971) found that in column and field studies, rhodamine WT did show sorptive tendencies.

Sulfo rhodamine B, also known as pontacyl brilliant pink, is more expensive than the other rhodamine dyes, and its toxicity appears to be slightly higher than that of rhodamine WT. It has not been used extensively as a ground-water tracer.

Blue fluorescent dyes have been used in increasing amounts in the past decade in textiles, paper, and other materials to enhance their white appearance. Water which has been contaminated by domestic waste can be used as a "natural" tracer if it contains detectable amounts of the brighteners. Glover (1972) described the use of optical brighteners in karst environments. Examples of the brighteners are amino G acid and photine CU. These two are the least sensitive of the dyes reviewed (Table 7-2), but the blue dyes have much lower background levels in uncontaminated water than do the green or orange dyes. Photine CU is significantly affected by temperature variations, and both dyes are affected by pH below 6.0. Amino G acid is fairly resistant to adsorption.

Toxicity studies on optical brighteners were reviewed by Akamatsu and Matsuo (1973). They concluded that the brighteners do not present any toxic hazard to man, even at excessive dosage levels.

7.8.5 Some Common Nonionized and Poorly Ionized Compounds

A number of chemical compounds will dissolve in water but will not ionize or will ionize only slightly under normal conditions of pH and Eh found in ground waters. Some of these compounds are relatively difficult to detect in small concentrations, others present a health hazard, and still others are present in moderate to large concentrations in natural waters, thus making the background effects difficult to deal with in most settings. A list of a few of these compounds is given in Table 7-4.

The use of these and similar compounds as injected tracers in ground water is limited to rather special cases. Of those listed, boric acid would probably act most conservatively over long distances of ground-water flow. Boric acid has been used successfully as a tracer in a geothermal system (Downs *et al.*, 1983). Large concentrations, 1,000 mg/l or more, would need to be used for injected tracers which, unfortunately, would pose difficult environmental questions if tracing were attempted in aquifers with potable water. From the standpoint of health concerns, sugars would be the most acceptable; however, they decompose rapidly in the subsurface and also tend to be sorbed on some materials. Alcohols such as ethanol would also tend to be sorbed on any solid organic matter which might be present. Another problem with the use of most of these compounds as tracers is that they would need to be introduced in moderately large concentrations which in turn would change the density and viscosity (particularly for glycerin) of the injected tracer mixture. Nevertheless, some of these compounds such as sugars may be useful for simulating the movement of other compounds which are also subject to rapid decomposition but which are too hazardous to inject directly into aquifers.

7.8.6 Gases

Numerous natural as well as artificially produced gases have been found in ground water. Some of the naturally produced gases can serve as tracers. Gas can also be injected into ground water where it dissolves and can serve as a tracer, but only a few examples of it being used for ground-water tracers are found in the literature. Gases of potential use in hydrogeologic studies are listed in Table 7-5.

Inert Radioactive Gases. Chemically inert but radioactive ^{133}Xe and ^{85}Kr appear to be suitable for many injected tracer applications (Robertson, 1969; and Wagner, 1977), provided legal restrictions can be overcome. Of the natural inert radioactive gases, ^{222}Rn is the most abundant. It is one of the daughter products from the spontaneous fission of ^{238}U. Radon is present in the subsurface, but owing to the short half-life (3.8 d) of ^{222}Rn, and the absence of parent uranium nuclides in the atmosphere, radon is virtually absent in surface water which has reached equilibrium with the atmosphere. Surveys of radon in surface streams and lakes have, therefore, been useful in detecting the locations of places where ground water enters surface waters (Rogers, 1958).

Inert Natural Gases. Because of their nonreactive and nontoxic nature, noble gases are potentially useful tracers. Helium is used widely as a tracer in industrial processes. It also has been used to a limited extent as a ground-water tracer (Carter *et al*, 1959). Neon, krypton, and xenon are other possible candidates for injected tracers because their natural concentrations are very low (Table 7-5). Although the gases do not undergo chemical reactions and do not participte in ion exchange, the heavier noble gases (krypton and xenon) do sorb to some extent on clay and organic material.

The very low natural concentrations of noble gases in ground water make them useful as tracers, particularly in determining ground-water velocities in regional aquifers. The solubility of the noble gases

Table 7-4 Some Simple Compounds Which are Soluble in Water

Name	Formula	Remarks
Silicic Acid	H_4SiO_4 (After combination with water)	Present in normal ground water in nonionized form in concentrations of between 4 and 100 mg/L. Low toxicity.
Boric Acid	H_3BO_3	Present in normal ground water in nonionized form in concentrations of 0.05 to 2.0 mg/L. Toxic to plants above 1 to 5 mg/L. Toxic to humans in higher concentrations.
Phosphoric Acid	H_3PO_4	Ionizes above pH of 6.0. Will form complexes with other dissolved constituents. Sorbs on or reacts with most aquifer materials. Natural concentrations mostly between 0.05 mg/L and 0.5 mg/L.
Acetic Acid	$C_2H_4O_2$	Moderately toxic in high concentrations. Water soluble. Natural concentrations are less than 0.1 mg/L in ground water.
Ethyl Alcohol (Ethanol)	C_2H_6O	Major component of alcoholic drinks. Water soluble. Natural concentrations are less than 0.05 mg/L in ground water.
Sugars		Major components of human and animal foods. Water soluble. Probably less than 0.2 mg/L in most ground water.
Sucrose	$C_{12}H_{22}O_{11}$	
Maltose	$C_{12}H_{22}O_{11}$	
Lactose	$C_{12}H_{22}O_{11}$	
Glucose	$C_6H_{12}O_6$	
Glycerol (Glycerin)	$C_3H_8O_4$	Water soluble. Low toxicity. Probably absent in natural ground water.

Table 7-5 Gases of Potential Use as Tracers

	Approximate Natural Background Assuming Equilibrium with Atmosphere at 20°C (mg gas/L water)	Maximum Amount in Solution Assuming 100% Gas at Pressure of 1 atm at 20°C (mg gas/L water)
Argon	0.57	60.6
Neon	1.7×10^{-4}	9.5
Helium	8.2×10^{-6}	1.5
Krypton	2.7×10^{-4}	234
Xenon	5.7×10^{-5}	658
Carbon Monoxide	6.0×10^{-6}	28
Nitrous Oxide	3.3×10^{-4}	1100

decreases with an increase in temperature. The natural concentrations of these gases in ground water are, therefore, an indication of surface temperatures at the time of infiltration of the water. This fact has been used to reconstruct the past movement of water in several aquifers (Sugisaki, 1969; Mazor, 1972; Andrews and Lee, 1979).

Fluorocarbons. Numerous artificial gases have been manufactured during the past decade and several of these gases have been released in sufficient volumes to produce measurable concentrations in the atmosphere on a worldwide scale. One of the most interesting groups of these gases are the fluorocarbons (Table 7-6). The gases generally pose a very low biological hazard, they are generally stable for periods measured in years, they do not react chemically with other materials, they can be detected in very low concentrations, and they sorb only slightly on most minerals. They do sorb strongly, however, on organic matter.

Fluorocarbons have two primary applications. First, because large amounts of fluorocarbons were not released into the atmosphere until the later 1940s and early 1950s, the presence of fluorocarbons in ground water indicates that the water was in contact with the atmosphere within the past 30 to 40 yr and that the ground water is very young (Thompson and Hayes, 1978). The second application of fluorocarbon compounds is for injected tracers (Thompson, Hayes, and Davis, 1974). Because detection limits are so low, large volumes of water can be labeled with the tracers at a rather modest cost. Despite the problem of sorption on natural material and especially on organics, initial tests have been quite encouraging.

Stable Isotopes. An isotope is a variation of an element produced by differences in the number of neutrons in the nucleus of that element. However, the difficulty in detecting small artificial variations of most isotopes against the natural background, the high cost of their analysis, and the expense of preparing isotopically enriched tracers, means that stable

Table 7-6 Properties of Fluorocarbon Compounds

Common Name	Chemical Formula	Boiling Point at 1 atm (°C)	Solubility in Water at 25°C (weight %)
Freon-11	CCl_3F	23.8	0.11
Freon-12	CCl_2F_2	−29.8	0.028
Freon-113	CCl_2F-$CClF_2$	47.6	0.017
--	$CBrClF_2$	−4.0	unknown
--	CBr_2F_2	24.5	unknown
--	$CBrI$-$CBrF_2$	47.3	unknown

isotopes are rarely used for artifically injected tracer studies in the field.

Research into the topic of stable isotopes of various elements in natural waters is progressing rapidly, and the potential usefulness of these isotopes to ground-water tracing will undoubtedly increase markedly in the near future.

The most common use of studies of 2H and 180 has been to trace the large-scale movement of ground water and to locate areas of recharge (Gat, 1971; Fritz and Fontes, 1980; and Ferronsky and Polyakov, 1982).

The two abundant isotopes of nitrogen (^{14}N and ^{15}N) can vary significantly in nature. Ammonia escaping as vapor from decomposing animal wastes, for example, will tend to remove the lighter (^{14}N) nitrogen and will leave behind a residue rich in heavy nitrogen. In contrast, many fertilizers with an ammonia base will be isotopically light. Natural soil nitrate will be somewhat between these two extremes. As a consequence, nitrogen isotopes have been useful in helping to determine the origin of unusually high amounts of nitrate in ground water. Also, the presence of more than about 5 mg/l of nitrate commonly is an indirect indication of contamination from chemical fertilizers and sewage.

The stable sulfur isotopes (^{32}S, ^{34}S, and ^{36}S) have been used to distinguish sulfate originating from natural dissolution of gypsum ($CaSO_4 \bullet 2H_2O$) from sulfate originating from an industrial spill of sulfuric acid (H_2SO_4).

Two stable isotopes of carbon (^{12}C and ^{13}C) and one unstable isotope (^{14}C) are used in hydrogeologic studies. Most isotopic studies of carbon in water have been centered on ^{14}C which will be discussed in a later portion of this chapter. Although not as commonly studied as ^{14}C, the ratio of the stable isotopes, $^{13}C/^{12}C$, is potentially useful in sorting out the origins of certain contaminants found in water. For example, methane (CH_4) originating from some deep geologic deposits is isotopically heavier then methane originating from near surface sources. This contrast forms the basis for identifying aquifers contaminated with methane from pipelines and from subsurface storage tanks.

Isotopes of other elements such as chlorine, strontium, and boron are related more to the determination of regional directions of ground-water flow than to problems of the identification of sources of contamination.

Radionuclides. Radioactive isotopes of various elements are collectively referred to as radionucl ides. In the early 1950's there was great enthusiasm for using radionuclides both as natural "environmental" tracers and as injected artificial tracers. The use of artificially injected radionuclides has all but ceased in many countries, including the United States. Most use of artificially introduced radioactive tracers is confined to carefully controlled laboratory experiments or to deep petroleum production zones which are devoid of potable water. However, the environmental use has been expanded greatly until it is now a major component of many hydrochemical studies.

A number of radionuclides are present in the atmosphere from natural and artificial sources. Many of these are carried into the subsurface by rain water. The most common hydrogeologic use of these radionuclides is to obtain some estimate of the average length of time ground water has been isolated from the atmosphere. This is complicated by dispersion in the aquifer and mixing in wells that sample several hydrologic zones. Nevertheless, it can usually be established that most or virtually all of the ground water is older than some given limiting value. In many situations we can say, based on atmospheric radionuclides, that the ground water was recharged more than 1,000 years ago or that, in another region, all the ground water in a given shallow aquifer is younger than 30 years.

Since the 1950s, atmospheric tritium, the radioactive isotope of hydrogen (^{3}H), has been dominated by tritium from the detonation of thermonuclear devices. Thermonuclear explosions had increased the concentration of tritium in local rainfall to more than 1,000 TU in the northern hemisphere by the early 1960s (Figure 7-14). As a result, ground water in the northern hemisphere which has more than about 5 TU is generally less than 30 years old. Very small amounts, 0.05 to 0.5 TU, can be produced by natural subsurface processes, so the presence of these low levels does not necessarily indicate water 40 to 60 years old or small amounts of more recent water mixed with very old water.

Figure 7-14 Average annual tritium concentration of rainfall and snow for Arizona, Colorado, New Mexico, and Utah.

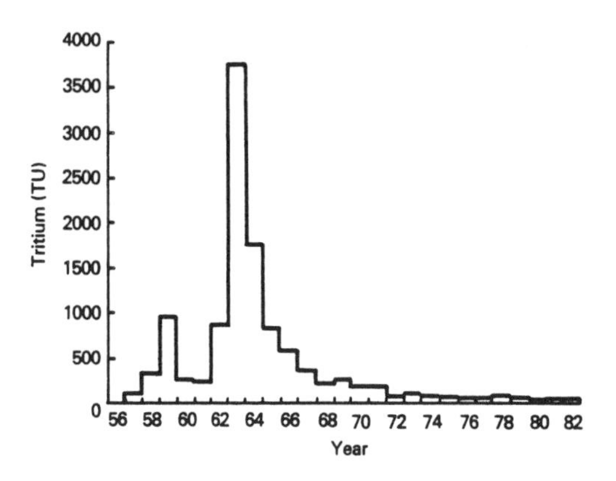

The radioactive isotope of carbon, ^{14}C, is also widely studied in ground water. In practice, the use of ^{14}C is rarely simple. Sources of old carbon, primarily from limestone and dolomite, will dilute the sample. A number of processes, such as the formation of CH_4 gas or the precipitation of carbonate minerals, will fractionate the isotopes and alter the apparent age. The complexity of the interpretation of ^{14}C "ages" of water is so great that it should be attempted only by hydrochemists specializing in isotope hydrology. Despite the complicated nature of ^{14}C studies, they are highly useful in determining the approximate residence time of old water (500 to 30,000 yr) in aquifers. In certain circumstances, this information cannot be obtained in any other way.

A number of radionuclides commonly used as tracers are shown in Table 7-7

Table 7-7 Commonly Used Radioactive Tracers for Ground-Water Studies

Radionuclide	Half-Life y = year, d = day, h = hour	Chemical Compound
^{2}H	12.3y	H_2O
^{32}P	14.3d	Na_2HPO_4
^{51}Cr	27.8d	EDTA-Cr and $CrCl_3$
^{60}Co	5.25y	EDTA-Co and $K_3Co\ (CN_6)$
^{82}Br	33.4h	NH_4Br, NaBr, LiBr
^{85}Kr	10.7y	Kr (gas)
^{131}I	8.1d	I and KI
^{198}Au	2.7d	$AuCl_3$

7.9 References

Akamatsu, K., and M. Matsuo. 1973. Safety of Optical Whitening Agents. Senryo to Yakuhin 18(2):2-11.

Aley, T., and M. W. Fletcher. 1976. Water Tracer's Cookbook. Journal of the Missouri Speleological Survey 16(3).

Allen, M. J., and S. M. Morrison. 1973. Bacterial Movement Through Fractured Bedrock. Ground Water 11(2):6-10.

Andrews, J. H., and D. J. Lee. 1979. Inert Gases in Groundwater from the Bunter Sandstone of England as Indicators of Age and Paleoclimatic Trends. Journal of Hydrology 41:233-252.

Arandjelovic, D. 1969. A Possible Way of Tracing Groundwater Flows in Karst. Geophysical Prospecting 17(4):404-418.

Arandjelovic, D. 1977. Determining Groundwater Flow in Karst Using "Geobomb." In: Karst Hydrology, edited by J. S. Tolson and F. L. Doyle. Memoirs of the 12th Congress of the Int. Assoc. Hydrogeologists, University of Alabama Press, Huntsville, AL.

Aulenbach, D. B., J. H. Bull, and B.C. Middlesworth. 1978. Use of Tracers to Confirm Ground-Water Flow. Ground Water 61(3):149-157.

Aulenbach, D. B., and N L. Clesceri. 1980. Monitoring for Land Application of Wastewater. Water, Air, and Soil Pollution 14:81-94.

Brown, M. C., and D.C. Ford. 1971. Quantitative Tracer Methods for Investigation of Karst Hydrology

Systems, with Reference to the Maligne Basin Area. Cave Research Group (Great Britain) 13(1):37-51.

Buss, D. F., and K. E. Bandt. 1981. An All-Teflon Bailer and an Air-Driven Pump for Evacuating Small-Diameter Ground-Water Wells. Ground Water 19(4):100-102.

California Department of Water Resources. 1968. Water Well Standards, State of California. Calif. Dept. Water Res. Bulletin 74:205.

Campbell, M. D., and J. H. Lehr. 1973. Water Well Technology. McGraw-Hill, New York, NY.

Carrera, J., and G. R. Walter. 1985. CONFLO, a New Numerical Model for Analyzing Convergent Flow Tracer Tests. Sandia Contractors Report (in preparation).

Carter, R. C.. 1959. Helium as a Ground-Water Tracer. Journal of Geophysical Research. 64:2433-2439.

Custodio, E. 1976. Trazadores y Tecnicas Radioisotopicas En Hidrologia Subterranea. In: Hidrologia Subterranea, Vol. 2, edited by E. Custodio and M. R. Llamas, Ediciones Omega, Barcelona, Spain.

Davis, J. T., E. Flotz, and W. S. Balkemore. 1970. Serratia Marcescens, a Pathogen of Increasing Clinical Importance. J. Am. Med. Assoc. 214:145-150.

Davis, S.N., D.J. Campbell, H.W. Bentley, and T.J. Flynn, 1985, Ground-Water Tracers: National Water Well Association, Worthington, OH.

Davis, S. N., G. M. Thompson, H.W. Bentley, and G. Stiles. 1980. Ground Water Tracers--a Short Review. Ground Water 18:14-23.

Downs, W. F., R. E. McAtee, and R.M. Capuano. 1983. Tracer Injection Tests in a Fracture Dominated Geothermal System. Am. Geophysical Union 61(18): 229.

Drew, D. P. 1968. A Study of the Limestone Hydrology of St. Dunstans Well and Ashwick Drainage Basins, Eastern Mendip Proc. Univ. Bristol Speleol. Soc. 11(3):257-276.

Drew, D. P., and D. I. Smith. 1969. Techniques for the Tracing of Subterranean Drainage. Br. Geomorphol. Res. Group Tech. Bulletin 2:36.

Dunn, J. A. 1963. New Method of Water Tracing. Journal of Eldon Pothole Club 5:5.

Ellis, J. 1980. A Convenient Parameter for Tracing Leachate from Sanitary Landfills. Water Research 14(9):1283-1287.

Ferronsky, V. I., and V. A. Polyakov. 1982. Environmental Isotopes in the Hydrosphere. John Wiley and Sons, Interscience Publications, New York, NY.

Fetter, C. W., Jr. 1981. Determination of the Direction of Ground Water Flow. Ground Water Monitoring Review 1(3):28-31.

Fried, J. J. 1975. Groundwater Pollution; Theory, Methodology Modeling, and Practical Rules. Elsevier Scientific Publishing Co.

Gann, E. E., and E. J. Harvey. 1975. Norman Creek: A Source of Recharge to Maramec Spring, Phelphs County, Missouri. Journal of Research of the U. S. Geological Survey 3(1):99-102.

Gaspar, E., and M. Oncescu. 1972. Radioactive Tracers in Hydrology. Elsevier Scientific Publishing Co.

Gat, J. R. 1971. Comments on the Stable Isotope Method in Regional Ground Water Investigations. Water Resources Research 7:980-993.

Gelhar, L. W. 1982. Analysis of Two-Well Tracer Tests with a Pulse Input. Rockwell International, Hanford, Washington. Report RHO-BW-CR-131P.

Glover, R. R. 1972. Optical Brighteners--A New Water Tracing Reagent. Transactions Cave Research Group, Great Britain 14(2):84-88.

Grisak, G. E., and J. F. Pickens. 1980. Solute Transport Through Fractured Media. The Effect of Matrix Diffusion. Water Resources Research 16(4): 719-730.

Grisak, G. E., J. F. Pickens and J. A. Cherry. 1979. Solute Transport Through Fractured Media. Column Study of Fractured Till. Manuscript submitted to Water Resources Research, July 1979.

Haas, J. L. 1959. Evaluation of Groundwater Tracing Methods Used in Speleology: Bulletin Natl. Spel. Soc. 21(2):67-76.

Hagedorn, C., D. T. Hansen and G. H. Simonson. 1978. Survival and Movement of Fecal Indicator Bacteria in Soil under Conditions of Saturated flow. J. Environ. Quality 17:55-59.

Halevy, E., and A. Nir. 1962. The Determination of Aquifer Parameters with the Aid of Radioactive Tracers. Jour. Geophysical Research 61:2403-2409.

Hubbard, E.F., F. A. Kilpatrick, L. A. Martens and J. F. Wilson, Jr. 1982. Measurement of Time of Travel and Dispersion in Streams by Dye Tracing. In: Techniques of Water Resources Investigations of the U.S. Geological Survey.

Johnson Division, UOP, Inc. 1972. Ground Water and Wells, 2nd ed. Edward E. Johnson, Co., St. Paul, MN.

Kaufman, W.J., and G. T. Orlob. 1956. Measuring Ground-Water Movements with Radioactive and

Chemical Tracers. Amer. Waterworks Association Journal 48:559-572.

Keely, J.F. 1984. Optimizing Pumping Strategies for Contaminant Studies and Remedial Actions. Ground Water Monitoring Review 4(3):63-74.

Keith, S.J., L. G. Wilson, H. R. Fitch, and D. M. Esposito. 1982. Sources of Spatial-Temporal Variability in Ground-Water Quality Data and Methods of Control: Case Study of the Cortaro Monitoring Program, Arizona. Second National Symposium on Aquifer Restoration and Ground-Water Monitoring, National Water Well Assoc., Worthington, OH.

Keswick, B.H., D. Wang, and C. P. Gerba. 1982. The Use of Microorganisms as Ground-Water Tracers; A Review. Ground Water 20(2):142-149.

Keys, W.S., and R. F. Brown. 1978. The Use of Temperature Logs to Trace the Movement of Injected Water. Ground Water 16(1):32-48.

Keys, W.S., and L. M. MacCary. 1971. Application of Borehole Geophysics to Water-Resources Investigations. In: Techniques of Investigations of the U.S. Geological Survey.

Klotz, D., H. Moser, and P. Trimborn. 1978. Single-Borehole Techniques; Present Status and Examples of Recent Applications. Isotope Hydrology, IAEA, Vienna, Part 1, pp. 159-179.

Knuttson, G. 1968. Tracers for Ground-Water Investigations. In: Ground Water Problems, edited by E. Eriksson, Y. Gustafsson and K. Nilsson. Pergamon Press, London.

Lange, A.L. 1972. Mapping Underground Streams Using Discrete Natural Noise Signals: A proposed method. Caves and Karst 14:41-44.

Lenda, A., and A. Zuber. 1970. Tracer Dispersion in Groundwater Experiments. Isotope Hydrology, Proc. Symp. Vienna, 1970, IAEA, pp. 619-641.

Lewis, D.C., G. J. Kriz, and R. H. Burgy. 1966. Tracer Dilution Sampling Technique to Determine Hydraulic Conductivity of Fractured Rock. Water Resources Research 2:533-542.

Mather, J.D., D. A. Gray, and D. G. Jenkins. 1969. The Use of Tracers to Investigate the Relationship Between Mining Subsidence and Groundwater Occurrence of Aberdare, South Wales. Journal of Hydrology 9:136-154.

Maurin, V., and J. Zotl. 1959. Die Untersuchung der Zusammenhange Unteirir-Discher Wasser Nit Besonderer Berucksichtigung der Karstver Baltnisse : Steierische Beitrage zur Hydrologie, Jahrgang. Graz, Austria.

Mayr, A. 1953. Bluten Pollen und Pflanzl Sporen Als Mittel Zur Untersuchung von Quellen und Karstwassen. Anz. Math-Natw. Kl. Ost. Ak. Wiss.

Mazor, E. 1972. Paleotemperatures and Other Hydrological Parameters Deduced from Noble Gases Dissolved in Ground Waters, Jordan Rift Valley, Israel. Geochimica et Cosmochimica Acta. 36: 1321-1336.

McLaughlin, M.J. 1982. A Review of the Use of Dyes as Soil Water Tracers. Water S.A., Water Research Commission, Pretoria, South Africa 8(4):196-201.

Murray, J.P., J. V. Rouse, and A. B. Carpenter. 1981. Groundwater Contamination by Sanitary Landfill Leachate and Domestic Wastewater in Carbonate Terrain: Principle Source Diagnosis, Chemical Transport Characteristics and Design Implications. Water Research 15(6):745-757.

Naymik, T.G., and M. E. Sievers. 1983. Ground-Water Tracer Experiment (II) at Sand Ridge State Forest, Illinois. Illinois State Water Survey Division, SWS Contract Report 334.

Pyle, B.H., and H. R. Thorpe. 1981. Evaluation of the Potential for Microbio logical Contamination of an Aquifer Using a Bacterial Tracer. Proceedings of Ground-Water Pollution Conference, 1979. Australian Water Resources Council Conference Series.

Rahe, T.M., C. Hagedorn, E. L. McCoy, and G. G. Kling. 1978. Transport of Antibiotic-Resistant Echerichia Coli Through Western Oregon Hill Slope Soils Under Conditions of Saturated Flow. J. Environ. Quality 7:487-494.

Repogle, J.A., L. E. Myers and K. J. Brust. 1966. Flow Measurements With Fluorescent Tracers. Journal of the Hydraulics Division ASCE 92:1-15.

Reynolds, E.R.C. 1966. The Percolation of Rainwater Through Soil Demonstrated by Fluorescent Dyes. Journal of Soil Science 17(1):127-132.

Robertson, J.B. 1969. Behavior of Xenon-133 Gas After Injection Underground. U.S. Geological Survey Open File Report ID022051.

Rogers, A.S. 1958. Physical Behavior and Geologic Control of Radon in Mountain Streams. U.S. Geological Survey Bulletin (1052E):187-211.

Sauty, J.P. 1978. Identification of Hydrodispersive Mass Transfer Parameters in Aquifers by Interpretation of Tracer Experiments in Radial Converging or Diverging Flow (in French). Journal of Hydrology 39:69-103.

Sauty, J.P. 1980. An Analysis of Hydrodispersive Transfer in Aquifers. Water Resources Research 16(1):145-158.

Schmidt, K.D. 1977. Water Quality Variations for Pumping Wells. Ground Water 15(2):130-137.

Schmotzer, J.K., W. A. Jester, and R. R. Parizek. 1973. Groundwater Tracing with Post Sampling Activation Analysis. Journal of Hydrology 20:217-236.

Simpson, E.S., S. P. Neuman, and G. M. Thompson. 1983. Field and Theoretical Investigations of Mass and Energy Transport in Subsurface Materials. Progress Report for the Nuclear Regulatory Commission by the Department of Hydrology and Water Resources, University of Arizona, Tucson, AZ.

Skoog, D.A., and D. M. West. 1980. Principles of Instrumental Analysis. Holt, Rinehart, and Winston, Philadelphia, PA.

Smart, P.L., and I.M.S. Laidlaw. 1977. An Evaluation of Some Fluorescent Dyes for Water Tracing. Water Resources Research 13(1):15-33.

Smart, P.L., and D.I. Smith. 1976. Water Tracing in Tropical Regions; The Use of Fluorometric Techniques in Jamaica. Journal of Hydrology 30: 179-195.

Sorey, M.L. 1971. Measurement of Vertical Ground-Water Velocity from Temperature Profiles in Wells. Water Resources Research 7(4):963-970.

Sugisaki, R. 1969. Measurement of Effective Flow Velocity of Groundwater by Means of Dissolved Gases. American Journal Science 259:144-153.

Tennyson, L.C., and C. D. Settergren. 1980. Percolate Water and Bromide Movement in the Root Zone of Effluent Irrigation Sites. Water Resources Bulletin 16(3):433-437.

Tester, J.W., R. L. Bivens, and R. M. Potter. 1982. Interwell Tracer Analysis of a Hydraulically Fractured Grantitic Geothermal Reservoir. Society of Petroleum Engineers Journal 8:537-554.

Theis, C.V. 1963. Hydrologic Phenomena Affecting the Use of Tracers in Timing Ground-Water Flow: Radioisotopes in Hydrology. International Atomic Energy Agency (Tokyo Symposium), Vienna, Austria.

Thompson, G.M., and J. M. Hayes. 1978. Trichlorofluoromethane in Ground Water. A Possible Tracer and Indicator of Ground-Water Age. Water Resources Research 15(3):546-554.

Thompson, G.M., J. M. Hayes, and S. N. Davis. 1974. Fluorocarbon Tracers in Hydrology. Geophysical Research Letters 1:177-180.

Todd, D.K. 1980. Groundwater Hydrology , 2nd ed. John Wiley and Sons, New York, NY.

Wagner, O.R. 1977. The Use of Tracers in Diagnosing Interwell Reservoir Heterogeneities. Jour. Petroleum Technology 11:1410-1416.

Wilson, J.F. 1968. Fluorometric Procedures for Dye Tracing. In: Techniques of Water-Resources Investigations of the U.S. Geological Survey.

Wilson, L.G. 1971. Investigations on the Subsurface Disposal of Waste Effluent at Inland Sites. Final report to Office of Saline Water, Grant u14-01-0001-1805. Water Resources Research Center, Tucson, AZ.

Wood, W.W., and G. G. Ehrlich. 1978. Use of Baker's Yeast to Trace Microbial Movement in Ground Water. Ground Water 16(6):398-403.

8. The Use of Models in Managing Ground-Water Protection Programs

8.1 The Utility of Models

8.1.1 Introduction

Mathematical models rely on the quantification of relationships between specific parameters and variables to simulate the effects of natural processes (Figures 8-1, 8-2). As such, mathematical models are abstract and provide little in the way of a directly observable link to reality. Despite this lack of intuitive grace, mathematical models can generate powerful insights into the functional dependencies between causes and effects in the real world. Large amounts of data can be generated quickly, and experimental modifications can be made with minimal effort, so that many possible situations can be studied in great detail for a given problem.

Figure 8-1 Typical ground-water contamination scenario. Several water-supply production wells are located downgradient of a contaminant source. The geology is complex.

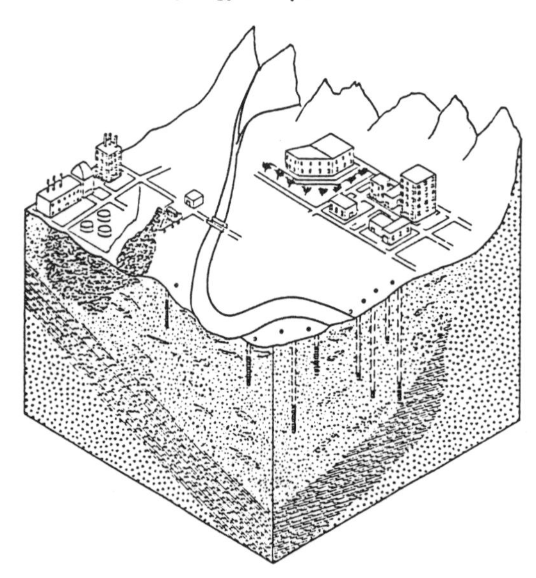

Figure 8-2 Possible contaminant transport model grid design for the situation shown in Figure 8-1.

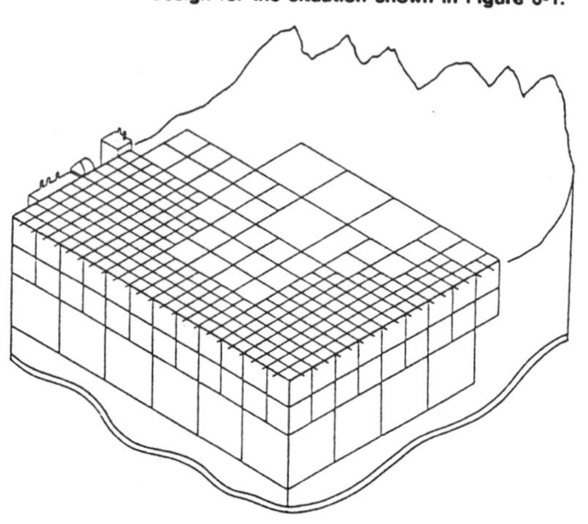

Values for natural process parameters would be specified at each node of the grid in performing simulations. The grid density is greatest at the source and at potential impact locations.

8.1.2 Management Applications

Mathematical models can and have been used to help organize the essential details of complex ground-water management problems so that reliable solutions are obtained (Holcomb Research Institute, 1976; Bachmat *et al.*, 1978; U.S. Congress, 1982; van der Heijde *et al.*, 1985). Some of the principal areas where mathematical models are now being used to assist in the management of ground-water protection programs are:

- o Appraising the physical extent, and chemical and biological quality, of ground-water reservoirs (e.g., for planning purposes)

- o Assessing the potential impact of domestic, agricultural, and industrial practices (e.g., for permit issuance)
- o Evaluating the probable outcome of remedial actions at waste sites, and aquifer restoration techniques generally
- o Providing health-effects exposure estimates.

The success of these efforts depends on the accuracy and efficiency with which the natural processes controlling the behavior of ground water, and the chemical and biological species it transports, are simulated (Boonstra and de Ridder, 1976; Mercer and Faust, 1981; Wang and Anderson, 1982). The accuracy and efficiency of the simulations, in turn, are heavily dependent on subjective judgments made by the modeler and management.

In the current philosophy of ground-water protection programs, the value of a ground-water resource is bounded by the most beneficial present and future uses to which it can be put (U.S. EPA, 1984). In most instances, physical appraisals of ground-water resources are conducted within a framework of technical and economic classification schemes. Classification of entire ground-water basins by potential yield is a typical first step (Domenico, 1972). After the initial identification and evaluation of a ground-water resource, strategies for its rational development need to be devised.

Development considerations include the need to protect vulnerable recharge areas, and the possibility of conjunctive use with available surface waters (Kazmann, 1972). Ground-water rights must be fairly administered to assure adequate supplies for domestic, agricultural, and industrial purposes.

Because basinwide or regional resource evaluations normally do not provide sufficient resolution for water allocation purposes, more detailed characterizations of the properties and behavior of an aquifer, or of a subdivision of an aquifer, are usually needed. Hence, subsequent classifications may involve local estimation of net annual recharge, rates of outflow, and the pumpage which can be sustained without undesirable effects.

The consequences of developments which might affect ground-water quality may be estimated initially by employing generalized classification schemes; for example, classifications based on regional hydrogeologic settings have been presented (Health, 1982; Aller *et al.*, 1985). Very detailed databases, however, must be created and molded into useful formats before decisions can be made on how best to protect and rehabilitate ground-water resources from site-specific incidents of natural and manmade contamination.

The latter are ordinary ground-water management functions which benefit from the use of mathematical models. There are other uses, however, which ought to be considered by management. The director of the International Ground Water Modeling Center discussed the role of modeling in the development of ground-water protection policies recently, noting its success in many policy formulation efforts in the Netherlands, the United States, and Israel. Nevertheless, he concluded that modeling was not widely relied upon for decision-making by managers. The primary obstacle has been an inability by modelers and program managers to communicate effectively (van der Heijde, 1985). The top executives of a leading high-tech ground-water contamination consulting firm made the same point clearly, going on to highlight the need for qualified personnel appreciative of the appropriateness, underlying assumptions, and limitations of specific models (Faust *et al.*, 1981). Because these views are widely held by technical professionals, it will be emphasized herein that mathematical models are useful only within the context of the assumptions and simplifications on which they are based. If managers are mindful of these factors, however, mathematical models can be a tremendous asset in the decision-making process.

8.1.3 Modeling Contamination Transport

Associated with most hazardous waste sites is a complex array of chemical wates and the potential for ground-water contamination. Since the hydrogeologic settings of these sites are usually quite complicated data acquisition and interpretation methods are needed which can examine to an unprecedented degree the physical, chemical, and biological processes which control the transport and fate of ground-water contaminants. The methods and tools that have been in use for large-scale characterizations (e.g., regional water quality studies) are applicable in concept to the specialized needs of hazardous waste site investigations; however, the transition to local-scale studies is not without scientific and economic consequences. In part, this stems from the highly variable nature of contaminant distributions at hazardous waste sites; but it also results from the limitations of the methods, tools, and theories used. Proper acknowledgement of the inherent limitations means that one must project the consequences of their use within the framework of the study at hand.

Assessments of the potential for contaminant transport require interdisciplinary analyses and interpretations. Integration of geologic, hydrologic, chemical, and biological approaches into an effective contaminant transport evaluation can only be possible if the data and concepts invoked are sound. The data must be accurate, precise, and appropriate for the intended problem scale. Just because a given paramèter (e.g., hydraulic conductivity) has been measured correctly at certain points with great reproducibility, is no guarantee that those estimates represent the volumes of aquifer material assigned to them by a modeler. The degree to which the data are representative, therefore, is not only relative to the physical scale of the problem, it is relative to the conceptual model to be used for interpretation efforts. It is crucial, then, to carefully define and qualify the conceptual model. In so doing, special attention should be given to the possible spatial and temporal variations of the data that will be collected.

To circumvent the impossibly large numbers of measurements and samples which would be needed to eliminate all uncertainties regarding the true relationships of parameters and variables, more comprehensive theories are constantly under development. The use of newly developed theories to help solve field problems, however, is often a frustrating exercise. Most theoretical advances call for some data which are not yet practically obtainable (e.g., chemical interaction coefficients, relative peremeabilities of immiscible solvents and water, and so on). The "state-of-the-art" in contaminant transport assessments is necessarily a compromise between the sophistication of "state-of-the-science" theories, the current limitations regarding the acquisition of specific data, and economics. In addition, the best attempts to obtain credible data still fall prey to natural and anthropogenic variabilities; and these lead to the need for considerable judgment on the part of the professional.

Despite these limitations, how well the problem is conceptualized remains the most serious concern in modeling efforts. For example, researchers recently produced dramatic evidence to show that extrapolations of two-dimensional model results to a truly three-dimensional problem lead to wildly inaccurate projections of the actual behavior of the system under study (Molz *et al.*, 1983). Therefore it is incumbent on model users to recognize the difference between an approximation and a misapplication. Models should never be used strictly on the basis of familiarity or convenience; an appropriate model should always be sought.

8.1.4 Categories of Models

The foregoing is not meant to imply that appropriate models exist for all ground-water problems, because a number of natural processes have yet to be fully understood. This is especially true for ground-water contaminant transport evaluations, where the chemical and biological processes are still poorly defined. For, although great advances have been made concerning the behavior of individual contaminants, studies of the interactions between contaminants are still in their infancy. Even the current understanding of physical processes lags behind what is needed, such as in the mechanics of multiphase flow and flow through fractured rock aquifers. Moreover, certain well-understood phenomena pose unresolved difficulties for mathematical formulations, such as the effects of partially penetrating wells in unconfined aquifers.

The technical-use categories of models are varied, but they can be grouped as follows (Bachmat *et al.*, 1978; van der Heijde *et al.*, 1985):

o Parameter identification models

o Prediction models

o Resource management models

o Data manipulation codes.

Parameter identification models are most often used to estimate the aquifer coefficients determining fluid flow and contaminant transport characteristics, like annual recharge (Puri, 1984), coefficients of permeability and storage (Shelton, 1982; Khan, 1986a and 1986b), and dispersivity (Guven *et al.*, 1984; Strecker and Chu, 1986). Prediction models are the most numerous kind of model, and abound because they are the primary tools for testing hypotheses about the problem one wishes to solve (Andersen *et al.*, 1984; Mercer and Faust, 1981; Krabbenhoft and Anderson, 1986).

Resource management models are combinations of predictive models, constraining functions (e.g., total pumpage allowed) and optimization routines for objective functions (e.g., optimization of wellfield operations for minimum cost or minimum drawdown/pumping lift). Very few of these are so well developed and fully supported that they may be considered practically useful, and there does not appear to be a significant drive to improve the situation (van der Heijde, 1984a and 1984b; van der Heijde *et al.*, 1985). Data manipulation codes also have received little attention until recently. They are now becoming increasingly popular, because they simplify data entry ("preprocessors") to other kinds of models and facilitate the production of graphic displays ("postprocessors") of the data outputs of other models (van der Heijde and Srinivasan, 1983; Srinivasan, 1984; Moses and Herman, 1986). Other software packages are available for routine and advanced statistics, specialized graphics, and database management needs (Brown, 1986).

8.2 Assumptions, Limitations, and Quality Control

The many natural processes that affect chemical transport from point to point in the subsurface can be arbitrarily divided into three categories: physical, chemical, and biological (Table 8-1). Conceptually, contaminant transport in the subsurface is an undivided phenomenon composed of these processes and their interactions. At this level the transport process may be gestalt: the sum of its parts, measured separately, may not equal the whole because of interactions between the parts. In the theoretical context, a collection of scientific laws and empirically derived relationships comprise the overall transport process. The universally shared premise that underlies theoretical expressions is that there are no interactions, measurable or otherwise.

Table 8-1 Natural processes that affect subsurface contaminant transport.

Physical Processes
Advection (porous media velocity)
Hydrodynamic dispersion
Molecular diffusion
Density stratification
Immiscible phase flow
Fractured media flow
Chemical Processes
Oxidation—reduction reactions
Radionuclide decay
Ion—exchange
Complexation
Co—solvation
Immiscible phase partitioning
Sorption
Biological Processes
Microbial population dynamics
Substrate utilization
Biotransformation
Adaption
Co—metabolism

Significant errors may result from the discrepancy between conceptual and theoretical approaches. Also the simplifications of theoretical expressions used to solve practical problems can cause substantial errors in the most careful analyses. Assumptions and simplifications, however, must often be made in order to obtain mathematically tractable solutions. Because of this, the magnitude of errors that arise from each assumption and simplification must be carefully evaluated. The phrase magnitude of errors is emphasized because highly accurate evaluations usually are not possible. Even rough approximations are rarely trivial exercises because they frequently demand estimates of some things which are as yet ill-defined.

8.2.1 Physical Processes

Until recently, ground-water scientists studied physical processes to a greater degree than chemical or biological processes. This bias resulted in large measure from the fact that, in the past, ground-water practitioners dealt mostly with questions of adequate water supplies. As quality considerations began to dominate ground-water issues, the need for studies of the chemical and biological factors, as well as more detailed representations of the physical factors became apparent.

There are two complimentary ways to view the physical processes involved in subsurface contaminant transport: the piezometric (pressure) viewpoint and the hydrodynamic viewpoint. Ground-water problems of yesterday could be addressed by the former, such as solving for the change in pressure head caused by pumping wells. Contamination problems of today also require detailed analyses of wellfield operations, for example, pump-and-treat plume removals; however, solutions depend principally on hydrodynamic evaluations, such as computing ground-water velocity (advection) distributions and dispersion estimates for migrating plumes.

8.2.1.1 Advection and dispersion

Ground-water velocity distributions can be approximated if the variations in hydraulic conductivity, porosity, and the strength and location of recharge and discharge can be estimated. While there are several field and laboratory methods for estimating hydraulic conductivity, these are not directly comparable because different volumes of aquifer material are affected by different tests. Laboratory permeameter tests, for example, obtain measurements from small core samples and thus give point value estimates. These tests are generally reliable for consolidated rock samples, such as sandstone, but can be highly unreliable for unconsolidated samples, such as sands, gravels, and clays. Pumping tests give estimates of hydraulic conductivity that are averages over the entire volume of aquifer subject to the pressure changes induced by pumping. These give repeatable results, but they are often difficult to interpret. Tracer tests are also used to estimate hydraulic conductivity in the field, but are difficult to conduct properly.

Regardless of the estimation technique used, the best that can be expected is order-of-magnitude estimates for hydraulic conductivity at the field scale appropriate for site-specific work. Conversely, porosity estimates that are accurate to better than a factor of two can be obtained. Estimation of the strength of nonpoint sources of recharge to an aquifer, such as infiltrating rainfall and leakage from other aquifers, is another order-of-magnitude effort. Similarly, nonpoint sources of discharge, such as losses to gaining streams, are difficult to quantify. Estimation of the strength of point sources of recharge or discharge (injection or pumping wells) can be highly accurate.

Consequently, it is not possible to generalize the quality of velocity distributions. They may be accurate to within a factor of two for very simple aquifers, but are more often accurate to an order-of-magnitude only. This situation has changed little over the past 20 years because better field and laboratory methods for characterizing velocity distributions have not been developed. This, however, is not the primary difficulty associated with defining the advective part of contaminant transport in the subsurface. The primary difficulty is that field tests for characterizing the physical parameters that control velocity distributions are not incorporated into contamination investigations on a routine basis. The causes seem to be: a perception that mathematical models can "back-out" an approximation of the velocity distribution (presumably eliminating the need for field tests); unfamiliarity with such methods by many practitioners; and a perception that field tests are too expensive. A more field-oriented approach is preferable because the non-uniqueness of modeling results has been amply demonstrated, and this leads to uncertain decisions regarding the design of remedies.

Dispersion estimates are predicted on velocity distribution estimates and their accuracy is therefore directly dependent on the accuracy of the estimated hydraulic conductivity distribution. Tracer tests have been the primary method used to determine dispersion coefficients until recently. Presently there are suggestions that any field method capable of generating a detailed understanding of the spatial variability of hydraulic conductivity, which in turn could give an accurate representation of the velocity distribution, may be used to derive estimates of dispersion coefficients. The manner in which data from field tests should be used to derive estimates of dispersion coefficients, however, is a controversial issue. There are both deterministic and stochastic schools of thought, and neither has been conclusively demonstrated in complex hydrogeological settings.

8.2.1.2 Complicating factors

Cert ain subtleties of the spatial variability of hydraulic conductivity must be understood because of its key role in the determination of velocity distributions and dispersion coefficients. Hydraulic conductivity is also known as the coefficient of permeability because it is comprised of fluid factors as well as the intrinsic permeability of the stratum in question. This means that a stratum of uniform intrinsic permeability (which depends strictly on the arrangement of its pores) may have a wide range of hydraulic conductivity because of differences in the density and viscosity of fluids that are present. The result is a dramatic downward shift in local flow directions near plumes that have as little as a one percent increase in density relative to uncontaminated water. Such density contrasts frequently occur at landfills and waste impoundments. It is often necessary to correct misimpressions of the direction of a plume because density considerations were not addressed.

Many solvents and oils are highly insoluble in water, and may be released to the subsurface in amounts sufficient to form a separate fluid phase. Because that fluid phase will probably have viscosity and density different from freshwater, it will flow at a rate and, possibly, in a direction different from that of the freshwater with which it is in contact. If an immiscible phase has a density approximately the same or less than that of ground water, this phase will not move down past the capillary fringe of the ground water. Instead, it will flow along the top of the capillary fringe in the direction of the maximum water-level elevation drop. If the density of an immiscible phase is substantially greater than the ground water, the immiscible phase will push its way into the ground water as a relatively coherent blob. The primary direction of its flow will then be down the dip of the first impermeable stratum encountered. There is a great need for better means of characterizing such behavior for site-specific applications. Currently, estimation methods are patterned after multiphase oil reservoir simulators. One of the key extensions needed is the ability to predict the transfer of trace levels of contaminants, such as xylenes from gasoline, from the immiscible fluid to ground water.

Anisotropy is a subtlety of hydraulic conductivity which relates to structural trends of the rock or sediments of which an aquifer is composed. Permeability and hydraulic conductivity are directionally dependent in anisotropic strata. When molten material from deep underground crystallizes to form granitic or basaltic rocks, for instance, it forms cleavage planes which may later becomes the preferred directions of permeability. Marine sediments accumulate to form sandstone, limestone, and shale sequences that have much less vertical than horizontal permeability. The seasonal differences in sediments that accumulate on lakebeds, and the stratification of grain sizes deposited by streams as they mature, give rise to similar vertical-to-horizontal anisotropy. Streams also cause anisotropy within the horizontal plane, by forming and reworking their sediments along a principal axis of movement. These structural variations in permeability would be of minimal concern except that ground water does not flow at right angles to water-level elevation contours under anisotropic conditions. Instead, flow proceeds along oblique angles, with the degree of deviation from a right-angle pathway proportional to the amount of anisotropy. This fact is all too often ignored and the causes again seem to be a reluctance to conduct the proper field tests, combined with an over-reliance on mathematical modeling.

If the pathways created by cleavage planes and fractures begin to dominate fluid flow through a subsurface stratum, the directions and rates of flow are no longer predictable by the equations used for porous rock and sediments. There have been a number of attempts to represent fractured flow as an equivalent porous medium, but these tend to give poor predictions when major fractures are present and when there are too few fractures to guarantee a minimum degree of interconnectedness. Other representations that have been studied are various dual porosity models, in which the bulk matrix of the rock has one porosity and the fracture system has another. Further development of the dual porosity approach is limited by the difficulty in determining a transfer function to relate the two different porosity schemes.

8.2.1.3 Considerations for predictive modeling

Equations for the combined advection-dispersion process are used to estimate the time during which a nonreactive contaminant will travel a specific distance, the pathway it will travel, and its concentration at any point. The accuracy of most predictions is only fair for typical applications, because of the complexity of the problems and the scarcity of site-specific hydrogeologic data. The lack of such data can be improved with much less effort than is commonly presumed, especially when costs of another round of chemical sampling are compared with the costs of additional borings, core retrievals, geophysical logging, or permeability testing.

Equations that assume a nonreactive contaminant have limited usefulness, because most contaminants react with other chemical constituents in subsurface waters and with subsurface solids in a manner that affects the rate at which they travel. Nevertheless, nonreactive advection-dispersion equations are often used to generate "worst-case" scenarios, on the presumption that the maximum transport velocity is obtained (equal to that of pure water). This may not be as useful as it first seems. Remedial action designs require detailed breakdowns of which contaminants will arrive at extraction wells and when; how long contaminants will continue their slow release from subsurface solids; and whether the contaminants will be transformed into other chemical species by chemical or biological forces. To address these points, special terms must be added to the advection-dispersion equations.

8.2.2 Chemical Processes

As difficult as the foregoing complications may be, predicting how chemical contaminants move through the subsurface is a relatively trivial matter when the contaminants behave as ideal, nonreactive substances. Unfortunately, such behavior is limited to a small group of chemicals. The actual situation is that most contaminants will, in a variety of ways, interact with their environment through biological or chemical processes.

This section focuses on the dominant chemical processes that may ultimately affect the transport behavior of a contaminant. As with the physical processes previously discussed, some of the knowledge of chemical processes has been translated into practical use in predictive models. However, the science has, in many instances, advanced well beyond what is commonly practiced. Furthermore, there is considerable evidence that suggests that numerous undefined processes affect chemical mobility. Most of the deviation from ideal nonreactive behavior of contaminants relates to their ability to change physical form by energetic interactions with other matter. The physical-chemical interactions may be grouped into: alterations in the chemical or electronic configuration of an element or molecule; alterations in nuclear composition; the establishment of new associations with other chemical species; and, interactions with solid surfaces.

8.2.2.1 Chemical/electronic alterations

The first of these possible changes is typified by oxidation-reduction or redox reactions. This class of reactions is especially important for inorganic compounds and metallic elements because the reactions often result in changes in solubility, complexing capacity, or sorptive behavior, which directly impact the mobility of the chemical. Redox reactions are reasonably well understood, but there are practical obstacles to applying the known science because it is difficult to determine the redox state of the aquifer zone of interest and to identify and quantify the redox-active reactants.

Hydrolysis, elimination, and substitution reactions that affect certain contaminants also fit into this classification. The chemistry of many organic contaminants has been well defined in surface water environments. The influence of unique aspects of the subsurface, not the least of which is long residence time, on such transformations of important organic pollutants is currently under investigation. There is also a need to investigate the feasibility of promoting in-situ abiotic transformations that may enhance the potential for biological mineralization of pollutants.

8.2.2.2 Nuclear alterations

Another chemical process interaction, which results in internal rearrangement of the nuclear structure of an element, is well understood. Radiodecay occurs by a variety of routes, but the rate at which it occurs is always directly proportional to the number of radioactive atoms present. This fact seems to make mathematical representation in contaminant transport models quite straightforward because it allows characterization of the process with a unique, well defined decay constant for each radionuclide.

A mistake that is often made when the decay constant is used in models involves the physical form of the reactant. If the decay constant is applied to the fluid concentrations and no other chemical interactions are allowed, then incorporation of the constant in the subroutine which computes fluid concentrations will not cause errors. If the situation being modeled involves chemical interactions such as precipitation, ion-exchange, or sorption, which temporarily remove the radionuclide from solution, then it is important to use a second subroutine to account for the non-solution phase decay of the radionuclide.

8.2.2.3 Chemical associations

The establishment of new associations with other chemical species is not as well understood. This category includes ion-exchange, complexation, and co-solvation. The lack of understanding derives from the nonspecific nature of these interactions, which are, in many instances, not characterized by the definite proportion of reactants to products (stoichiometry) typical of redox reactions. While the general principles and driving mechanisms by which these interactions occur are known, the complex subsurface matrix in which they occur provides many possible outcomes and renders predictions uncertain. Ion-exchange and complexation reactions heavily influence the mobility of metals and other ionic species in the subsurface in a reasonably predictable fashion. Their influence on organic contaminant transport, however, is not well understood. Based on studies of pesticides and other complex organic molecules, natural organic matter (such as humic and fulvic materials) can complex and thereby enhance the apparent solubility and mobility of synthetic organic chemicals. Research is needed to define the magnitude of such interactions, not only with naturally occurring organic molecules but also with man-made organics present in contaminated environments. Research is also needed to determine if these complexes are stable and liable to transport through the subsurface. Examination of the degree to which synthetic organic chemicals complex toxic metals is also necessary. There is no theoretical objection to such interactions, and there is ample evidence that metals are moving through the subsurface at many waste sites.

Co-solvation occurs when another solvent is in the aqueous phase at concentrations that enhance the solubility of a given contaminant. This occurs in agricultural uses, for example, where highly insoluble pesticides and herbicides are mixed with organic solvents to increase their solubility in water prior to field application. There is every reason to expect similar behavior at hazardous waste sites, where a variety of solvents are typically available. At present, prediction of the extent of the solubility increases that might occur at disposal sites in the complex mixture of water and organic solvents is essentially impossible. Researchers have started examining co-solvation as an influence on pollutant transport, by working on relatively simple mixed solvent systems. This research will be extremely useful, even if the results are limited to a qualitative appreciation for the magnitude of the effects.

At the extreme, organic solvents in the subsurface may result in a phase separate from the aqueous phase. In addition to movement of this separate phase through the subsurface, contaminant mobility that involves partitioning of organic contaminants between the organic and aqueous phases must also be considered. The contaminants will move with the organic phase and will, depending on aqueous phase concentrations, be released into the aqueous phase to a degree roughly proportional to their octanol-water partition coefficients. An entire range of effects is possible, from increasing to slowing the mobility of the chemical in the subsurface relative to its migration rate in the absence of the organic phase. The equilibrium partitioning process increases the total volume of ground water affected by contaminants, by releasing a portion of the organic phase constituents into adjacent waters. It may also interfere with transformation processes by affecting pollutant availability for reaction, or by acting as a biocidal agent to the native microflora.

8.2.2.4 Surface interactions

Of those interactions that involve organic chemicals in the environment, none has been as exhaustively studied as sorption. Sorption studies relate, in terms of a sorption isotherm, the amount of contaminant in solution to the amount associated with the solids.

Most often the sorption term in transport models is estimated for simplicity from the assumption that the response is linear. This approximation can produce serious mass balance errors. Typically, the contaminant mass in the solution phase is under-estimated and contaminant retardation is thereby over-estimated. In practical applications, this means that the contaminant can be detected at a monitoring well long before it is anticipated. To resolve the discrepancy between predicted and actual transport, most practitioners arbitrarily adjust some other poorly-characterized model parameter, for example, dispersion. This leads to the creation of a model that does not present various natural process influences in proper perspective. The predictions from such models are likely to be qualitatively, as well as quantitatively, incorrect. More widespread consideration should be given to accurate representation of non-linear sorption, particularly in transport modeling at contaminated sites.

The time dependency of the sorption process is a related phenomenon that has also been largely ignored in practical applications of sorption theory. Most models assume that sorption is instantaneous and completely reversible. A growing body of evidence argues to the contrary, not only for large organic molecules in high-carbon soils and sediments, but also for solvent molecules in low-carbon aquifer materials. Additionally, there must be some subtle interplay between sorption kinetics and ground-water flow rates which gains significance in pump-and-treat remediation efforts, where flow rates are routinely substantially increased. Constant pumpage at moderate-to-high flow rates may not allow contaminants that are sorbed to solids sufficient times of release to increase solution concentrations to maximum (equilibrium) levels prior to their removal from the aquifer. Hence, treatment costs may rise substantially due to the prolonged pumping required to remove all of the contaminants and due to the lowered efficiency of treatment of the less contaminated pumped waters.

Evidence from Superfund sites and ongoing research activities suggests that contaminant association with a solid surface does not preclude mobility. In many instances, especially in glacial tills that contain a wide distribution of particle sizes, fine aquifer materials have accumulated in the bottom of monitoring wells. Iron-based colloids have been identified in ground water downgradient from a site contaminanted with domestic wastewater. If contaminants can associate with these fine particles, their mobility through the subsurface could be markedly enhanced. To determine the significance of particle transport to pollutant movement, studies must be performed at such contaminanted sites.

Although knowledge about chemical processes that function in the subsurface has been significantly expanded in recent years, this information is only slowly finding its way into practical interpretations of pollutant transport at contaminated sites. Evidence from field sites suggests that much remains to be learned about these processes.

8.2.3 Biological Processes

Many contaminants that enter the subsurface environment are biologically reactive. Under appropriate circumstances they can be completely degraded to harmless products. Under other circumstances, however, they can be transformed to new substances that are more mobile or more toxic than the original contaminant. Quantitative predictions of the fate of biologically reactive substances are at present very primitive, particularly compared to other processes that affect pollutant transport and fate. This situation resulted from the ground-water community's choice of an inappropriate conceptualization of the active processes: subsurface biotransformations were presumed to be similar to biotransformations known to occur in surface water bodies. Only very recently has detailed field work revealed the inadequacy of the traditional view.

8.2.3.1 Surface water model analogy

As little as five years ago ground-water scientists considered aquifers and soils below the zone of plant roots to be essentially devoid of organisms capable of transforming contaminants. As a result, there was no reason to include terms for biotransformations in transport models. Recent studies have shown that water-table aquifers harbor appreciable numbers of metabolically active microorganisms, and that these microorganisms frequently can degrade organic contaminants. It became necessary to consider biotransformation in transport models. Unfortunately, many ground-water scientists adopted the conceptual model most frequently used to describe biotransformations in surface waters.

The presence of the contaminant was assumed to have no effect on microorganism populations that degrade it. It was also assumed that contaminant concentration does not influence transformation kinetics, and that the capacity to transform the contaminant is uniformly distributed throughout the body of water under study. These assumptions are often appropriate for surface waters: contaminant concentration is usually too low and the residence time too short to allow adaption of the microbial community to the contaminant, and the organisms that are naturally pre-adapted to the contaminant are mixed throughout the water body by turbulence. Consequently, utilization kinetics can conveniently be described by simple first-order decay constants. In surface waters these constants are usually obtained by monitoring contaminant disappearance in water samples.

8.2.3.2 Ground-water biotransformations

These circumstances rarely apply to biotransformation in ground water. Contaminant residence time is usually long, at least weeks or months, and frequently years or decades. Further, contaminant concentrations that are high enough to be of environmental concern are often high enough to elicit adaption of the microbial community. For example, the U.S. Environmental Protection Agency's Maximum Contaminant Level (MCL) for benzene is 5 ug/L. This is very close to the concentration of alkylbenzenes required to elicity adaption to this class of organic compounds in soils. As a result, the biotransformation rate of a contaminant in the subsurface environment is not a constant, but increases after exposure to the contaminant in an unpredictable way. Careful field work has shown that the transformation rate in aquifers of typical organic contaminants, such as alkylbenzenes, can vary as much as two orders of magnitude over a meter vertically and a few meters horizontally. This surprising variability in transformation rate is not related in any simple way to system geology or hydrology.

It is difficult to determine first-order rate constants in subsurface material. Most microbes in subsurface material are firmly attached to solid surfaces; usually less than one percent of the total population is truly planktonic. As a result, the microbes in a ground-water sample grossly underrepresent the total microbial population in the aquifer. Thus, contaminant disappearance kinetics in a ground-water sample do not represent the behavior of the material in the aquifer. It is therefore necessary to do microcosm studies with samples representative of the entire aquifer system - a formidable technical challenge.

8.2.3.3 A Ground-Water Model

These concerns have prompted re-examination of assumptions about biotransformation implicit or explicit in a particular modeling approach, with the realization that no one qualitative description of biotransformation can be universally applicable. Field experience has shown that the relationships that describe the biological fate of contaminants actually change within aquifers in response to geochemical constraints on microbial physiology. Rather than describing biotransformation with a continuous function applicable at all points in the aquifer, it may be more realistic to examine key geochemical parameters and to use that information to identify the relationship for biotransformation that applies at any particular point. These key parameters could include the contaminant concentration, oxygen or other electron acceptor concentration, redox state, pH, toxicity of the contaminant or co-occurring materials, and temperature. One such model has been evaluated in the field.

The model described an alkylbenzene and polynuclear aromatic hydrocarbon plume in a shallow water-table aquifer. Microcosm studies showed that organisms in the aquifer had adapted to these contaminants, and would degrade them very rapidly when oxygen was available. As a result of this adaption, the hydrocarbon biodegradation rate was not controlled by any inherent property of the organisms. Rather, physical transport processes such as diffusion and dispersion seemed to dominate by controlling oxygen availability to the plume.

Because the biotransformation rate was controlled by physical processes, the actual model was very simple. Oxygen and hydrocarbon transport were simulated as conservative solutes using the U.S. Geological Survey method-of-characteristics code. A subroutine examined oxygen and hydrocarbon concentrations at each node and generated new concentrations based on oxidative metabolism stoichiometry. When the model was projected forward in time it illustrated an important property of many such plumes. The plume grew with time until the rate of admixture of oxygen balanced the rate of release of hydrocarbons from the source. Afterward, the extent of the plume was at steady-state.

The body of field experience which can be drawn upon to properly assign laws for biotransformation is growing rapidly. Transport-limited kinetics may commonly apply to releases of petroleum hydrocarbons and other easily degradable materials such as ethanol or acetone in oxygenated ground water. On the other hand, materials that can support a fermentation, in which an exogenous electron acceptor is not required, may follow first-order kinetics. Unfortunately, many important biotransformations in ground water are still mysteries. The reductive dehalogenation of small halogenated hydrocarbons such as trichloroethene and 1,1,1-trichloroethane is a good example. In such cases transformation kinetics of the compound are controlled by transformation kinetics of a second compound, the primary substrate that supports the metabolism of the active microorganisms. These complex interactions are poorly understood and cannot be described quantitatively at the present time. However, this is an area of active research, and hopefully the appropriate relationships may soon be determined.

Rapid field methods to determine if adaption has occurred at a site are needed. Tools to predict whether adaption can be expected, and to estimate the time required for adaption if it does occur, are also needed. For systems that are limited by transport processes, field methods to estimate the aquifer processes that control mixing, such as transverse dispersion and exchange processes across the water table, are required. For systems that are limited by the intrinsic biotransformation rate, new laboratory test methods (possibly, improved microcosms) that will provide reliable estimates of the kinetic parameters are required.

In addition to being sufficiently accurate and precise, these new methods should provide estimates that are truly representative of the hydrologic unit being simulated. Because contaminants typically have long residence times in aquifers, slow transformation rates can have environmental significance. The test methods should therefore be sufficiently sensitive to measure transformation rates that are significant in the hydrologic context being simulated. Finally, there is a need for models that go beyond simple prediction of contaminant concentrations at points in the aquifer, and forecast the concentrations produced by production wells.

8.2.4 Analytical and Numerical Models

One of the more subtly involved decisions which must be made is whether to use an analytical model or a numerical model to solve a particular problem. Analytical models provide exact solutions, but many simplifying assumptions must be made for the solutions to be tractable; this places a burden on the user to test and justify the underlying assumptions and simplifications (Javendel *et al.*, 1984). For example, the Theis equation is an analytical expression which is used to predict the piezometric head changes for pumping or injection wells in confined aquifers (Freeze and Cherry, 1979; Todd, 1980):

$$s = [Q/(4\pi T)] \times [-0.5772 - \ln(u) + u - (u^2/(2 \times 2!)) + (u^3/(3 \times 3!)) - (u^4/(4 \times 4!)) \ldots]$$

where:

s = the change in piezometric head

Q = the flowrate of the well
T = the transmissivity of the aquifer
u = $(r^2 S) / (4 T t)$;
r = the radial distance from the well
S = the storage coefficient of the aquifer
t = the length of time the well has been operating.

Here the principal assumptions are (Lohman, 1972):

- o The aquifer is homogenous and isotropic
- o The aquifer is of infinite areal extent, relative to the effects of the well (no boundaries)
- o The well is screened over the entire saturated thickness of the aquifer
- o The saturated thickness of the aquifer does not vary as a result of the operation of the well
- o The well has an infinitesimal diameter so that waters in storage in the casing represent an insignificant volume
- o Water is removed from or injected into the aquifer with an instantaneous change in the piezometric head.

Evaluation of the infinite Taylor series representing the well function integral can be accomplished graphically using type curves (Walton, 1962; Lohman, 1972). Alternatively, a simplification can be made so that the Theis equation is directly solvable (Cooper and Jacob, 1946). This is done by dropping all terms in the Taylor series with powers greater than one, and is strictly valid for cases where "u" has a value less than 0.01 (e.g., Figure 8-3). Physically, this corresponds to a limitation on the predictive power of the modified Theis equation; head changes predicted at locations far from the well are inaccurate, except for long durations of pumpage (i.e., approaching equilibrium or steady-state conditions).

Numerical models are much less burdened by these assumptions and are therefore inherently capable of addressing more complicated problems, but they require significantly more data and their solutions are inexact (numerical approximations). For example, the assumptions of homogeneity and isotropicity are unnecessary due to the ability to assign point (nodal) values of transmissivity and storage. Likewise, the capacity to incorporate complex boundary conditions obviates the need for the "infinite areal extent" assumption. There are, however, difficult choices facing the user of numerical models; i.e. time steps, spatial grid designs, and ways to avoid truncation errors and numerical oscillations must be chosen (Remson *et al.*, 1971; Javendel *et al.*, 1984). These choices, if improperly made, may result in errors unlikely to be made with analytical approaches (e.g., mass imbalances, incorrect velocity distributions, and grid-orientation effects).

8.2.5 Quality Control

These latter points signify a greater need for quality control measures when contemplating the use of numerical models. Three levels of quality control have been suggested previously (Huyakorn *et al.*, 1984):

1) Validation of the model's mathematics by comparison of its output with known analytical solutions to specific problems,
2) Verification of the general framework of the model by successful simulation of observed field data, and
3) Benchmarking of the model's efficiency in solving problems by comparison with other models.

These levels of quality control address the soundness and utility of the model alone, and do not treat questions of its application to a specific problem. Hence, at least two additional levels of quality control appear justified:

4) Critical review of the problem conceptualization to ensure that the modeling effort considers all physical, chemical, and biological processes which may affect the problem, and
5) Evaluation of the specifics of the application; e.g., appropriateness of the boundary conditions, grid design, time steps, etc.

Validation of the mathematical framework of a numerical model is deceptively simple. The usual approach for ground-water flow models involves a comparison of drawdowns predicted by the Theis analytic solution to those obtained by using the model, such as depicted in Figure 8-4. The "deceptive" part is the foreknowledge that the Theis solution can treat only a very simplified situation as compared with the scope of situations addressable by the numerical model. In other words, analytical solutions cannot test most of the capabilities of the numerical model in a meaningful way; this is particularly true with regard to simulation of complex aquifer boundaries and irregular chemical distributions.

Field verification of a numerical model consists of first calibrating the model using one set of historical records (e.g., pumping rates and water levels from a certain year), and then attempting to predict the next set of historical records. In the calibration phase, the aquifer coefficients and other model parameters are adjusted to achieve the best match between model outputs and known data; in the predictive phase, no adjustments are made (excepting actual changes in pumping rates, etc.). Presuming that the aquifer coefficients and other parameters were known with sufficient accuracy, a mismatch means that either the model is not correctly formulated or that it does not treat all of the important phenomena affecting the situation being simulated (e.g., does not allow for

Figure 8-3 **Example of plots prepared with the Jacob's approximation of the Theis analytical solution to well hydraulics in an artesian aquifer.**

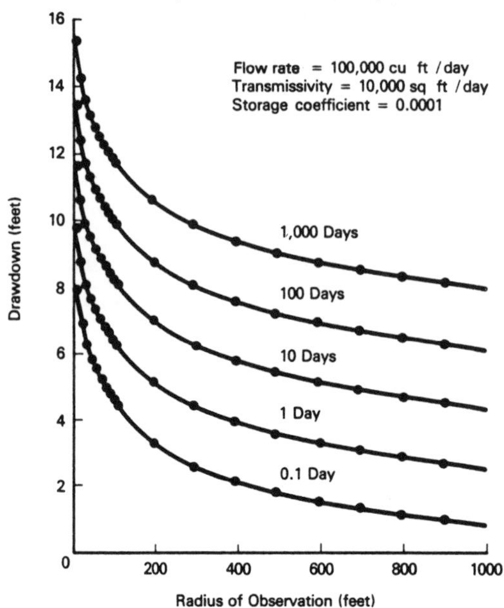

Figure 8-4 **Mathematical validation of a numerical method of estimating drawdown, by comparison with an analytical solution.**

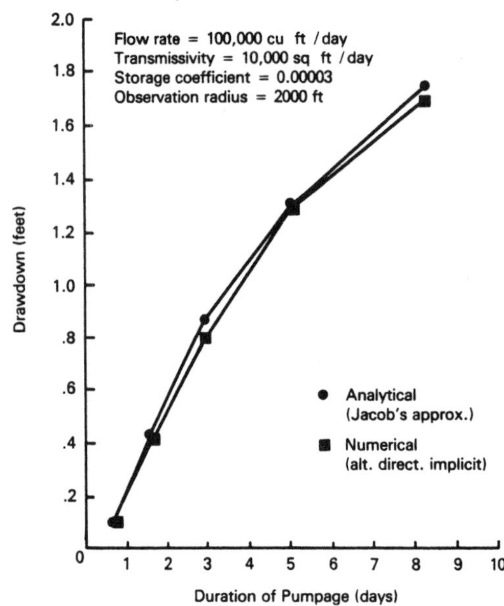

leakage between two aquifers when this is actually occurring).

Field verification exercises usually lead to additional data gathering efforts, because existing data for the calibration procedure are often insufficient to provide unique estimates of key parameters. This means that a "black box" solution may be obtained, which may be good only for the records used in the calibration. For this reason, the blind prediction phase is an essential check on the uniqueness of the parameter values used. In this regard, field verification of models using datasets from controlled research experiments may be much more achievable practically.

Benchmarking routines to compare the efficiency of different models in solving the same problem have only recently become available (Ross *et al.*, 1982; Huyakorn *et al.*, 1984). Much more needs to be done in this area, because some unfair perceptions continue to persist regarding the ostensibly greater utility of certain modeling techniques. For example, it has been said many times that finite element models (FEMs) have an inherent advantage over finite difference models (FDMs) in terms of the ability to incorporate irregular boundaries (Mercer and Faust, 1981); the number of points (nodes) which must be used by FEMs is considerably less due to the flexible nodal spacings that are allowed. Benchmarking routines, however, show that the much longer computer time required to evaluate FEM nodes causes there to be little, if any, cost advantage for simulations of comparable accuracy.

8.3 Applications in Practical Settings

8.3.1 Stereotypical Applications

As stated in preceding sections, models are simplifications of reality that may or may not faithfully simulate the actual situation. Typically, attempts are made to mimic the effects of hydrogeologic, chemical, and biological processes in practical applications of models. These almost always involve idealizations of known or suspected features of the problem on hand. For example, the stratification of alluvial, fluvial, and glacial deposits may be assumed to occur in uniformly thick layers, despite the great variability of stratum thicknesses found in actual settings. Large blocks of each stratum are assumed to be homogenous. Sources of chemical input are commonly assumed to have released contaminants at constant rates over the seasons and years of operational changes that the sources were active. The areal distribution of rainfall and the actual schedules of pumpage from production wells are also artificially homogenized in most modeling exercises.

All these idealizations are made necessary by a lack of the appropriate historical records and field-derived parameter estimates, and all reduce the reliability of predictions made with models. The degree of usefulness of a model is therefore directly dependent on the subjective judgments that must be made in data collection and preparation efforts prior to attempting mathematical simulations. This is true not only in a quantitative sense, but also in a qualitative sense because it is the data gathering phase of a project that begets the conceptualization on which the model will be based.

8.3.2 Real-World Applications

To illustrate this point, the highlights of two very different contamination problems will be described. The first involves a relatively limited contamination incident arising from a very small source and having few contaminants. The second involves a major contamination incident arising from the operation of a chemical reprocessing facility that handled dozens of different contaminants in large amounts. The common theme that is shared by the two cases, as should also apply to virtually all cases, is one of seeking to define the relative influences of natural processes affecting contaminant transport in order to optimize the assessment and remediation of the problem. It is the validity of the conceptual model of what is happening at these sites that is most important, not the application of a particular mathematical model.

8.3.2.1 Field example no. 1

The Lakewood Water District in Lakewood, Washington, operates a number of wells for drinking water supply purposes. Some of the wells operated by the District, such as the two primary wells at its Ponder's Corner site (Figures 8-5, 8-6), have been contaminated (USEPA-Region 10, 1981) with low levels of volatile organic chemicals (VOCs). During the course of the investigations at the Ponder's Corner site, a number of cost-saving sampling alternatives were chosen. These related principally to the field use of a portable gas chromatograph (Organic Vapor Analyzer) for the screening of water samples and soil extracts taken while drilling monitoring wells, and to the use of selective analyses (volatiles only) of ground-water samples when initial results showed only a narrow group of contaminants to be present. The lowered analytical costs, in part, allowed for increased expenditures for geotechnical characterization of the site (Wolf and Boateng, 1983). The geotechnical efforts, particularly the pump tests which were conducted, led to a realization that the source of the contaminants was to be found regionally downgradient (Keely and Wolf, 1983). The pumping strength of the water-supply wells, when operating, was sufficient to pull contaminants over 400 feet back against the regional flow direction. Because most contaminant sources are found upgradient of the wells they affect, this behavior was somewhat unexpected. A unique feature of the field investigation was the taking of ground-water samples from the pumping wells concurrent with drawdown measurements obtained during pump tests (Keely, 1982).

Figure 8-5 Location map for Lakewood Water District wells contaminated with volatile organic chemicals.

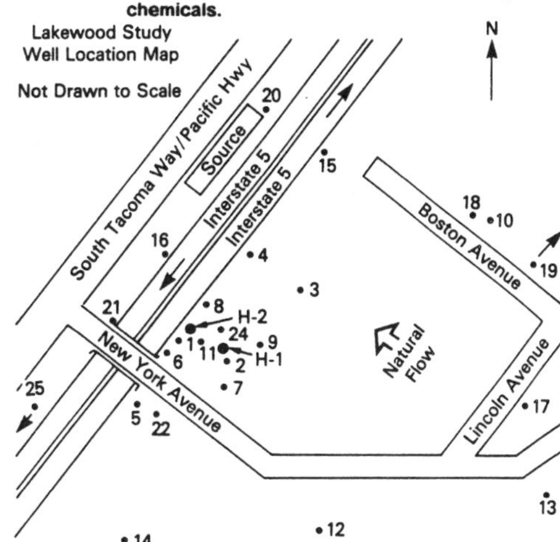

The pump tests yielded estimates of local transmissivity and storage coefficients. It also confirmed the presence of a major aquifer boundary nearby; a buried glacial till drumlin just west of the site parallels the general direction (north) of regional flow. The pump tests clearly showed some anisotropy of the sediments as well; drawdown contours produced an elliptical cone of drawdown, the major axis of which was aligned with the regional flow to the north. This information resulted in modifications to the original plans, which called for drilling and constructing several monitoring wells west of the site. Instead, more monitoring wells were drilled along the north-south axis. Chemical analysis of the samples taken concurrent with drawdown measurements formed a time-series of contaminant concentrations that provided a clue to where the contaminant source was located (Keely, 1982). The time-series showed that the well nearest the downgradient edge of the well field was exposed to increasing contaminant levels as pumping continued, whereas the upgradient pumping well remained largely unaffected (Keely and Wolf, 1983).

The hydrogeologic parameter estimates obtained from the pump tests strengthened the conceptualization of contaminants being drawn back against the regional flow because the capture zones of the pumping wells were sufficiently distorted by the local anisotropy to more than encompass the contaminant source.

Without considering the anisotropic bias along the regional flow path, the estimated boundaries of the capture zone for either of the two wells marginally reached the distance to the contaminant source. The

Figure 8-6 Geologic logs for Lakewood Water District wells contaminated with volatile organic chemicals.

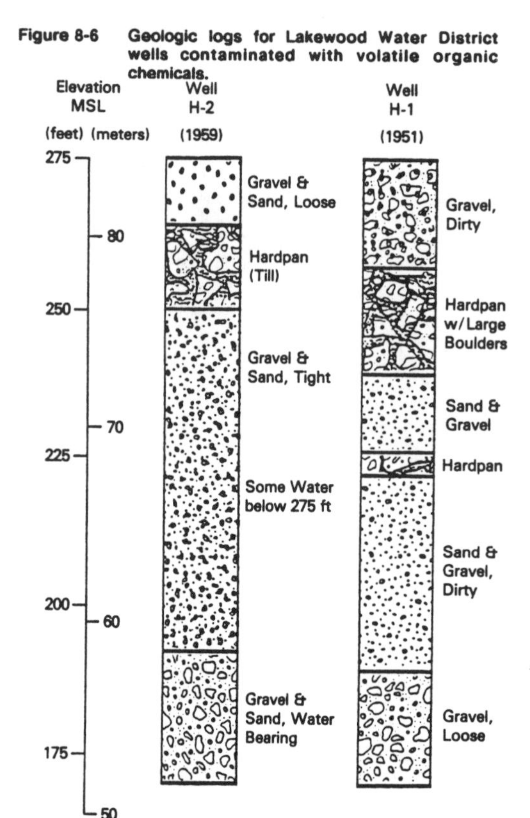

Figure 8-7 Schematic illustrating the mechanism by which a downgradient source may contaminate a production well, and by which a second well may isolate the source through hydraulic interference.

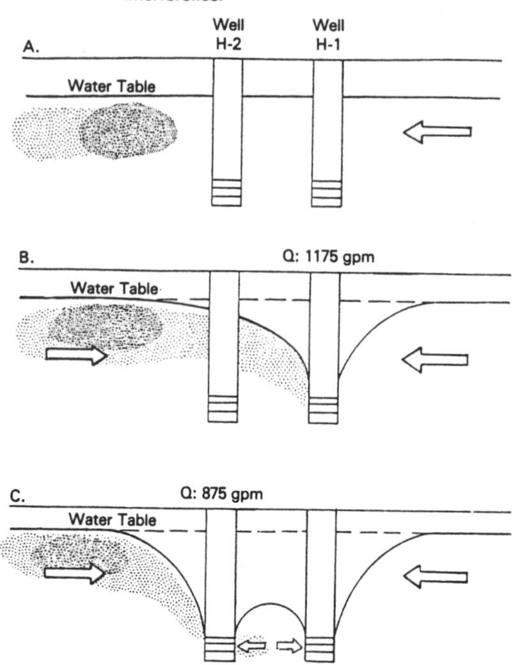

mechanism by which the two wells became contaminated seemed to be understood from a hydraulic point of view (Figure 8-7), but the chemical information did not seem to provide a consistent picture.

The source of contamination, a septic tank at a dry-cleaning facility, was found to have received large amounts of tetrachloroethylene and trichloroethylene, but no known amounts of cis- or trans-dichloroethylene; whereas the contaminated wells had relatively high concentrations of dichloroethylene. Initially it was thought that other sources might also be present and would explain the high concentrations of dichloroethylene. However, it soon became clear that recent research results regarding the potential for biotransformation of tetrachloroethylene and trichloroethylene (Wilson and McNabb, 1981) would more satisfactorily explain the observations.

Simulations of this kind of problem could be adequately performed only by contaminant transport models capable of incorporating the effects of the pumping wells on the regional flow field. More sophisticated approximations would also require the ability to account for the anisotropic and hexterogeneous character of the site, the retardation of the VOCs by sorption, and their possible biotransformations. Given the higly localized nature of the contaminant source and limited extent of the plume, however, there was insufficient justification for pursuing such efforts. The resolution of the problem was possible by relatively simple source removal techniques (excavation of the septic tank and elimination of discharges).

8.3.2.2 Field example no. 2.

Similar experience with special use of geotechnical methods and state-of-the-art research findings occurred at the 20-acre Chem-Dyne solvent reprocessing site in Hamilton, Ohio (Figure 8-8). During operation of the site (1974-1980), poor waste handling practices such as on-site spillage of a wide variety of industrial chemicals and solvents, direct discharge of liquid wastes to a stormwater drain beneath the site, and mixing of imcompatible wastes were engaged in routinely. These caused extensive soil and ground-water contamination, massive fish kills in the Great Miami River, and major on-site fires and explosions, respectively. The stockpiling of liquid and solid wastes resulted in thousands of badly

Figure 8-8 Location map for Chem-Dyne Superfund Site.

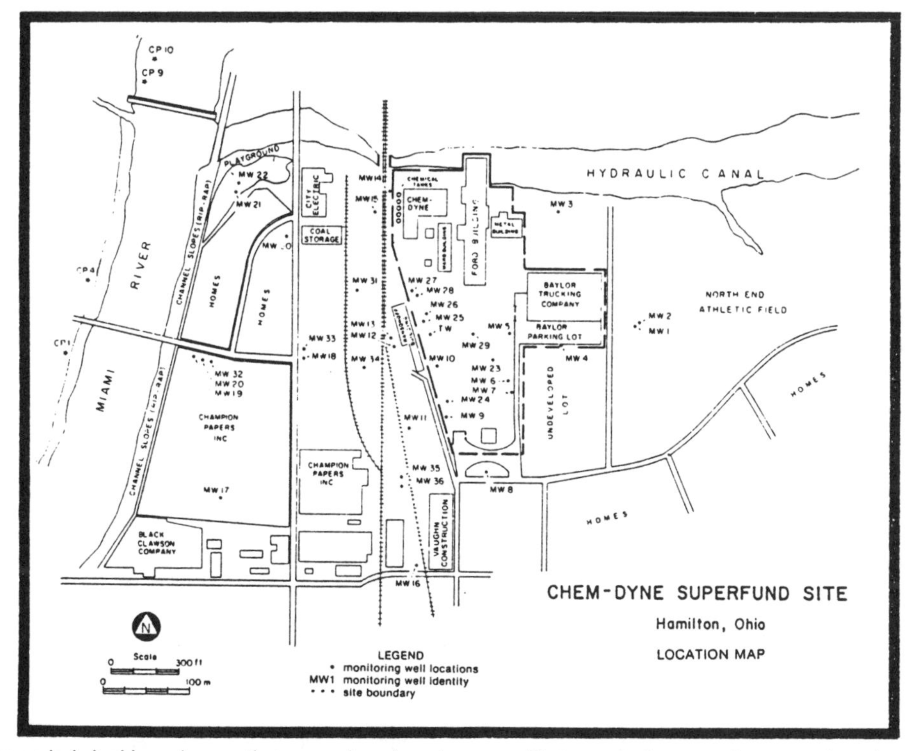

corroded leaking drums that posed a long-term threat to the environment (CH_2M-Hill, 1984).

The seriousness of the ground-water contamination problem became evident during the initial site survey (1980-1981), which included the construction and sampling of over 20 shallow monitoring wells (Ecology and Environment, 1981). The initial survey indicated that the contaminant problem was much more limited than was later shown to be the case (Roy F. Weston Inc., 1983, Ch_2M-Hill, 1984). A good portion of the improvement in delineating the plume was brought about by an improved understanding of the natural processes controlling transport of contaminants at the site.

The initial site survey indicated that ground-water flow was generally to the west of the site, toward the Great Miami River, but that a shallow trough paralleled the river itself as a result of weak and temporary stream influences. The study concluded that contaminants would be discharged from the aquifer into the river (Ecology and Environment, 1981). That study also concluded that the source was limited to highly contaminated surface soils, and that removal of the uppermost three feet of the soil would essentially eliminate the source.

That conclusion was, however, based on faulty soil sampling procedures. The soil samples that were taken were not preserved in air-tight containers, so that most of the VOCs leaked out prior to analysis. That the uppermost soil samples showed high VOC levels is probably explained by the co-occurrence of viscous oils and other organic chemicals that may have served to entrap the VOCs. The more fiscous and highly retarded chemicals did not migrate far enough into the vertical profile to exert a similar influence on samples collected at depths greater than a few feet.

Subsequent studies of the site corrected these misinterpretations by producing data from proper soil samplings and by incorporating much more detailed characterization of the fluvial sediments and the natural flow system. In those studies vertical profile characterizations were obtained from each new borehole drilled by continuous split-spoon samples of subsurface solids; and clusters of vertically-separated monitoring wells were constructed. The split-spoon samples helped to confirm the general locations of intefingered clay lenses and clearly showed the high degree of heterogeneity of the sediments (Figures 8-9 and 8-10). While an extensive network of shallow wells confirmed earlier indications of general ground-water flow toward the river (Figure 8-11), the clusters of vertically separated wells revealed that dramatic downward gradients existed adjacent to the Great Miami River (Figure 8-12). This finding indicated that the migrating plume would not be discharged to the river, but would instead flow under the river.

The presence of major industrial wells on the other side of the river supported this conclusion. The plume would be drawn to greater depths in the aquifer by the locally severe downward gradient, but whether the industrial wells would actually capture the plume could not be determined. That determination would require careful evaluation of the hydrogeologic features beneath the river; something that has not been attempted because of the onset of remedial actions designed to stop the plume from reaching the river.

The field characterization efforts, however, did include the performance of a major pump test so that the hydrogeologic characteristics of the contaminated portion of the aquifer could be estimated. The pump test was difficult to arrange, because the pumping well had to be drilled onsite for reasons of potential liability and lack of property access elsewhere. The drillers were considerably slowed in their work by the need to don air-tanks when particularly contaminated subsoils were encountered because the emission of volatile fumes from the borehole presented unacceptable health risks. Since the waters which would be pumped were expected to be contaminated, it was necessary to construct 10 large temporary holding tanks (100,000 gallons each) onsite to impound the waters for testing and possible treatment prior to being discharged to the local sewer system (CH_2M-Hill, 1984).

The costs and difficulty of preparing for and conducting the test were worth the effort, however. The water levels in thirty-six monitoring wells were observed during the test and yielded a very detailed picture of transmissivity variations (Figure 8-13), which has been used to help explain the unusual configuration of the plume (Figure 8-14) and which were used to guide the design of a pump-and-treat system. Storage coefficients were also estimated; and though the short duration of the test (14 hours) did not allow for definitive estimates to be obtained, it was clear that qualitative confirmation of the generally non-artesian (water-table) nature of the aquifer beneath the site was confirmed. An automated data acquisition system (computer controlled pressure transducer) was used to monitor the water levels and provide real-time drawdown plots of 19 of the 36 wells (Table 8-2), greatly enhancing the information obtained with only minimal manpower requirements. The benefits from conducting the pump test cannot be overemphasized; qualitative confirmation of lithologic information and semi-quantitative estimation of crucial parameters were obtained.

Finally, the distribution patterns of contaminant species that emerged from the investigations at Chem-Dyne were made understandable by considering research results and theories regarding chemical and microbiological influences. Once again there seemed to be evidence of transformation of tetrachloroethene (Figure 8-15) to less halogenated daughter products such as trichloroethene (Figure 8-16), dichloroethene (Figure 8-17), and vinyl chloride/monochloroethene (Figure 8-18). The relative rates of movement of these contaminants, as well as other common solvents like benzene (Figure 8-19) and chloroform (Figure 8-20), generally conformed to predictions based on sorption principles. The remediation efforts also made use of these contaminant transport theories in estimating the capacity of the treatment system needed and the length of time necessary to remove residuals from the aquifer solids (CH_2M-Hill, 1984).

During the latter stages of negotiations with the Potentially Responsible Parties (PRPs), government contractors prepared mathematical models of the flow system and contaminant transport at Chem-Dyne (GeoTrans, 1984). These were used to estimate the possible direction and rate of migration of the plume in the absence of remediation, the mass of contaminants removed during various remedial options, and the effects of sorption and dispersion on those estimates. Because of the wide range of sorption properties associated with the variety of VOCs found in significant concentrations it was necessary to select values of retardation constants that represented the likely upper- and lower-limits of sorptive effects. It was also necessary to estimate or assume the values of other parameters known to affect transport processes, such as dispersion coefficients.

While the developers of the models would be the first to acknowledge the large uncertainties associated with those modeling efforts due to lack of information about the actual history of chemical inputs and other important data, there was agreement between the government and PRP technical experts that the modeling efforts had been very helpful in assessing the magnitude of the problem and in determining minimal requirements for remediation. Consequently, modeling efforts will continue at Chem-Dyne. Data generated during the remediation phase will be used to refine models in an ongoing process so that the effectiveness of the remedial action can be evaluated properly.

8.3.3 Practical Concerns

In many ways, there may be too much confidence among those not directly involved in ground-water quality research regarding current abilities to predict

Figure 8-9 Chem-Dyne geologic cross-section along NNW-SSE axis.

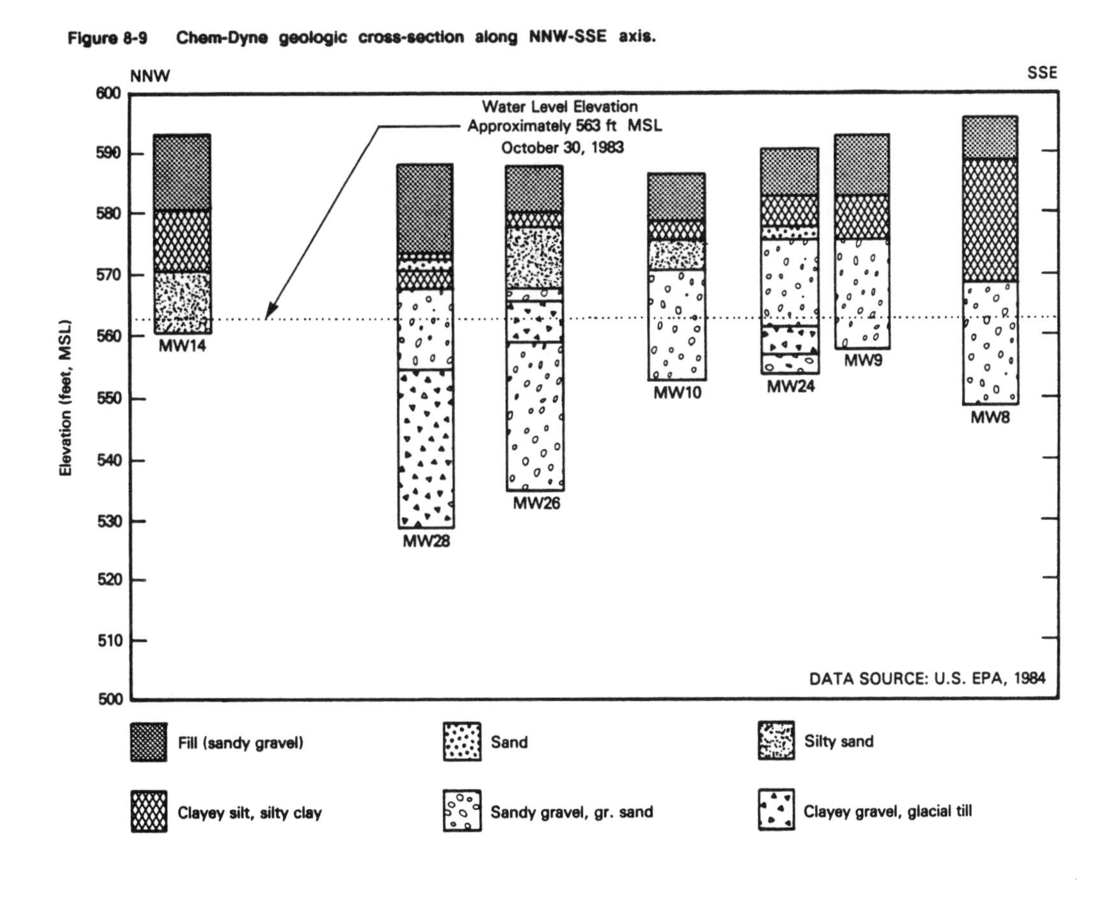

Figure 8-10 Chem-Dyne geologic cross-section along WSW-ENE axis.

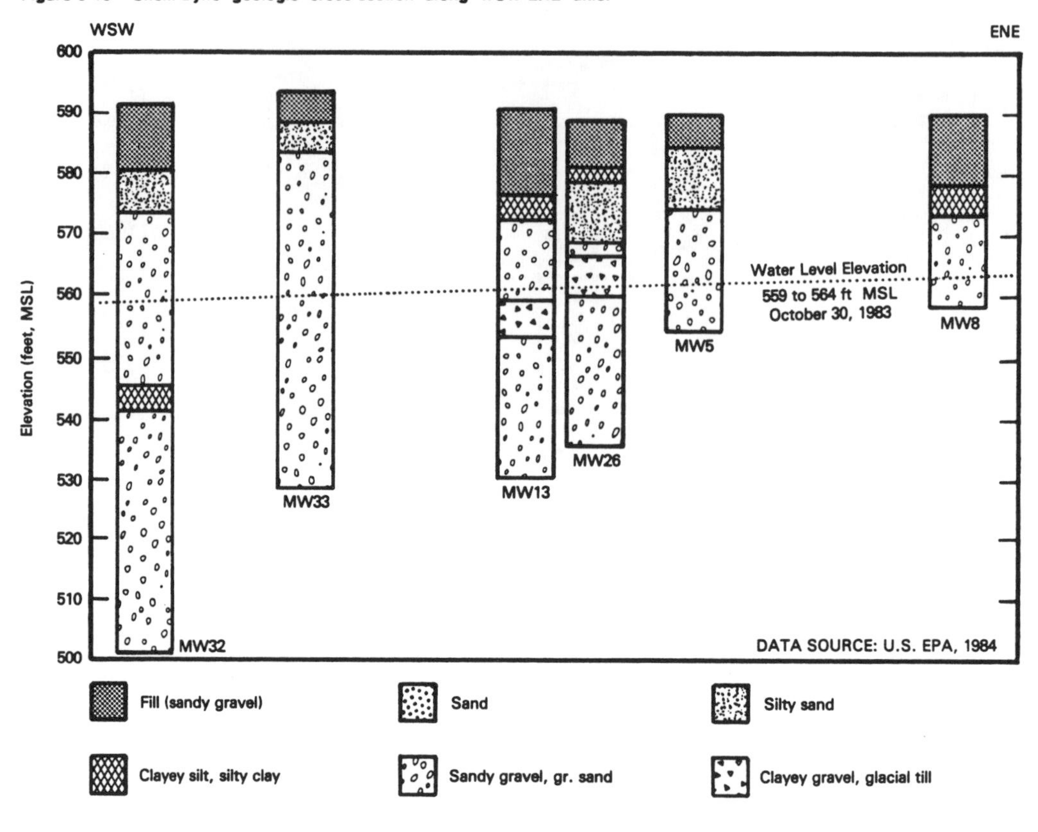

Figure 8-11 Shallow well ground-water contour map for Chem-Dyne. Flow is generally to the river (west) and down the valley (southwest).

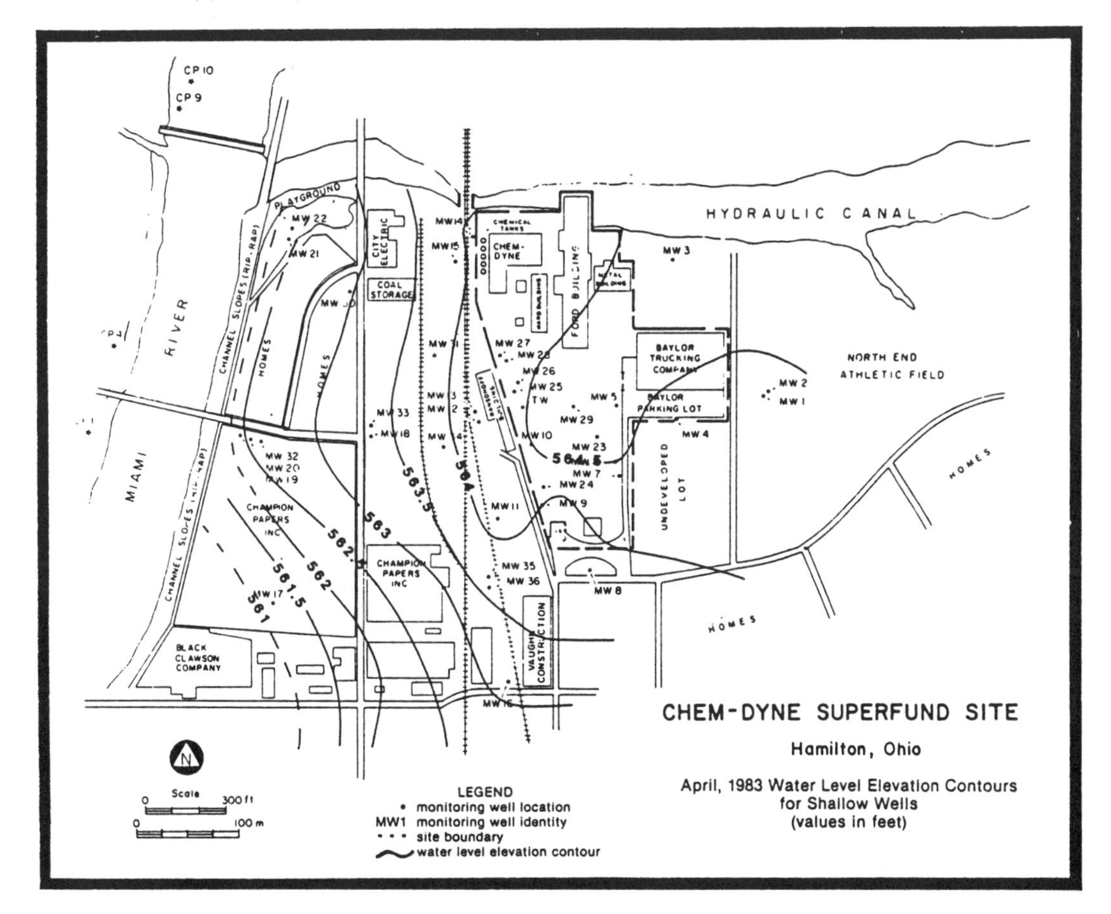

Figure 8-12 Typical arrangement of clustered, vertically-separated wells installed adjacent to Chem-Dyne and the Great Miami River.

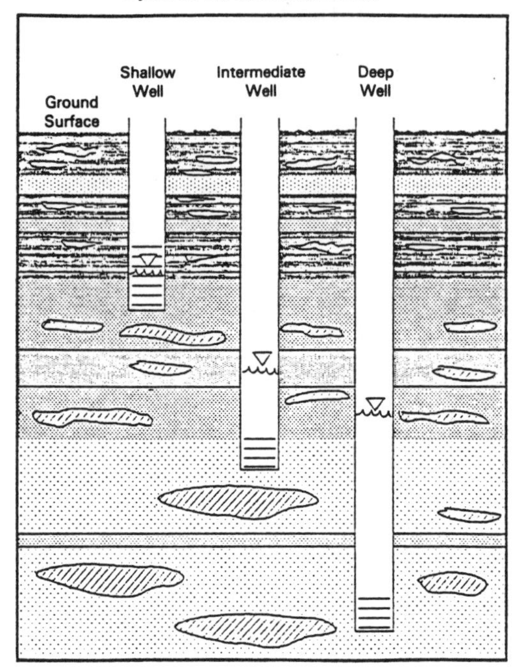

transport and fate of contaminants in the subsurface. The discussions in the preceding sections should place in proper perspective the admittedly remarkable advances that have been made in recent years by illustrating the practical and conceptual uncertainties that remain unresolved. Continuing research efforts will eventually resolve these uncertainties, but those efforts will be considerably slower if existing results are not routinely incorporated into practical situations. Research results must be tested in real-world settings because there is no alternative mechanism for validating them. Just as importantly, there are economic arguments for incorporating research findings and state-of-the-art techniques into routine contaminant investigations and remediations.

Additional effort devoted to site-specific characterizations of natural process parameters, rather than relying almost exclusively on chemical analyses of ground-water samples, can significantly improve the quality and cost-effectiveness of remedial actions at such sites. To underscore this point, condensed summaries are provided of the principal activities, benefits, and shortcomings of three possible site characterization approaches: conventional (Table 8-3), state-of-the-art (Table 8-4), and state-of-the-science (Table 8-5). To further illustrate this, a qualitative assessment of desired trade-offs between characterization and clean-up costs is presented in Figure 8-21.

As illustrated there, some investments in specialized equipment and personnel will be necessary to make transitions to more sophisticated approaches, but those investments should be more than paid back in reduced clean-up costs. The maximum return on increased investments is expected for the state-of-the-art approach, and will diminish as the state-of-the-science approach is reached because highly specialized equipment and personnel are not widely available. It is vitally important that this philosophy be considered, because the probable benefits in lowered total costs, health risks, and time for effective remediations can be substantial.

8.4 Liabilities, Costs, and Recommendations for Managers

There are many texts available that describe the derivation of the theories underlying mathematical models, the technical applications of models, and related technical topics (e.g., data collection and parameter estimation techniques). Few texts treat the nontechnical issues that managers face when evaluating the possible uses of models, such as potential liabilities, costs, and communications between the modeler and management. These are, however, important considerations because many modeling efforts fail as a consequence of insufficient attention to them. This section is therefore directed to those issues.

8.4.1 Potential Liabilities

Some of the liabilities attending the use of mathematical models relate to the degree to which predictive models are relied on to set conditions for permitting or banning specific practices or products. If a model is incapable of treating specific applications properly, substantially incorrect decisions may be made. Depending on the application, unacceptable environmental effects may begin to accumulate long before the nature of the problem is recognized. Conversely, unjustified restrictions may be imposed on the regulated community. Inappropriate or inadequate models may also cause the 're-opening clause' of a negotiated settlement agreement to be invoked when, for instance, compliance requirements that were guided by model predictions of expected plume behavior are not met.

Certain liabilities relate to the use of proprietary codes in legal settings, where the inner workings of a model may be subject to disclosure in the interests of justice. The desire for confidentiality by the model developer would likely be subordinate to the public right to full information regarding actions predicated on modeling results. The mechanisms for protection of proprietary rights do not currently extend beyond extracted promises of confidentiality by reviewers or

Figure 8-13 Estimates of transmissivity obtained from shallow and deep wells during Chem-Dyne pump test.

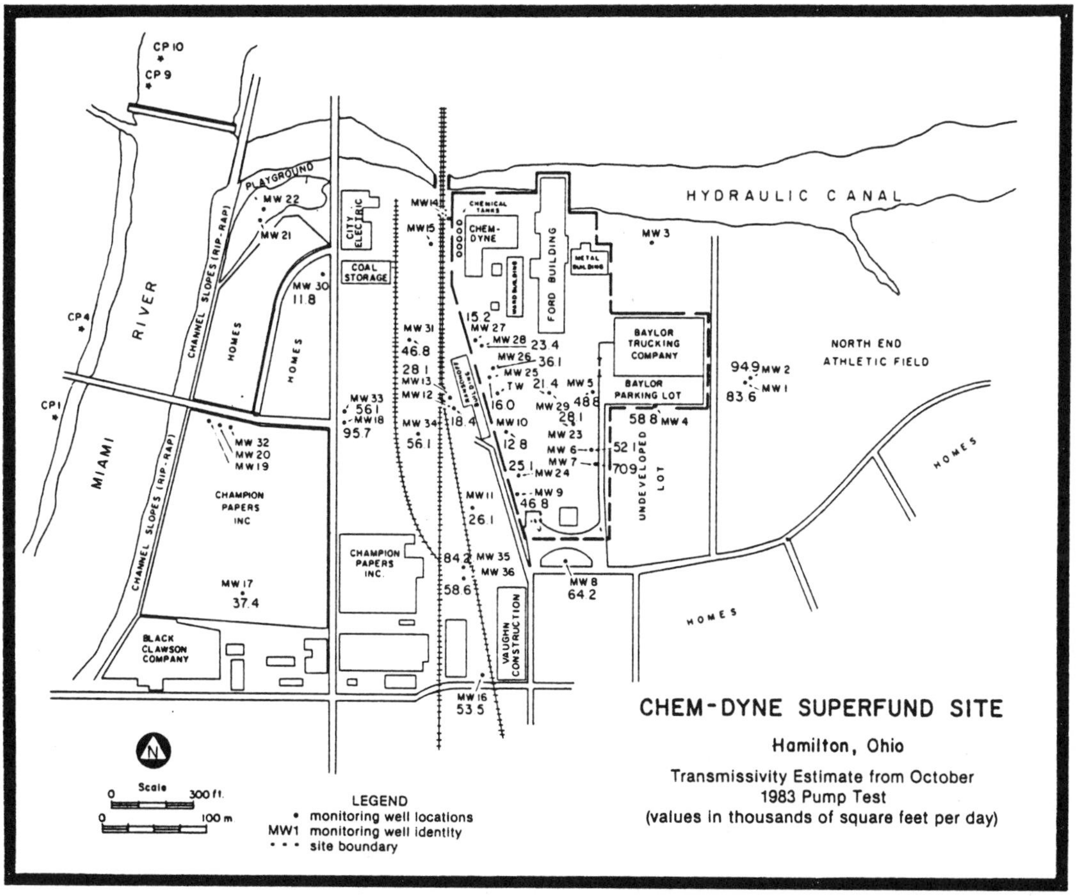

CHEM-DYNE SUPERFUND SITE

Hamilton, Ohio

Transmissivity Estimate from October 1983 Pump Test

(values in thousands of square feet per day)

LEGEND

• monitoring well locations

MW1 monitoring well identity

• • • site boundary

Figure 8-14 Distribution of total volatile organic chemical contamination in shallow wells at Chem-Dyne during October, 1983 sampling.

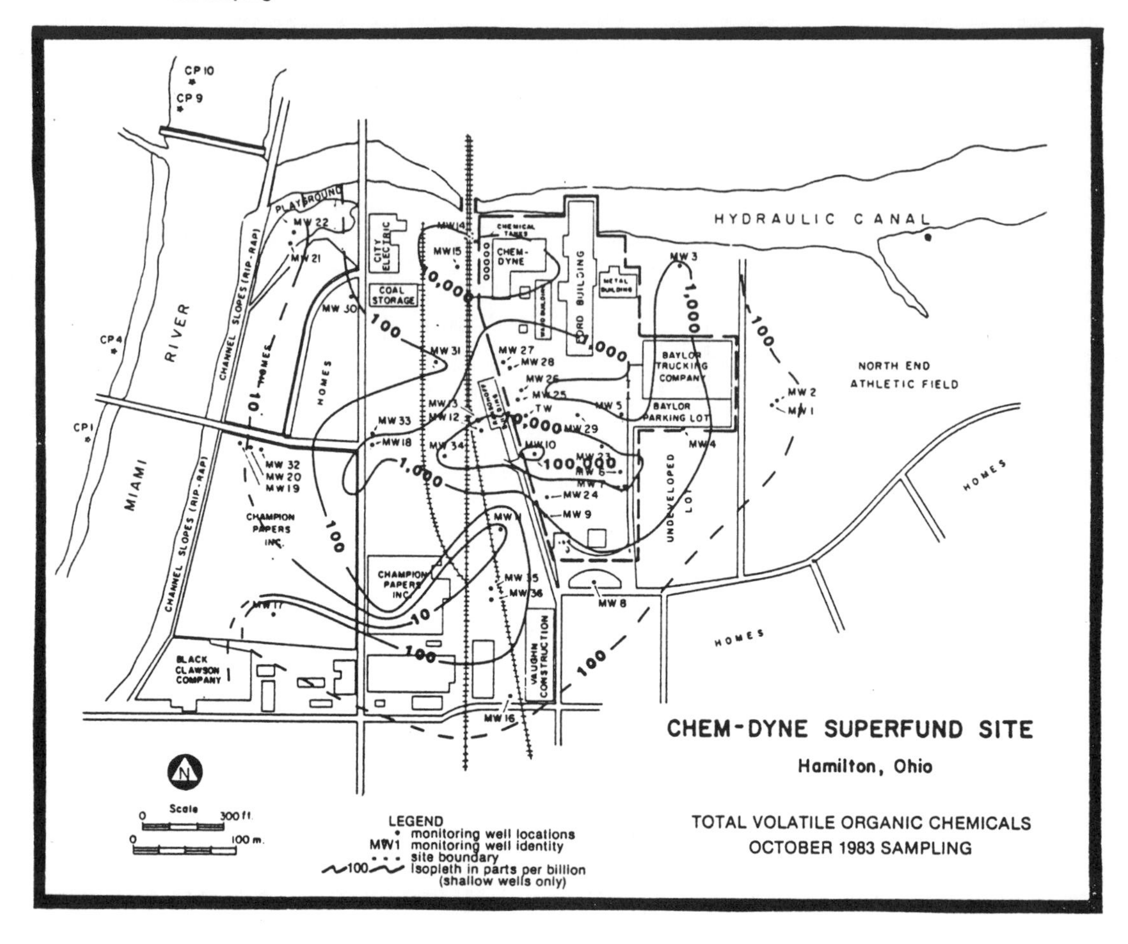

Table 8-2 Chem-Dyne Pump Test Observation Network

Observation Well Number	Radial Distance (ft)	Initial Water Level (ft, MSL)	Method of Measurement (type, field unit)
MW1	957	563.68	Manual, electric probe
MW2	965	563.74	Manual, electric probe
MW3	848	563.96	Automatic, float-type
MW4	537	563.27	Manual, electric probe
MW5	313	563.31	Manual, electric probe
MW6	420	564.40	Manual, electric probe
MW7	480	563.30	Manual, electric probe
MW8	740	563.01	Manual, electric probe
MW9	487	563.08	Manual, electric probe
MW10	186	563.29	Automatic, pressure transducer
MW11	502	562.90	Automatic, pressure transducer
MW12	232	562.39	Automatic, pressure transducer
MW13	232	563.19	Automatic, pressure transducer
MW14	701	–	Dry – no data collected
MW15	611	563.10	Automatic, pressure transducer
MW16	1275	562.47	Manual, electric probe
MW17	1518	560.03	Manual, electric probe
MW18	692	562.67	Manual, electric probe
MW19	1204	559.80	Automatic, float-type
MW20	1225	562.10	Automatic, float-type
MW21	1259	561.29	Manual, electric probe
MW22	1261	559.95	Automatic, pressure transducer
MW23	298	563.07	Automatic, pressure transducer
MW24	398	563.07	Automatic, pressure transducer
MW25	53	563.04	Automatic, pressure transducer
MW26	62	562.96	Automatic, pressure transducer
MW27	272	563.13	Automatic, pressure transducer
MW28	248	562.99	Automatic, pressure transducer
MW29	167	563.23	Automatic, pressure transducer
MW30	993	561.25	Manual, electric probe
MW31	465	562.78	Automatic, float-type
MW32	1236	559.56	Automatic, pressure transducer
MW33	690	562.06	Automatic, pressure transducer
MW34	454	562.29	Automatic, pressure transducer
MW35	651	562.93	Automatic, pressure transducer
MW36	696	562.69	Manual, electric probe
Pumping Well	0 (reference point)	562.97	Automatic, pressure transducer

other interested parties. Hence, a developer of proprietary codes still assumes some risk of exposure of innovative techniques, even if the code is not pirated outright.

Yet other liabilities may arise as the result of misapplication of models or the application of models later found to be faulty. Frequently, the choices of boundary and initial conditions for a given application are hotly contested; misapplications of this kind are undoubtedly responsible for many of the reservations expressed by would-be model users. It has also happened many times in the past that a widely used and highly regarded model code was found to contain errors that affected its ability to faithfully simulate situations for which it was designed. The best way to minimize these liabilities is to adopt strict quality control procedures for each application.

8.4.2 Economic Considerations

The nominal costs of the support staff, computing facilities, and specialized graphics' production equipment associated with numerical modeling efforts can be high. In addition, quality control activities can result in substantial costs; the determining factor in controlling these costs is the degree to which a manager must be certain of the characteristics of the model and the accuracy of its output.

As a general rule, costs are greatest for personnel, moderate for hardware, and minimal for software. The exception to this ordering relates to the combination of software and hardware purchased. An optimally outfitted business computer (e.g., VAX 11/785 or IBM 3031) costs about $100,000; but it can rapidly pay for itself in terms of dramatically increased speed and computational power. A well complimented personal computer (e.g., IBM-PC/AT or DEC Rainbow) may cost $10,000; but the significantly slower speed and

Figure 8-15 Distribution of tetrachloroethane in shallow wells at Chem-Dyne during October, 1983 sampling.

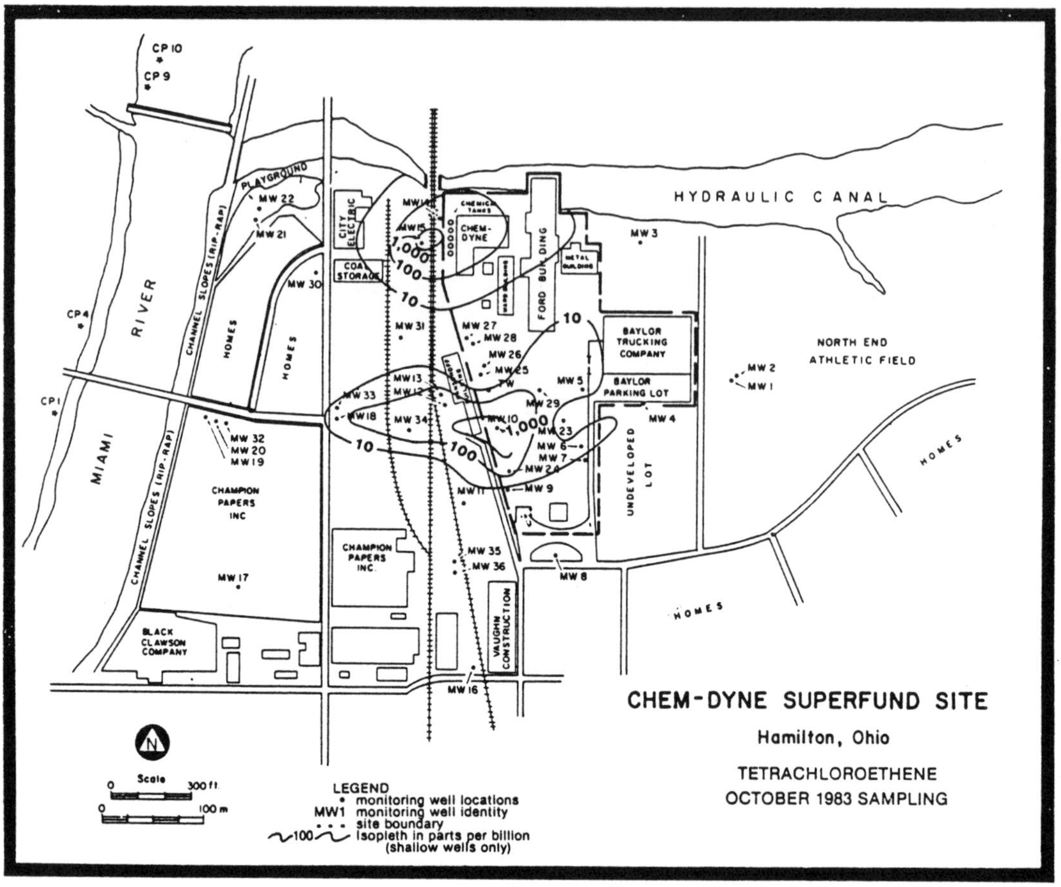

CP 10
CP 9
PLAYGROUND
MW 22
MW 21
HYDRAULIC CANAL
MW 14
MW 15
CITY ELECTRIC
CHEMICAL TANKS
CHEM-DYNE
FORD BUILDING
METAL BUILDING
MW 3
1,000
100
10
COAL STORAGE
MW 30
CP 4
RIVER
CHANNEL SLOPES (RIP-RAP)
HOMES
HOMES
MW 31
MW 27
MW 28
MW 26
MW 25
BAYLOR TRUCKING COMPANY
NORTH END ATHLETIC FIELD
MW 2
MW 1
MW 5
BAYLOR PARKING LOT
CP 1
MW 13
MW 12
MW 33
MW 18
MW 34
MW 29
MW 10
1,000
MW 23
MW 4
MW 32
MW 20
MW 19
10
100
MW 6
MW 7
MW 24
UNDEVELOPED LOT
HOMES
MIAMI
CHAMPION PAPERS INC
MW 11
MW 9
CHANNEL SLOPES (RIP-RAP)
CHAMPION PAPERS INC.
MW 35
MW 36
MW 8
MW 17
HOMES
BLACK CLAWSON COMPANY
VAUGHN CONSTRUCTION
MW 16

CHEM-DYNE SUPERFUND SITE
Hamilton, Ohio
TETRACHLOROETHENE
OCTOBER 1983 SAMPLING

N
Scale
0 300 ft
0 100 m

LEGEND
• monitoring well locations
MW1 monitoring well identity
• • • site boundary
~100~ isopleth in parts per billion (shallow wells only)

Figure 8-16 Distribution of trichloroethane in shallow wells at Chem-Dyne during October, 1983 sampling.

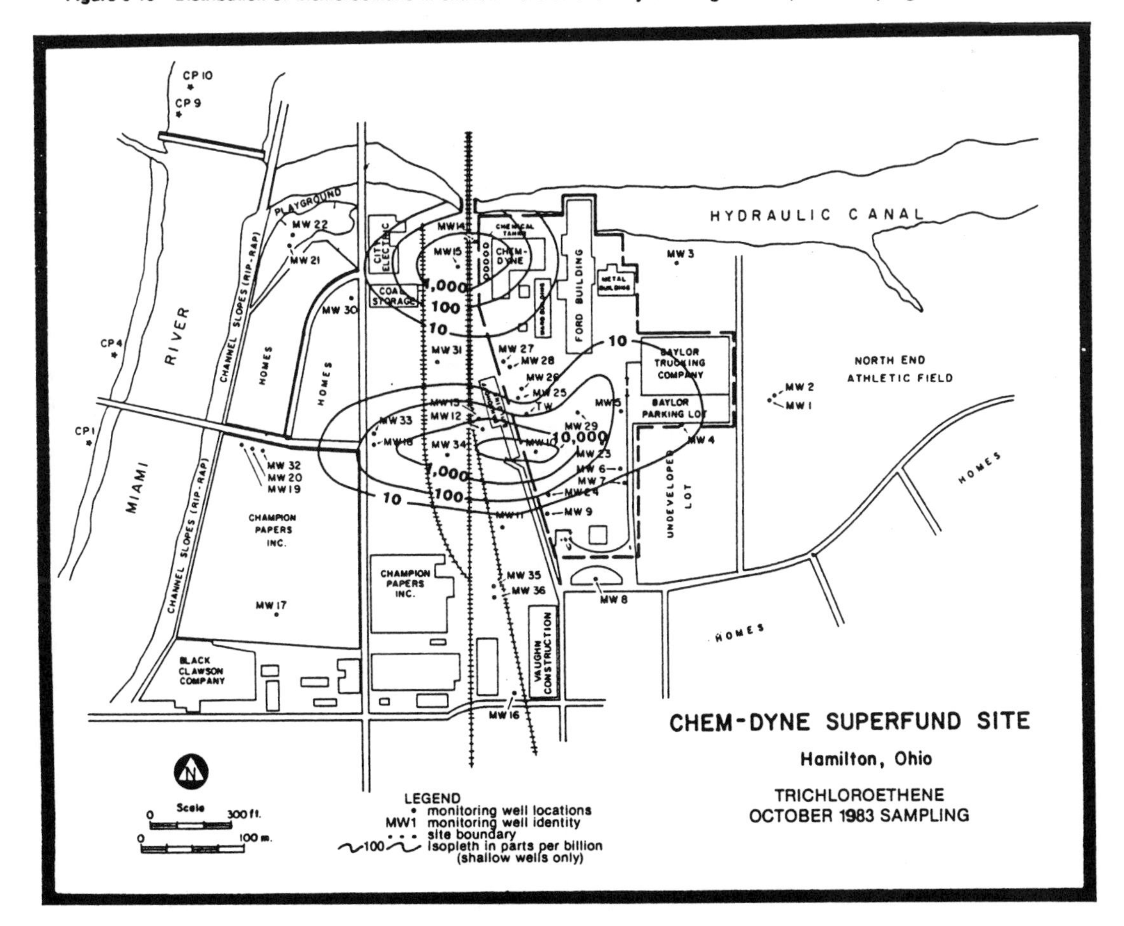

Figure 8-17 Distribution of trans-dichloroethene in shallow wells at Chem-Dyne during October, 1983 sampling.

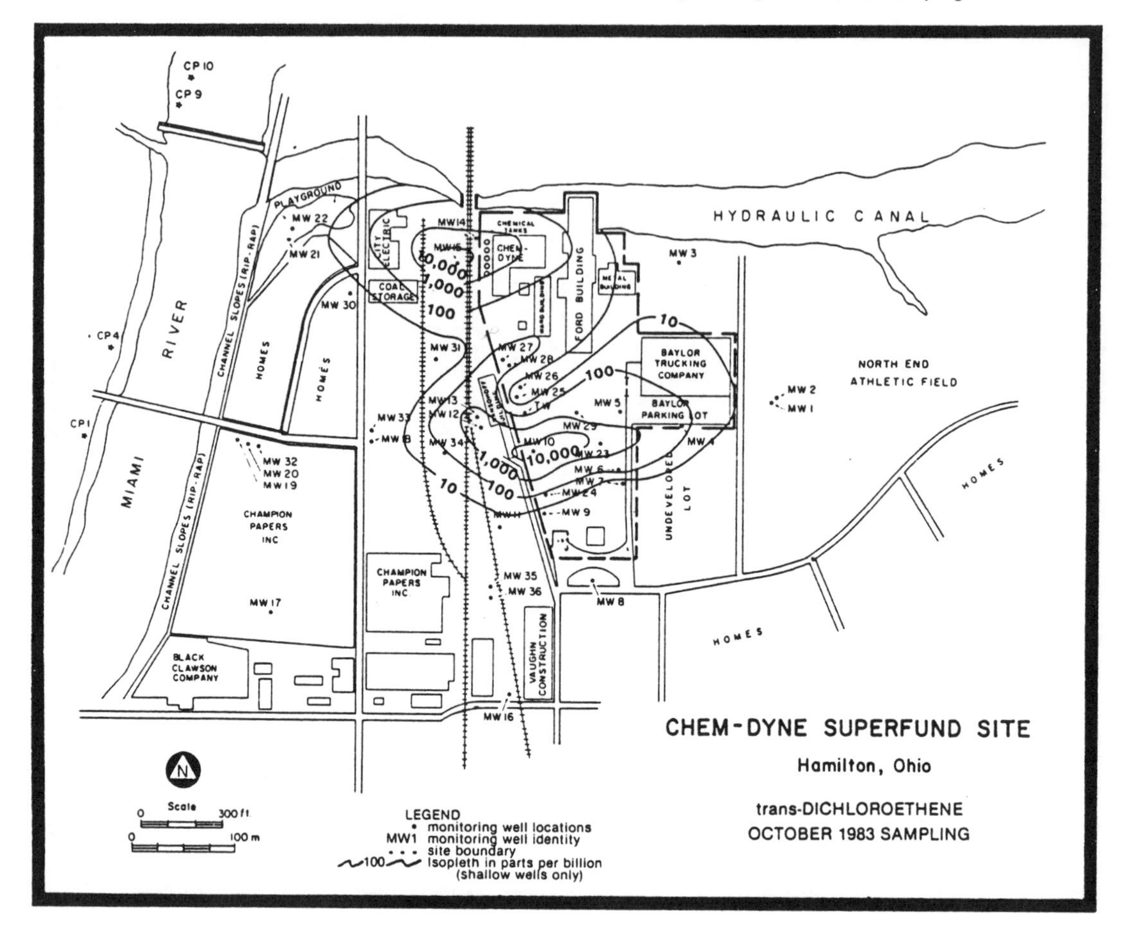

Figure 8-18 Distribution of vinyl chloride in shallow wells at Chem-Dyne during October, 1983 sampling.

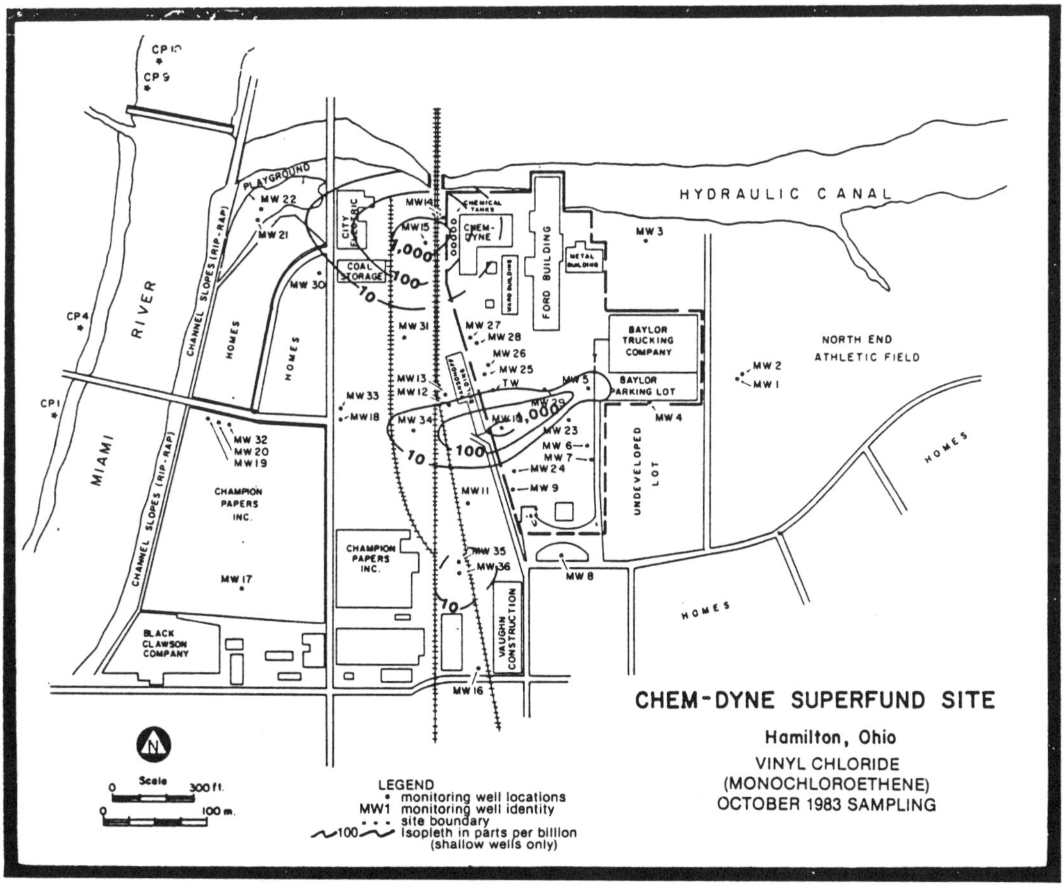

Figure 8-19 Distribution of benzene in shallow wells at Chem-Dyne during October, 1983 sampling.

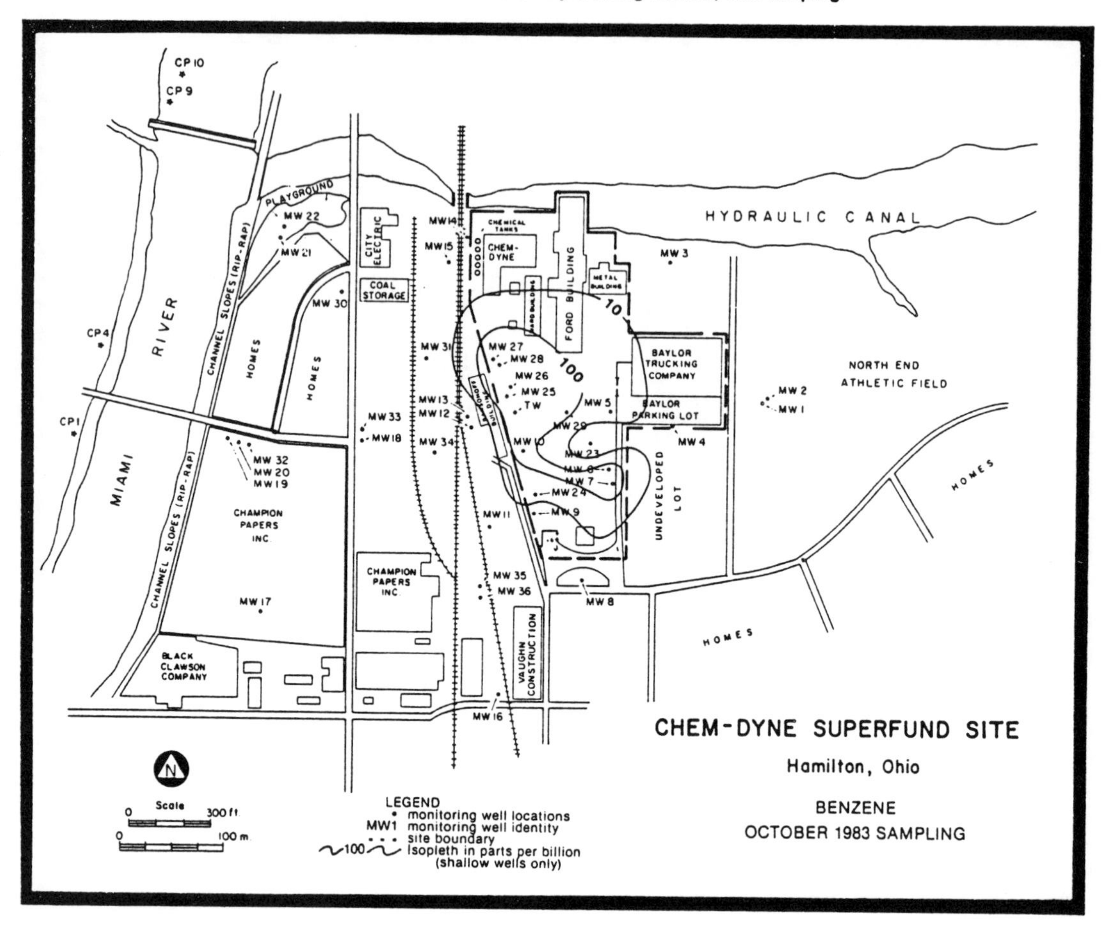

Figure 8-20 Distribution of chloroform in shallow wells at Chem-Dyne during October, 1983 sampling.

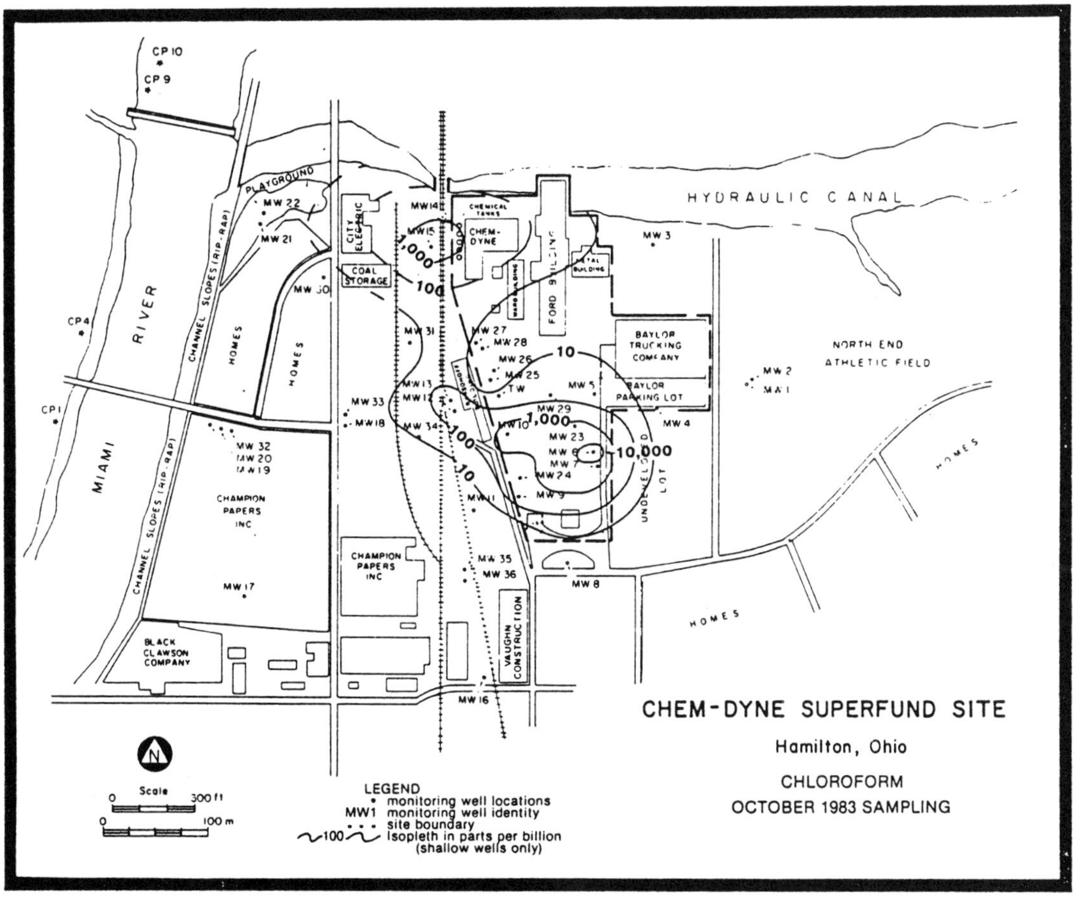

CHEM-DYNE SUPERFUND SITE

Hamilton, Ohio

CHLOROFORM

OCTOBER 1983 SAMPLING

LEGEND

• monitoring well locations

MW1 monitoring well identity

• • • site boundary

~100~ Isopleth in parts per billion (shallow wells only)

Scale 0 300 ft

0 100 m

Table 8-3 Conventional Approach to Site Characterization Efforts

Actions Typically Taken
- Install a few dozen shallow monitoring wells
- Sample and analyze numerous times for 129+ pollutants
- Define geology primarily by driller's log and cuttings
- Evaluate hydrology with water level maps only
- Possibly obtain soil and core samples (chemical extractions)

Benefits
- Rapid screening of problem
- Moderate costs involved
- Field and lab techniques standardized
- Data anlaysis relatively straightforward
- Tentative identification of remedial options possible

Shortcomings
- True extent of problem often misunderstood
- Selected remedial alternative may not be appropriate
- Optimization of remedial actions not possible
- Clean-up costs unpredictable and excessive
- Verification of compliance uncertain and difficult

Table 8-4 State-of-the-Art Approach to Site Characterization Efforts

Recommended Actions
- Install depth-specific well clusters
- Sample and analyze for 129+ pollutants initially
- Analyze selected contaminants in subsequent samplings
- Define geology by extensive coring/split-spoon samples
- Evaluate hydrology with well clusters and geohydraulic tests
- Perform limited tests on solids (grain size, clay content)
- Conduct geophysical surveys (resistivity soundings, etc.)

Benefits
- Conceptual understanding of problem more complete
- Better prospect for optimization of remedial actions
- Predictability of remediation effectiveness increased
- Clean-up costs lowered; estimates improved
- Verification of compliance soundly based, more certain

Shortcomings
- Characterization costs somewhat higher
- Detailed understanding of problem still difficult
- Full optimization of remedial actions not likely
- Field tests may create secondary problems
- Demand for specialists increased

limited computational power may infer hidden costs in terms of the inability to perform specific tasks. For example, highly desirable statistical packages like SAS and SPSS are unavailable or available only with reduced capabilities for personal computers; many of the most sophisticated mathematical models are available in their fully-capable form only on business computers.

Figure 8-22 gives a brief comparison of typical costs for software for different levels of computing power. Obviously, the software for less capable computers is cheaper, but the programs are not equivalent; so managers need to thoroughly think through what level is appropriate. If the decisions to be made are to be based on very little data, it may not make sense to insist on the most elegant software and hardware. If the intended use involves substantial amounts of data and sophisticated analyses are desired, it would be unwise to opt for the least expensive combination.

Based on experience and observation, there does seem to be an increasing drive away from both ends of the spectrum and toward the middle; that is, the use of powerful personal computers is increasing rapidly, whereas the use of small programmable calculators and large business computers alike is

Table 8-5 State-of-the-Science Approach to Site Characterization Efforts

Idealized Approach
- **Assume "State-of-the-Art Approach" as starting point**
- **Conducter tracer tests and borehole geophysical surveys**
- **Determine % organic carbon, exchange capacity, etc. of solids**
- **Measure redox potential, pH, dissolved oxygen, etc. of fluids**
- **Evaluate soprtion-desorption behavior using select cores**
- **Identify bacteria and assess potential for biotransformation**

Benefits
- **Thorough conceptual understanding of problem obtained**
- **Full optimization of remedial actions possible**
- **Predictability of remediation effectiveness maximized**
- **Clean-up costs lowered significantly; estimates reliable**
- **Verification of compliance assured**

Shortcomings
- **Characterization costs significantly higher**
- **Few previous field applications of advanced theories**
- **Field and laboratory techniques not yet standardized**
- **Availability of specialized equipment low**
- **Demand for specialists dramatically increased**

Figure 8-21 General relationship between site characterization costs and clean-up costs as a function of the characterization approach.

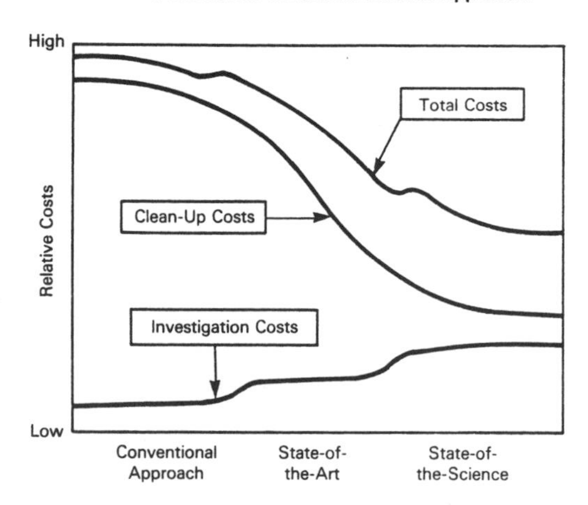

Figure 8-22 Average price per category for ground-water models from the International Ground Water Modeling Center.

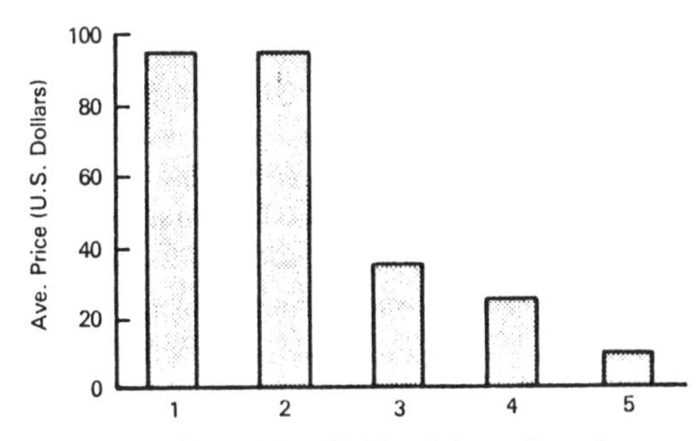

Categories
1 Mainframe / business computer models
2 Personal computer versions of mainframe models
3 Original IBM-PC and compatibles' models
4 Handheld microcomputer models (e.g., Sharp PC1500)
5 Programmable calculator models (e.g., HP41-CV)
Prices include software and all available documentation, reports, etc.

declining. In part, this stems from the significant improvements in the computing power and quality of printed outputs obtainable from personal computers. In part, it is due to the improved telecommunications capabilities of personal computers, which are now able to emulate the interactive terminals of large business computers so that vast computational power can be accessed and the results retrieved with no more than a phone call. Most importantly for ground-water managers, many of the mathematical models and data packages have been "down-sized" from mainframe computers to personal computers; many more are being written directly for this market.

Since it is expected that most managers will want to explore this situation a bit more, Figure 8-23 has been prepared to provide some idea of the costs of available software and hardware for personal computers.

The technical considerations discussed in previous sections indicate that the desired accuracy of the modeling effort directly affects the total costs of mathematical simulations. Thus managers will want to determine the incremental benefits gained by increased expenditures for more involved mathematical modeling efforts. There are many economic theories which can be helpful in determining the incremental benefits gained per increased level of investment. The most straightforward of these are the cost-benefit approaches commonly used to evaluate the economic desirability of water resource projects. There are two generalized approaches in common practice: the "benefit/cost ratio" method and the "net benefit" method.

The benefit/cost ratio method involves tallying the economic value of all benefits and dividing that sum by the total costs involved in generating those benefits (i.e., B/C = ?). A ratio greater than one is required for the project to be considered viable, though there may be sociopolitical reasons for proceeding with projects that do not meet this criterion. Consider the example of a project that is about to get underway and has gained considerable social or political momentum when the initial cost estimates begin to prove to be too low. Not proceeding or substantially altering the work may be economically wise; however, such a decision may be viewed as a breach of faith by the public. Regardless of how this kind of situation evolves, it is not uncommon for certain costs to be forgiven or subsidized, which muddies the picture for incremental benefits or trade-off analyses.

The "net benefit" method involves determining the arithmetic difference of the total benefits and total costs (i.e., B-C = ?). Here the obvious criterion is that the proposed work results in a situation where total benefits exceed total costs. This approach is most often adopted by profit-making enterprises, because they seek to maximize the difference as a source of income. The ratio method, by contrast, has long been used by government agencies and other non-profit organizations because they seek to show the simple viability of their efforts irrespective of the costs involved.

In a very real sense, then, these two general economic assessment methods stem from different philosophies. They share many common difficulties and limitations, however. For example, there is a need to predict the present worth of future costs and to amortize benefits over the life of a project. The mechanics of such calculations are well known, but they necessarily involve substantial uncertainties. For example, the present worth of a series of equal payments for equipment or software can be computed by (White *et al.*, 1984):

$$P = A \times ((1 + i)[n] - 1)/(i \times (1 + i)[n]) \qquad (8\text{-}2)$$

where:

P = present worth
A = series payment each interest period
i = interest rate per period
n = number of interest periods.

Note, however, that the interest rate must be estimated; this has fluctuated widely in the past two decades as a result of inflationary and recessionary periods in our economy. The significance of this is that a small difference in the interest rate results in

Figure 8-23 Price ranges for IBM-PC ground-water models available from various sources.

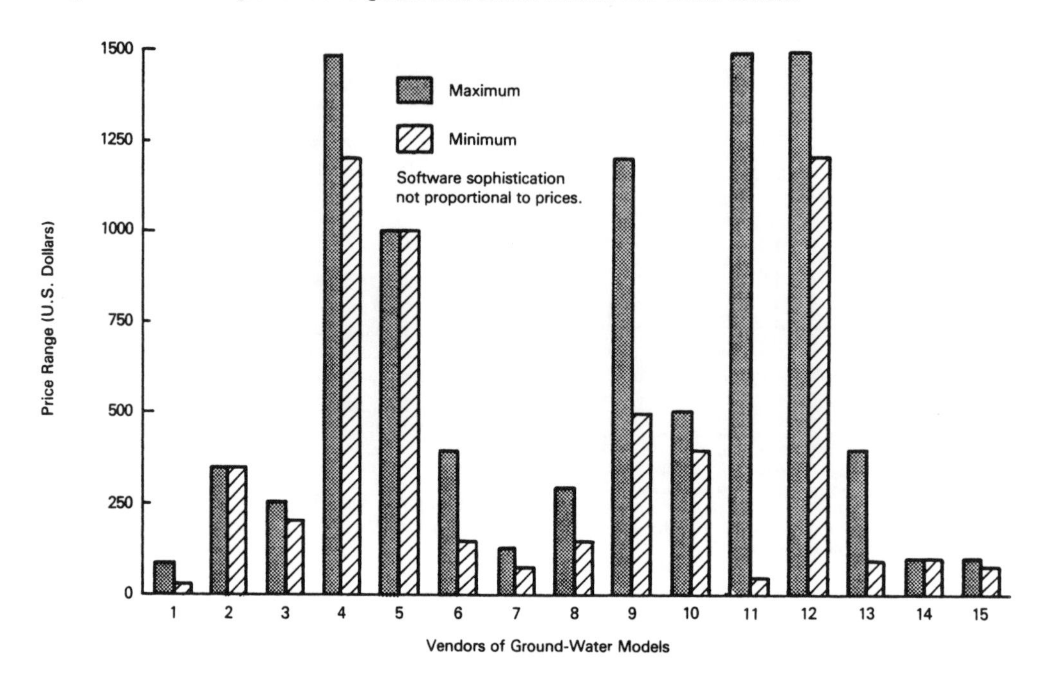

Vendors
1 International Ground Water Modeling Center
2 Computapipe Co.
3 Data Services, Inc.
4 GeoTrans, Inc.
5 Hydrosoft, Inc.
6 In Situ, Inc.
7 Irrisco Co.
8 Koch and Assoc.
9 KRS Enterprises, Inc.
10 Michael P. Spinks Co.
11 RockWare, Inc.
12 Solutech Corp.
13 Thomas A. Prickett & Assoc.
14 James S. Ulrick Co.
15 Watershed Research, Inc.

tremendous differences in the present worth estimate because of the exponential nature of the equation.

It is also possible to compute the future worth of a present investment, to calculate the percentage of worth annually acquired through single payments or serial investments, and so on. One should be aware that these methods of calculating costs belong to the general family of "single-objective" or "mutually-exclusive alternative" analyses which presuppose that the cost of two actions is obtained by simple addition of their singly-computed costs. In other words, the efforts being evaluated are presumed to have no interactions. For some aspects of ground-water modeling efforts, this assumption may not be valid; e.g., one may not be able to specify software and hardware costs independently. In addition, these methods rely on the "expected value concept", wherein the expected value of an alternative is viewed as the single product of its effects and the probability of their occurrence. This means that high-risk, low-probability alternatives and low-risk, high-probability alternatives have the same expected value.

To overcome these difficulties it is necessary to use methods which can incorporate functional dependencies between various alternatives and which do not rely on the expected value concept, such as multi-objective decision theories (Asbeck and Haimes, 1984; Haimes and Hall, 1974; Haimes, 1981). A conceivable use would be the estimation of lowered health risks associated with various remedial action alternatives at a hazardous waste site. In such a case the output of a contaminant transport model would be used to provide certain inputs (i.e., water levels, contaminant concentrations, etc.) to a health effects model, and it would convert these into the inputs for the multiobjective decision model (e.g., probability of additional cancers per level of contaminant). The primary difficulty with these approaches to cost-benefit analyses is in clearly formulating the overall probabilities of the alternatives, so that the objectives which are to be satisfied may be ranked in order of importance. A related difficulty is the need to specify the functional form of the inputs (e.g., the "population distribution function" of pumpage rates or contaminant levels). Historical records about the inputs may be insufficient to allow their functional forms to be determined.

Another problem compounding the cost-benefit analysis of mathematical modeling efforts relates to the need to place an economic value on intangibles. For example, the increased productivity a manager might expect as a result of rapid machine calculations replacing hand calculations may not be as definable in terms of the improved quality of judgments made as it is in terms of time released for other duties. Similarly, the estimation of improved ground-water quality protection benefits may necessitate some valuation of the human life and suffering saved (rather nebulous quantities). Hence, there is often room for considerable "adjustment" of the values of costs and benefits. This flexibility can be used inappropriately to improve otherwise unsatisfactory economic evaluations. Lehr (1986) offers a scathing indictment of the Tennessee Valley Authority for what he described as an "extreme injustice", perpetrated by TVA in the form of hydroelectric projects which have "incredibly large costs and negative cost benefit ratios".

Finally, some costs and benefits may be incorrectly evaluated because the data on which they are based are probabilistic and this goes unrecognized. For instance, we often know the key parameters affecting ground-water computations (i.e., hydraulic conductivity) only to within an order of magnitude due to data collection limitations. In these situations great caution must be exercised. On the one hand, excessive expenditures may be made to ensure that the model "accurately" simulates observed (though inadequate) data. On the other, the artistic beauty of computer generated results sometimes generates its own sense of what is "right", regardless of apparent clashes with common sense. The reason the basic data are uncertain is very important. Costs are not uncertain just because of lack of information about future interest rates; many times expectations are not realized because of societal and technological changes. Miller (1980) noted that EPA overestimated the cost of compliance with its proposed standard for vinyl chloride exposure by 200 times the actual costs.

8.4.3 Managerial Considerations

The return on investments made to use mathematical models rests principally with the training and experience of the technical support staff applying the model to a problem, and on the degree of communication between those persons and management. In discussing the potential uses of computer modeling for ground-water protection efforts, Faust and others (1981) summarized by noting that "the final worth of modeling applications depends on the people who apply the models". Managers should be aware that a fair degree of specialized training and experience are necessary to develop and apply mathematical models, and relatively few technical support staff can be expected to have such skills presently (van der Heijde *et al.*, 1985). This is due in part to the need for familiarity with a number of scientific disciplines, so that the model may be structured to faithfully simulate real-world problems.

What levels of training and experience are necessary to apply mathematical models properly? Do we need "Rennaissance" specialists or can interdisciplinary teams be effective? The answers to these questions are not clear-cut. From experience it is easy to see that the more informed an individual is, the more effective he or she can be. It is doubtful, however, that any individual can master each discipline with the same depth of understanding that specialists in those fields have. What is clear is that some working knowledge of many sciences is necessary so that appropriate questions may be put to specialists, and so that some sense of integration of the various disciplines can evolve. In practice this means that ground-water modelers have a great need to become involved in continuing education efforts. Managers should expect and encourage this because the benefits to be gained are tremendous, and the costs of not doing so may be equally large.

An ability to communicate effectively with management is essential also. Just as is the case with statistical analyses, an ill-posed problem yields answers to the wrong questions ("I know you heard what I said, but did you understand what I meant?"). Some of the questions managers should ask technical support staff, and vice versa, to ensure that the solution being developed is appropriate to the actual problems are listed in Table 8-6 through 8-8. Table 8-6 consists of "screening level" questions. Table 8-7 addresses the need for correct conceptualizations, and Table 8-8 is comprised of sociopolitical concerns.

On another level of communication, managers should appreciate how difficult it will be to explain the results of complicated models to non-technical audiences such as in public meetings and courts of law. Many scientists find it a trying exercise to discuss the details of their labors without the convenience of the jargon of their discipline. Some of the more useful means of overcoming this limitation involve the production of highly simplified audio-visual aids, but this necessarily involves a great deal of work. The efforts which will be required to sell purportedly self-explanatory graphs from computer simulations may rival the efforts spent on producing the simulations initially.

Table 8-6 Screening-Level Questions for Mathematical Modeling Efforts

General Problem Definition

- What are the key issues; quantity, quality, or both?
- What are the controlling geologic, hydrologic, chemical, and biological features?
- Are there reliable data (proper field scale, quality controlled, etc.) for preliminary assessments?
- Do we have the model(s) needed for appropriate simulations?

Initial Responses Needed

- What is the time-frame for action (imminent or long-term)?
- What actions, if taken now, can significantly delay the projected impacts?
- To what degree can mathematical simulations yield meaningful results for the action alternatives, given available data?
- What other techniques or information (generic models, past experience, etc.) would be useful for initial estimates?

Strategies for Further Study

- Are the critical data gaps identified; if not, how well can simulations determine the specific data needs?
- What are the trade-offs between additional data and increased certainty of the simulations?
- How much additional manpower and resources are necessary for further modeling efforts?
- How long will it take to produce useful simulations, including quality control and error-estimation efforts?

Table 8-7 Conceptualization Questions for Mathematical Modeling Efforts

Assumptions and Limitations

- What are the assumptions made, and do they cast doubt on the model's projections for this problem?
- What are the model's limitations regarding the natural processes controlling the problem; can the full spectrum of probable conditions be addressed?
- How far in space and time can the results of the model simulations be extrapolated?
- Where are the weak spots in the application, and can these be further minimized or eliminated?

Input Parameters and Boundary Conditions

- How reliable are the estimates of the input parameters; are they quantified within accepted statistical bounds?
- What are the boundary conditions, and why are they appropriate to this problem?
- Have the initial conditions with which the model is calibrated been checked for accuracy and internal consistency?
- Are the spatial grid design(s) and time-steps of the model optimized for this problem?

Quality Control and Error Estimation

- Have these models been mathematically validated against other solutions to this kind of problem?
- Has anyone field verified these models before, by direct applications or simulation of controlled experiments?
- How do these models compare with others in terms of computational efficiency, and ease of use or modification?
- What special measures are being taken to estimate the overall errors of the simulations?

Table 8-8 Sociopolitical Questions for Mathematical Modeling Efforts

Demographic Considerations

- Is there a larger population endangered by the problem than we are able to provide sufficient responses to?
- Is it possible to present the model's results in both nontechnical and technial formats, to reach all audiences?
- What role can modeling play in public information efforts?
- How prepared are we to respond to criticism of the model(s)?

Political Constraints

- Are there nontechnical barriers to using this model, such as "tainted by association" with a controversy elsewhere?
- Do we have the cooperation of all involved parties in obtaining the necessary data and implementing the solution?
- Are similar technical efforts for this problem being undertaken by friend or foe?
- Can the results of the model simulations be turned against us; are the results ambiguous or equivocal?

Legal Concerns

- Will the present schedule allow all regulatory requirements to be met in a timely manner?
- If we are dependent on others for key inputs to the model(s), how do we recoup losses stemming from their nonperformance?
- What liabilities are incurred for projections which later turn out to be misinterpretations originating in the model?
- Do any of the issues relying on the application of the model(s) require the advice of attorneys?

8.5 References

Aller, L., T. Bennett, J.H. Lehr, and R.J. Petty. 1985. DRASTIC: A Standardized System for Evaluating Ground Water Pollution Potential Using Hydrogeologic Settings. EPA-600/2-85-018, U.S. Environmental Protection Agency, Robert S. Kerr Environmental Research Laboratory, Ada, OK.

Andersen, P.F., C.R. Faust and J.W. Mercer. 1984. Analysis of Conceptual Designs for Remedial Measures at Lipari Landfill. Ground Water 22(2): 176-190.

Asbeck, E., and Y.Y. Haimes. 1984. The Partitioned Multiobjective Risk Method. Large Scale Systems 13(38).

Bachmat, Y., B. Andrews, D. Holtz, and S. Sebastian. 1978. Utilization of Numerical Groundwater Models for Water Resource Management. EPA-600/ 8-78-012, U.S. Environmental Protection Agency, Robert S. Kerr Environmental Research Laboratory, Ada, OK.

Boonstra, J., and N.A. de Ridder. 1981. Numerical Modeling of Groundwater Basins. ILRI Publication No. 29, International Institute for Land Reclamation and Improvement, Wageningen, The Netherlands.

Brown, J. 1986. 1986 Environmental Software Review. Pollution Engineering 18(1):18-28.

Daubert, J.T., and R.A. Young. 1982. Ground-Water Development in Western River Basins: Large Economic Gains with Unseen Costs. Ground Water 20(1):80-86.

Domenico, P.A. 1972. Concepts and Models in Groundwater Hydrology. McGrawHill, New York, NY

Faust, C.R., L.R. Silka, and J.W. Mercer. 1981. Computer Modeling and Ground-Water Protection. Ground Water 19(4):362-365.

Freeze, R.A., and J.A. Cherry. 1979. Groundwater. Prentice Hall, Englewood Cliffs, NJ.

Graves, B. 1986. Ground Water Software - Trimming the Confusion. Ground Water Monitoring Review 6(1):44-53.

Haimes, Y.Y., ed. 1981. Risk/Benefit Analysis in Water Resources Planning and Management. Plenum Publishers, New York, NY.

Haimes, Y.Y., and W.A. Hall. 1974. Multiobjectives in Water Resources Systems Analysis: The Surrogate Worth Trade-off Method. Water Resources Research 10:615-624.

Heath, R.C. 1982. Classification of Ground-Water Systems of the United States. Ground Water 20(4):393-401.

Holcolm Research Institute. 1976. Environmental Modeling and Decision Making. Praeger Publishers, New York, NY.

Huyakorn, P.S., A.G. Kretschek, R.W. Broome, J.W. Mercer, and B.H. Lester. 1984. Testing and Validation of Models for Simulating Solute Transport in Ground Water: Development, Evaluation, and Comparison of Benchmark Techniques. IGWMC Report No. GWMI 84-13, International Ground Water Modeling Center, Holcolm Research Institute, Butler University.

International Ground Water Modeling Center. 1986. Price List of Publications and Services Available from IGWMC (January 1986). International Ground Water Modeling Center, Holcolm Research Institute, Butler University.

Javendel, I., C. Doughty, and C.F. Tsang. 1984. Groundwater Transport: Handbook of Mathematical Models. AGU Water Resources Monograph No. 10., American Geophysical Union, Washington, DC.

Kazmann, R.G. 1972. Modern Hydrology. 2nd ed. Harper & Row Publishers, New York, NY.

Khan, I.A. 1986a. Inverse Problem in Ground Water: Model Development. Ground Water 24(1):32-38.

Khan, I.A. 1986b. Inverse Problem in Ground Water: Model Application. Ground Water 24(1):39-48.

Krabbenhoft, D.P., and M.P. Anderson. 1986. Use of a Numerical Ground-Water Flow Model for Hypothesis Testing. Ground Water 24(1):49-55.

Lehr, J.H. 1986. The Myth of TVA. Ground Water 24(1):2-3.

Lohman, S.W. 1972. Ground-Water Hydraulics. U.S. Geological Survey Professional Paper 708, U.S. Government Printing Office, Washington, D C.

Mercer, J.W., and C.R. Faust. 1981. Ground-Water Modeling. National Water Well Association, Worthington, OH.

Miller, S. 1980. Cost-Benefit Analyses. Environmental Science and Technology 14(12):1415-1417.

Molz, F.J., O. Guven, and J.G. Melville. 1983. An Examination of Scale Dependent Dispersion Coefficients. Ground Water 21(6):715-725.

Moses, C.O., and J.S. Herman. 1986. Computer Notes - WATIN - A Computer Program for Generating Input Files for WATEQF. Ground Water 24(1):83-89.

Puri, S. 1984. Aquifer Studies Using Flow Simulations. Ground Water 22(5):538-543.

Remson, I., G.M. Hornberger, and F.J. Molz. 1971. Numerical Methods in Subsurface Hydrology. John Wiley and Sons, New York, NY.

Ross, B., J.W. Mercer, S.D. Thomas, and B.H. Lester. 1982. Benchmark Problems for Repository Siting Models. U.S. NRC Publication No. NUREG/CR-3097, U.S. Nuclear Regulatory Commission, Washington, DC.

Shelton, M.L. 1982. Ground-Water Management in Basalts. Ground Water 20(1):86-93.

Srinivasan, P. 1984. PIG - A Graphic Interactive Preprocessor for Ground-Water Models. IGWMC Report No. GWMI 84-15, International Ground Water Modeling Center, Holcolm Research Institute, Butler University.

Strecker, E.W., and W. Chu. 1986. Parameter Identification of a Ground-Water Contaminant Transport Model. Ground Water 24(1):56-62.

Todd, D.K. 1980. Groundwater Hydrology. 2nd ed. John Wiley and Sons. New York.

U.S. Congress. 1982. Use of Models for Water Resources Management, Planning, and Policy. Office of Technology Assessment U.S. Government Printing Office, Washington, DC.

U.S. Environmental Protection Agency. 1984. Ground-Water Protection Strategy. Office of Ground-Water Protection, Washington, DC.

van der Heijde, P.K.M. 1985. The Role of Modeling in Development of Ground-Water Protection Policies. Ground Water Modeling Newsletter 4(2).

van der Heijde, P.K.M., Y. Bachmat, J. Bredehoeft, B. Andrews, D. Holtz, and S. Sebastian. 1985. Groundwater Management: The Use of Numerical Models. 2nd ed. AGU Water Resources Monograph No. 5, American Geophysical Union, Washington, DC.

van der Heijde, P.K.M. 1984a. Availability and Applicability of Numerical Models for Ground Water Resources Management. IGWMC Report No. GWMI 84-14, International Ground Water Modeling Center, Holcolm Research Institute, Butler University.

van der Heijde, P.K.M. 1984b. Utilization of Models as Analytic Tools for Groundwater Management. IGWMC Report No. GWMI 84-19, International Ground Water Modeling Center, Holcolm Research Institute, Butler University.

van der Heijde, P.K.M., and P. Srinivasan. 1983. Aspects of the Use of Graphic Techniques in Ground Water Modeling. IGWMC Report No. GWMI 83-11, International Ground Water Modeling Center, Holcolm Research Institute, Butler University.

Wang, H.F., and M.P. Anderson. 1982. Introduction to Groundwater Modeling: Finite Difference and Finite Element Methods. W.H. Freeman and Company, San Francisco, CA.

Warren, J., H.P. Mapp, D.E. Ray, D.D. Kletke, and C. Wang. 1982. Economics of Declining Water Supplies of the Ogallala Aquifer. Ground Water 20(1):73-80.

White, J.A., M.H. Agee, and K.E. Case. 1984. Principles of Engineering Economic Analysis. 2nd ed. John Wiley and Sons, New York, NY.

9. Basic Geology

Geology, the study of the earth, includes the investigation of earth materials, the processes that act on these materials, the products that are formed, the history of the earth, and the origin and development of life forms. There are several subfields of geology. Physical geology deals with all aspects of the earth and includes most earth science specialties. Historical geology is the study of the origin of the earth, continents and ocean basins, and life forms, while economic geology is an applied approach involved in the search and exploitation of mineral resources, such as metallic ores, fuels, and water. Structural geology deals with the various structures of the earth and the forces that produce them. Geophysics is the examination of the physical properties of the earth and includes the study of earthquakes and methods to evaluate the subsurface.

From the perspective of ground water, all of the subfields of geology are used, some more than others. Probably the most difficult concept to comprehend by individuals with little or no geological training is the complexity of the subsurface, which is hidden from view and, at least presently, cannot be adequately sampled. In geologic or hydrogeologic studies, it is best to always keep in mind a fundamental principle of geology. The present is the key to the past. That is, the processes occurring today are the same processes that occurred throughout the geologic past -- only the magnitude changes from one time to the next.

Consider, for example, the channel and flood plain of a modern day river or stream. The watercourse constantly meanders from one side of the flood plain to the other, eroding the banks and carrying the sediments farther downstream. The channel changes in size and position, giving rise to deposits of differing grain size and, perhaps, composition. The changes may be abrupt or gradual, both vertically and horizontally, as is evident from an examination of the walls of a gravel pit or the bluffs along a river. Because of the dynamic nature of streams and deltas, one will find a geologic situation that is perplexing not only to the individual involved in a ground-water investigation, but to the geologist as well. Each change in grain size will cause differences in permeability and ground-water velocity, while changes in mineral composition can lead to variances in water quality. At the other end of the depositional spectrum are deposits collected in lakes, seas, and the oceans, which are likely to be much more widespread and uniform in thickness, grain size, and composition.

As one walks from the sandy beach of a lake into the water, the sediments become finer and more widely distributed as the action of waves and currents sort the material brought into the lake by streams. Farther from shore, the bottom of the lake may consist of mud, which is a mixture of silt, clay, and organic matter. In some situations the earthy mud grades laterally into a lime ooze or mud. In geologic time, these sediments become lithified or changed into rock -- the sand to sandstone, the mud to shale, and the limey mud to limestone. It is important to note, however, that the sand, mud, and lime were all deposited at the same time, although with lithification each sediment type produced a different sedimentary rock.

9.1 Geologic Maps and Cross-Sections

Geologists use a number of techniques to graphically represent surface and subsurface conditions. These include surficial geologic maps, columnar sections, cross-sections of the subsurface, maps that show the configuration of the surface of a geologic unit, such as the bedrock beneath glacial deposits, maps that indicate the thickness or grain size of a particular unit, a variety of contour maps, and a whole host of others.

A surficial geologic map depicts the geographic extent of formations and their structure. Columnar sections describe the vertical distribution of rock units, their lithology, and thickness. Geologic cross-sections attempt to illustrate the subsurface distribution of rock units between points of control, such as outcrops or well bores. An isopach map shows the geographic range in thickness of a unit; these maps are based largely or entirely on well logs.

Whatever the graphic techniques, it must be remembered that these maps represent only best guesses and may be based on scanty data. In reality, they are interpretations, presumably based on scientific thought, a knowledge of depositional characteristics of rock units, and a data base that provides some control. They are not exact because the features they attempt to show are complex, nearly always hidden from view, and difficult to sample.

All things considered, graphical representations are exceedingly useful, if not essential, to subsurface studies. On the other hand, a particular drawing that is prepared for one purpose may not be adequate for another even though the same units are involved. This is largely due to scale and generalizations.

A combination topographic and geologic map of a glaciated area is shown in Figure 9-1. The upland area is mantled by glacial till (Qgm) and the surficial material covering the relatively flat flood plain has been mapped as alluvium (Qal). Beneath the alluvial cover are other deposits of glacial origin that consist of glacial till, outwash, and local lake deposits. A water well drillers log of a boring at point A states, "This well is just like all of the others in the valley," and that the upper 70 feet of the valley fill consists of a "mixture of clay, sand, silt, and boulders." This is underlain by 30 feet of "water sand," which is the aquifer. The aquifer overlies "slate, jingle rock, and coal." The terminology may be quaint, but it is nonetheless a vocabulary that must be interpreted. Examination of the local geology, as evidenced by strata that crop out along the hillsides, indicate that the bedrock or older material that underlies the glacial drift consists of shale, sandstone cemented by calcite, and lignite, which is an immature coal. These are the geologic terms, at least in this area, for "slate, jingle rock, and coal," respectively.

For generalized purposes, it is possible to use the driller's log to construct a cross section across or along the stream valley (Figure 9-2). In this case, one would assume for the sake of simplicity the existence of an aquifer that is rather uniform in composition and thickness. A second generation cross section, shown in Figure 9-3, is based on several bore hole logs described by a geologist who collected samples as the holes were being drilled. Notice in this figure that the subsurface appears to be much more complex, consisting of several isolated permeable units that are incorporated within the fine grained glacial deposits that fill the valley. In addition, the aquifer does not consist of a uniform thickness of sand, but rather a unit that ranges from 30 to 105 feet in thickness and from sand to a mixture of sand and gravel. The water-bearing characteristics of these units are all different. This cross section too is quite generalized, which becomes evident as one examines an actual log of one of the holes (Table 9-1).

Figure 9-1 **Generalized geologic map of a glaciated area along the Souris River Valley in central North Dakota.**

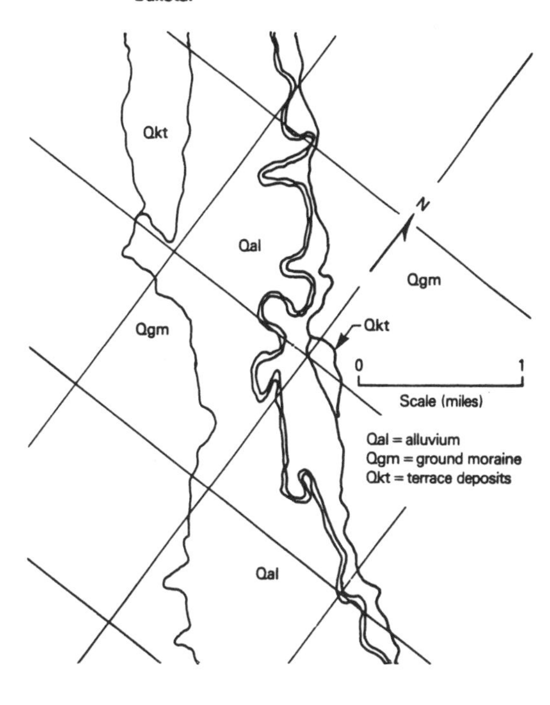

Figure 9-2 **Generalized geologic cross section of the Souris River Valley based on driller's log.**

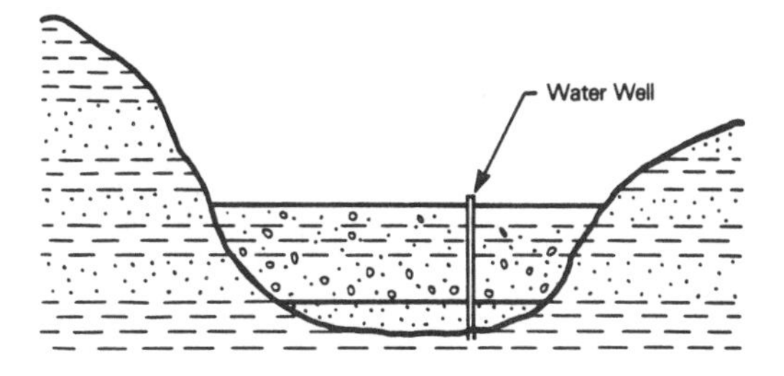

Table 9-1 Geologist's Log of a Test Hole, Souris River Valley, North Dakota

Sample Description and Drilling Condition	Depth (ft)
Topsoil, silty clay, black	0-1
Clay, silty, yellow brown, poorly consolidated	1-5
Clay, silty, yellow grey, soft, moderately compacted	5-10
Clay, silty, as above, silty layers, soft	10-15
Silty, clayey, gray, soft, uniform drilling	15-20
Clay, silty, some fine to medium sand, gray	20-30
Clay, gray to black, soft, very tight	30-40
Clay, as above, gravelly near top	40-50
Clay, as above, no gravel	50-60
Clay, as above, very silty in spots, gray	60-70
Clay and silt, very easy drilling	70-80
Clay, as above to gravel, fine to coarse, sandy, thin clay layers, taking lots of water	80-90
Gravel, as above, some clay near top, very rough drilling, mixed three bags of mud, lots of lignite chips	90-100
Gravel, as above, cobbles and boulders	100-120
Gravel, as above, to sand, fine to coarse, lots of lignite, much easier drilling	120-130
Clay, gravelly and rocky, rough drilling, poor sample return	130-140
Sandy clay, gravelly and rocky, rough drilling, poor sample return (till)	140-150
Sandy clay, as above, poor sample return	150-160
Clay, sandy, gray, soft, plastic, noncalcareous	160-170
Clay, sandy, as above, tight, uniform drilling	170-180
Clay, as above, much less sand, gray, soft, tight, plastic	180-190
Clay, as above, no sand, good sample return	190-200
Clay, as above	200-210

In addition to showing more accurately the composition of the subsurface, logs can also provide some interesting clues concerning the relative permeabilities of the water-bearing units. Referring to Table 9-2, the depth interval ranging from 62 to 92 feet, a generalized log of well 1 contains the remark "losing water" and in well 5, at depths of 80 to 120 feet, is the notation, "3 bags of bentonite." In the first case this means that the material being penetrated by the drill bit from 62 to 92 feet was more permeable than the annulus of the cutting-filled bore hole. The water, pumped down the hole through the drill pipe to remove the cuttings, found it easier to move out into the formation than to flow back up the hole. The remark is a good indication of a permeability that is higher than that present in those sections where water was not being lost.

In the case of well 5, the material was so permeable that much of the drilling fluid was moving into the formation and there was no return of the cuttings. To regain circulation, bentonite, or to use the field term, mud, was added to the drilling fluid to seal the permeable zone. Even though the geologist described

Table 9-2 Generalized Geologic Logs of Five Test Holes, Souris River Valley, North Dakota

Material	Depth (ft)
Test Hole 1	
Fill	0-3
Silt, olive-gray	3-14
Sand, fine-medium	14-21
Silt, sandy, gray	21-25
Clay, gray	25-29
Sand, fine-coarse	29-47
Clay, gray	47-62
Gravel, fine to coarse, losing water	62-92
Silt, sandy, gray	92-100
Observation well depth 80 feet	
Test Hole 2	
Fill	0-2
Clay, silty and sandy, gray	2-17
Clay, very sandy, gray	17-19
Sand, fine-medium	19-60
Sand, fine-coarse with gravel	60-80
Gravel, coarse, 2 bags bentonite and bran	80-100
Observation well depth 88 feet	
Test Hole 3	
Silt, yellow	0-5
Clay, silty, black	5-15
Sand, fine to coarse	15-29
Clay, silty, gray	29-65
Sand, medium-coarse, some gravel	65-69
Gravel, sandy, taking water	69-88
Sand, fine to medium, abundant chips of lignite	88-170
Observation well depth 84 feet	
Test Hole 4	
Fill	0-5
Silt, brown	5-12
Sand, fine-medium	12-28
Clay, silty and sandy, gray	28-37
Sand, fine	37-49
Clay, dark gray	49-55
Sand, fine	55-61
Clay, sandy, gray	61-66
Sand, fine-coarse, some gravel	66-103
Silt, gray	103-120
Observation well depth 96 feet	
Test Hole 5	
Clay, silty, brown	0-10
Silt, clayey, gray	10-80
Gravel, fine-coarse, sandy, taking lots of water 3 bags bentonite	80-120
Sand, fine to coarse, gravelly	120-130
Clay, gravelly and rocky (till)	130-150
Sand, fine, Fort Union Group	150-180
Observation well depth 100 feet	

Figure 9-3 Geologic cross section of the Souris River Valley based on detailed logs of test holes.

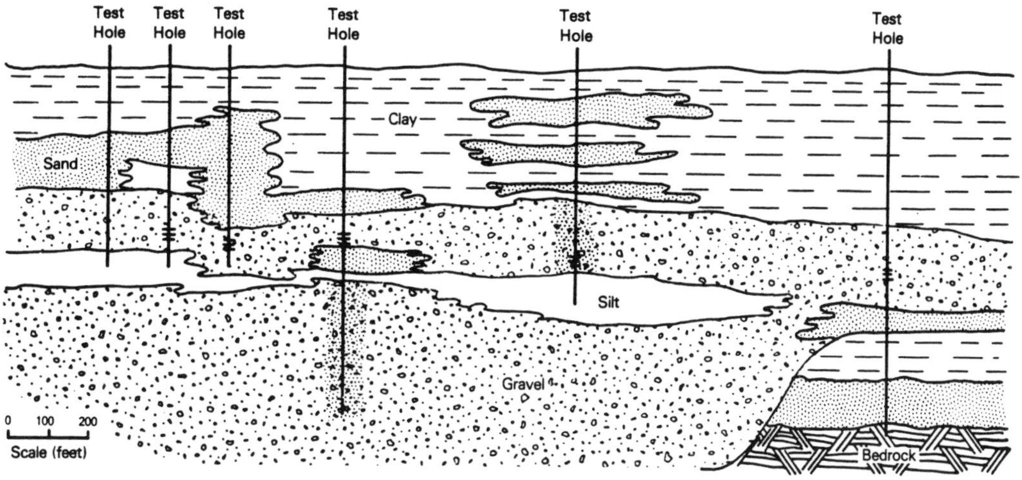

Test Hole
Test Hole
Test Hole
Test Hole
Test Hole
Test Hole
Clay
Sand
Silt
Gravel
Bedrock
0 100 200
Scale (feet)

Figure 9-4 Schematic of general features of the Columbia Plateau region (from Heath, 1984).

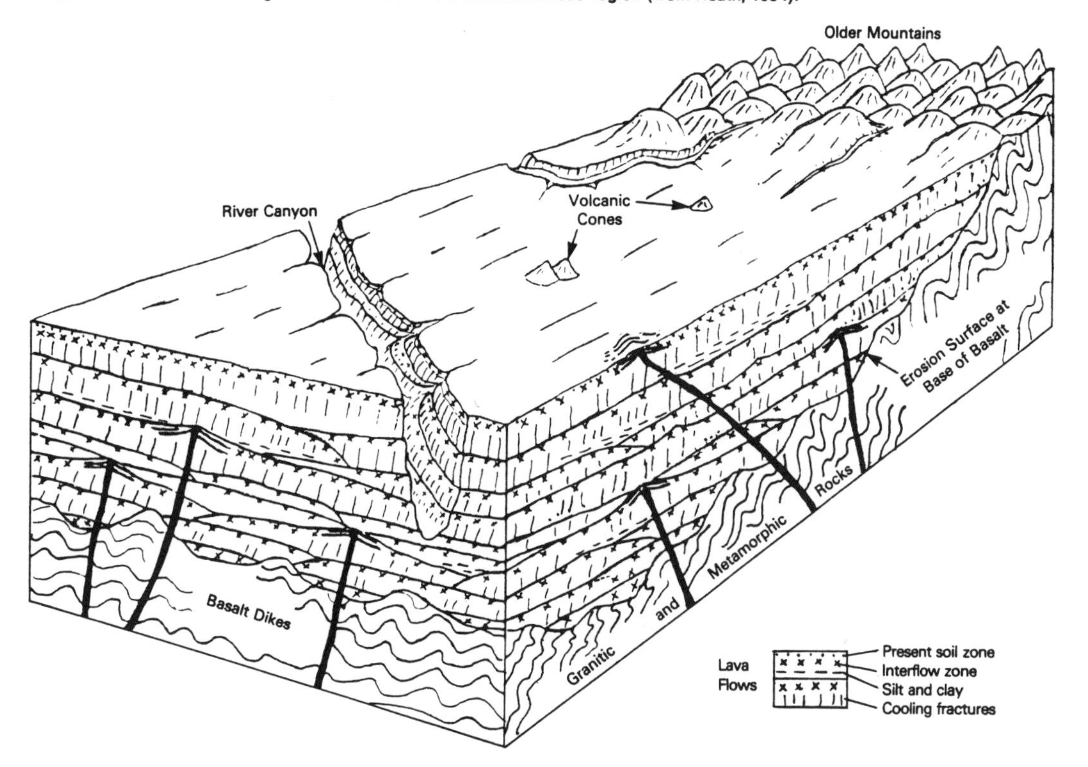

the aquifer materials from both zones similarly, the section in well 5 is more permeable than the one in well 1, which in turn is more permeable than the other coarse-grained units penetrated where there was no fluid loss.

The three most important points to be remembered here are, first, that graphical representations of the surface or subsurface geology are merely guesses of what might actually occur, and even these depend to some extent on the original intended usage. Second, the subsurface is far more complex than is usually anticipated, particularly in regard to unconsolidated deposits. Finally, evaluating the original data such as well logs might lead to a better evaluation of the subsurface, an evaluation that far surpasses the use of generalized lithologic logs alone.

9.2 Ground Water in Igneous and Metamorphic Rocks

Nearly all of the porosity and permeability of igneous and metamorphic rocks is the result of secondary openings such as fractures, faults, and the dissolution of certain minerals. A few notable exceptions include large lava tunnels present in some flows, interflow or coarse sedimentary layers between individual lava flows, and deposits of selected pyroclastic materials (Figure 9-4).

Because the openings in igneous and metamorphic rocks are quite small volumetrically, rocks of this type are poor suppliers of ground water. The supplies that are available commonly drain rapidly after a period of recharge by infiltration of precipitation. In addition they are subject to contamination from the surface where these rocks crop out.

The width, spacing, and depth of fractures range widely, as do their origins. the surface to 0.003 inches at a depth of 200 feet, while spacing increased from 5 to 10 feet near the surface to 15 to 35 feet at depth in the Front Range of the Rocky Mountains. He also reported that porosity decreased from below 300 feet or so, but there are many recorded exceptions. Exfoliation fractures in the crystalline rocks of the Piedmont near Atlanta, Georgia range from 1 to 8 inches in width (Cressler and others, 1983).

The difficulty of evaluating water and contaminant movement in fractured rocks is that the actual direction of movement may not be in the direction of decreasing head, but rather in some different though related direction. The problem is further compounded by the difficulty in locating the fractures. Because of these characteristics, evaluation of water availability, direction of movement, and velocity is exceedingly difficult. As a general rule, at least in the eastern part of the United States, LeGrand pointed out that well yields, and therefore fractures, permeability, and porosity, are greater in valleys and broad ravines than on flat uplands, which in turn are higher than on hill slopes and hill crests (Figure 9-5). The reason this occurs in parts of North Carolina is because stream valleys have formed along fracture zones.

Unless some special circumstance exists, such as where rocks crop out at the surface, water obtained from igneous and metamorphic rocks is nearly always of excellent chemical quality. Dissolved solids are commonly less than 100 mg/l. Water from metamorphosed carbonate rocks may have moderate to high concentrations of hardness.

9.3 Ground Water in Sedimentary Rocks

Usable supplies of ground water can be obtained from all types of sedimentary rocks, but the fine grained strata such as shale and siltstone may only provide a few gallons per day and even this can be highly mineralized. Even though fine grained rocks may have relatively high porosities, the primary permeability is very low. On the other hand, shale is likely to contain a great number of joints that are both closely spaced and extend to considerable depths. Therefore, rather than being impermeable, as many individuals imply, they can be quite transmissive. This is of considerable importance in waste disposal schemes because insufficient attention might be paid during engineering design to the potential for flow through fractures. In addition, the leachate that is formed as water infiltrates through waste might be small in quantity but highly mineralized. Because of the low bulk permeability, it would be difficult to pump out the contaminated water or even to properly locate monitoring wells.

From another perspective, fine grained sedimentary rocks, owing to their high porosity, can store huge quantities of water. Some of this water can be released to adjacent aquifers when a head difference is developed due to pumping. No doubt fine grained confining units provide, on a regional scale, a great deal of water to aquifer systems. The porosity, however, decreases with depth because of compaction brought about by the weight of overlying sediments.

The porosity of sandstones range from less than 1 percent to a maximum of about 30 percent. This is a function of sorting, grain shape, and cementation. Cementation can be variable both in space and time and on outcrops can differ greatly from that in the subsurface.

Fractures also play an important role in the movement of fluids through sandstones and transmissivities may be as much as two orders of magnitude greater in a fractured rock than in an unfractured part of the same geologic formation.

Sandstone units that were deposited in a marine or near-marine environment can be very widespread, covering tens of thousands of square miles, such as the St. Peter sandstone of Cambrian age. Those

Figure 9-5 Schematic of general features of the Piedmont and Blue Ridge region (from Heath, 1984).

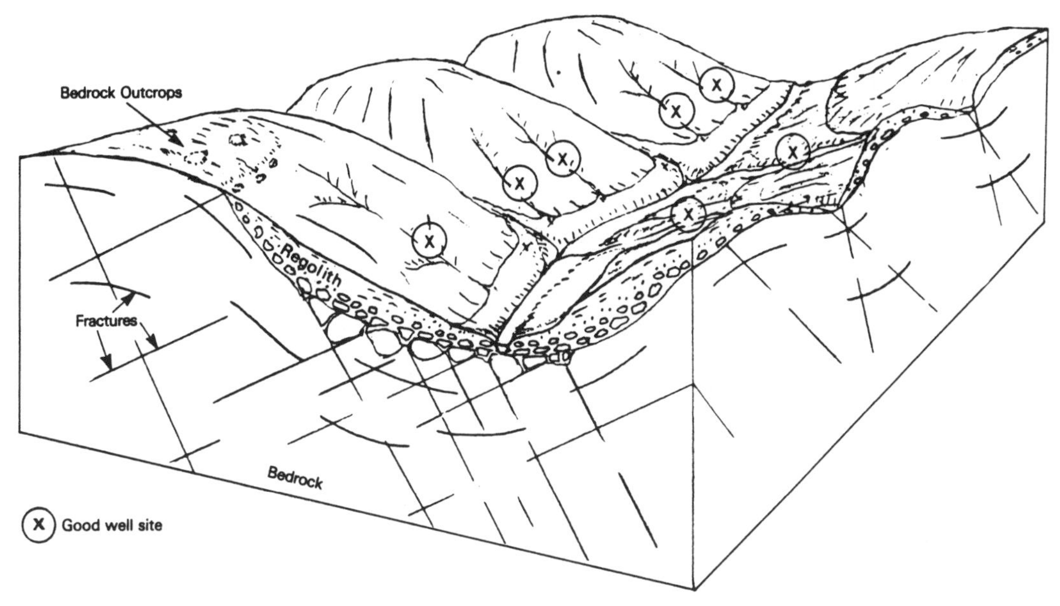

Figure 9-6 Schematic of general features of the Gulf Coastal Plain (from Heath, 1984).

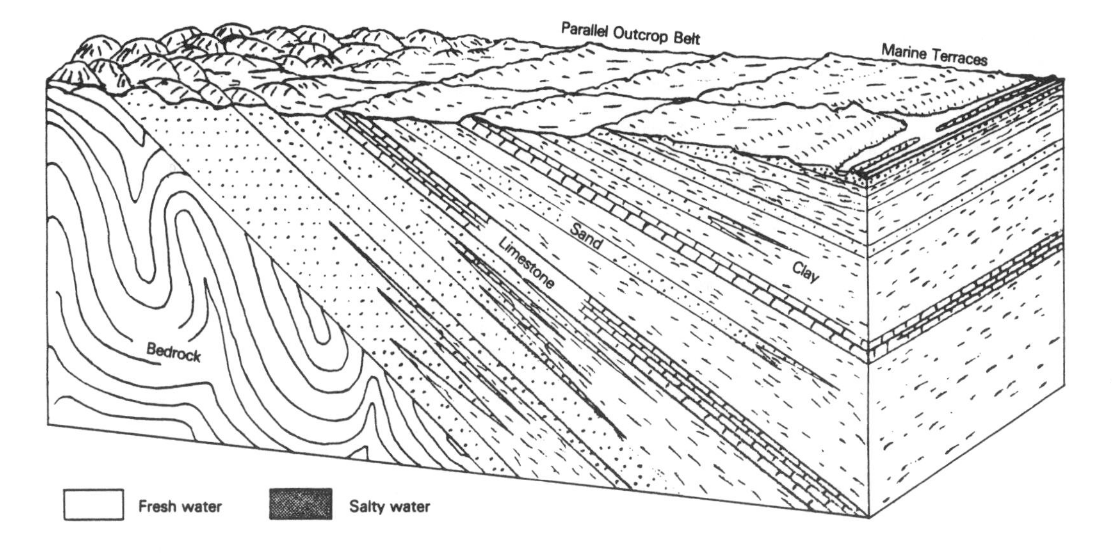

representing ancient alluvial channel fills, deltas, and related environments of deposition are more likely to be discontinuous and erratic in thickness. Individual units are exceedingly difficult to trace in the subsurface. Regional ground-water flow and storage may be strongly influenced by the geologic structure (Figures 9-6 and 9-7).

Carbonate rocks are formed in many different environments and the original porosity and permeability are modified rapidly after burial. Some special carbonate rocks, such as coquina and some breccias, may remain very porous and permeable, but these are the exception.

It is the presence of fractures and other secondary openings that develop high yielding carbonate aquifers. One important aspect is the change from calcite to dolomite ($CaMgCO_3$), which results in a volumetric reduction of 13 percent and the creation of considerable pore space. Of particular importance and also concern in many of the carbonate regions of the world is the dissolution of carbonates along fractures and bedding planes by circulating ground water. This is the manner in which caves and sinkholes are formed. As dissolution progresses upward in a cave, the overlying rocks may collapse to form a sinkhole that contains water if the cavity extends below the water table. Regions in which there has been extensive dissolution of carbonates leading to the formation of caves, underground rivers, and sinkholes, are called karst. Notable examples include parts of Missouri, Indiana, and Kentucky (Figure 9-8).

Karst areas are particularly troublesome even though they can provide large quantities of water to wells and springs because they are easily contaminated, it is often difficult to trace the contaminant, the water can flow very rapidly, and there is no filtering action to degrade the waste. Not uncommonly a well owner may be unaware that he is consuming unsafe water.

9.4 Ground Water in Unconsolidated Sediments

Unconsolidated sediments accumulate in many different environments, all of which leave their mark on the characteristics of the deposit. Some are thick and areally extensive, as the alluvial fill in the Basin and Range Province; others are exceedingly long and narrow, such as the alluvial deposits along streams and rivers; and others may cover only a few hundred square feet, for example, some glacial forms. In addition to serving as major aquifers, unconsolidated sediments are also important as sources of raw materials for construction.

Although closely related to sorting, the porosities of unconsolidated materials range from less than 1 to more than 90 percent, the latter representing uncompacted mud. Permeabilities also range widely. Cementing of some type and degree is probably universal, but not obvious, with silt and clay being the predominant form.

Most unconsolidated sediments owe their emplacement to running water and consequently, some sorting is expected. On the other hand, water as an agent of transportation will vary in both volume and velocity, which is climate dependent, and this will leave an imprint on the sediments. It is to be expected that stream related material, which most unconsolidated material is, will be variable in extent, thickness, and grain size (Figure 9-9). Other than this, one can draw no general guidelines; therefore, it is essential to develop some knowledge of the resulting stratigraphy that is characteristic of the most common environments of deposition. The water-bearing properties of glacial drift, of course, are exceedingly variable, but stratified drift is more uniform and better sorted than glacial till (Figure 9-10).

9.5 Relationship Between Geology, Climate, and Ground-Water Quality

The availability of ground-water supplies and their chemical quality are closely related to precipitation. As a general rule, the least mineralized water, both in streams and underground, occurs in areas of the greatest amount of rainfall. Inland, precipitation decreases, water supplies diminish, and quality deteriorates. Water-bearing rocks exert a strong influence on ground-water quality and thus, the solubility of the rocks may override the role of precipitation.

Where precipitation exceeds 40 inches per year, shallow ground water usually contains less than 500 mg/l and commonly less than 250 mg/l of dissolved solids. Where precipitation ranges between 20 and 40 inches, dissolved solids may range between 400 and 1,000 mg/l, and in drier regions dissolved solids commonly exceed 1,000 mg/l.

The dissolved solids concentration of ground water increases toward the interior of the continent. The increase is closely related to precipitation and the solubility of the aquifer framework. The least mineralized ground water is found in a broad belt that extends southward from the New England States along the Atlantic Coast to Florida, and then continues to parallel much of the Gulf Coast. Similarly, along the Pacific Coast from Washington to central California the mineral content is also very low. Throughout this belt, dissolved solids concentrations are generally less than 250 mg/l and commonly less than 100 mg/l (Figure 9-11).

The Appalachian region consists of a sequence of strata that range from nearly flat-lying to complexly folded and faulted. Likewise, ground-water quality in this region is also highly variable, being generally

Figure 9-7 Schematic of general features of the Colorado Plateau and Wyoming Basin region (from Heath, 1984).

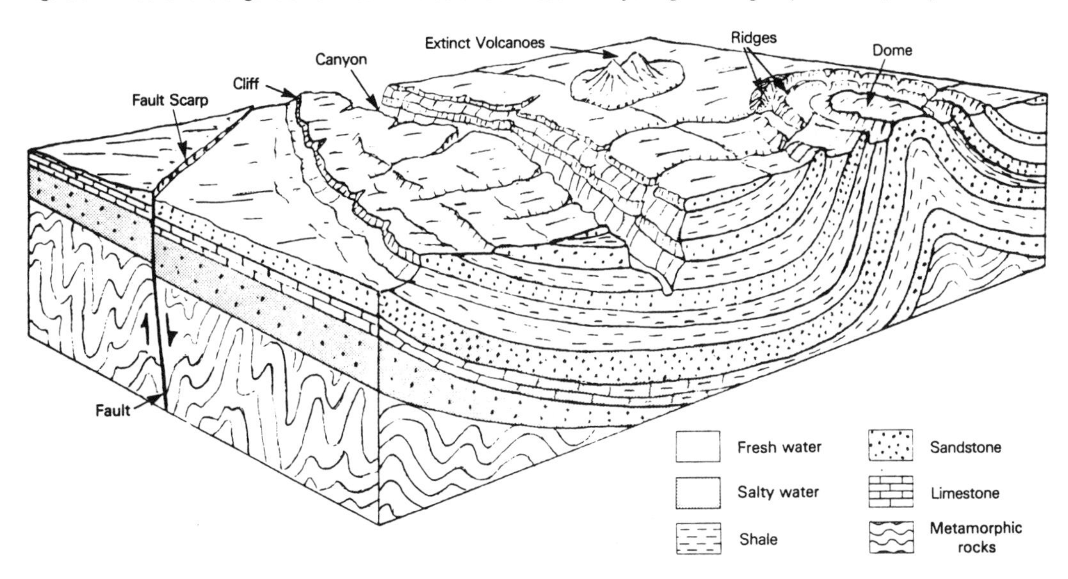

Figure 9-8 Schematic of general features of the Nonglaciated Central region (from Heath, 1984).

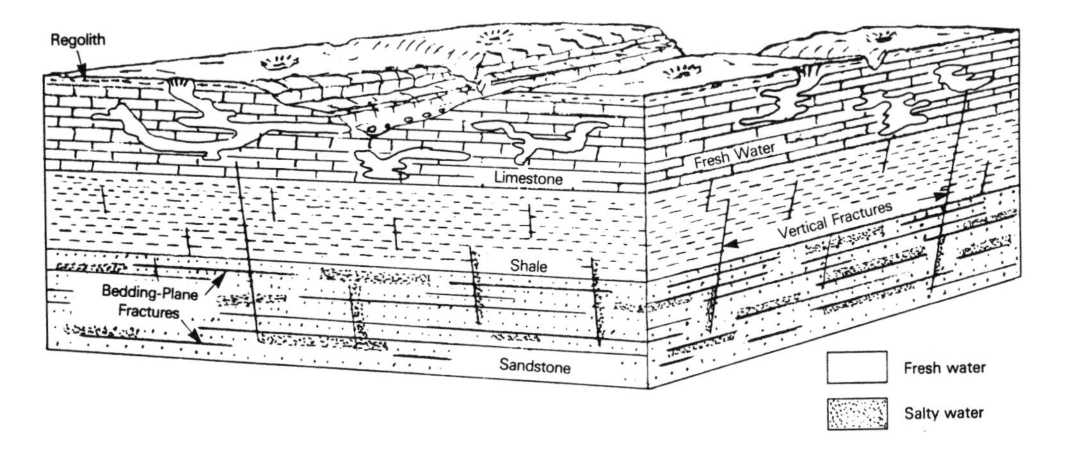

Figure 9-9 Schematic of general features of the High Plains region (from Heath, 1984).

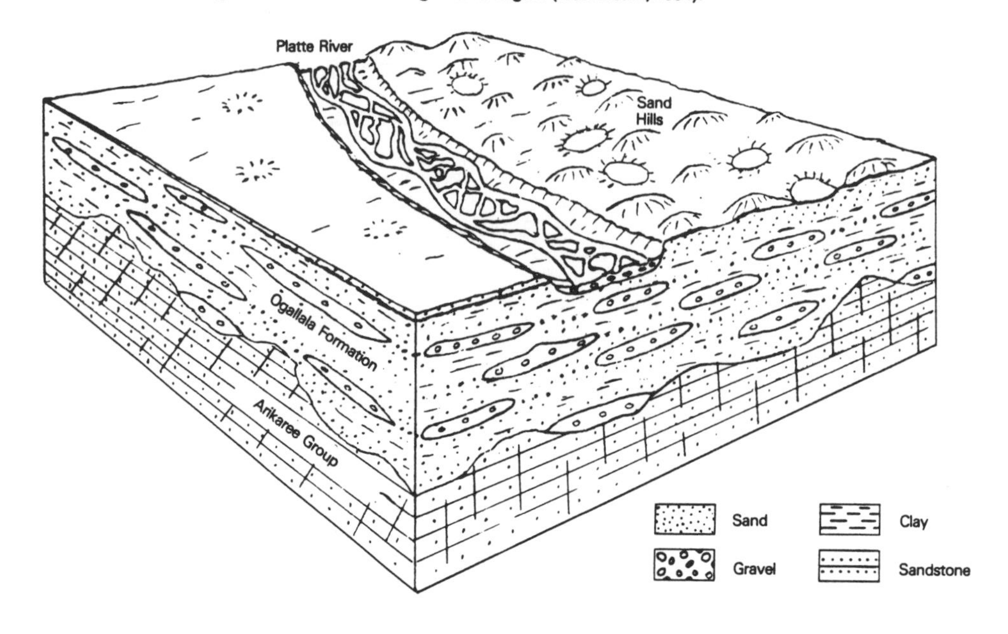

Figure 9-10 Schematic of general features of the Glaciated Central region (from Heath, 1984).

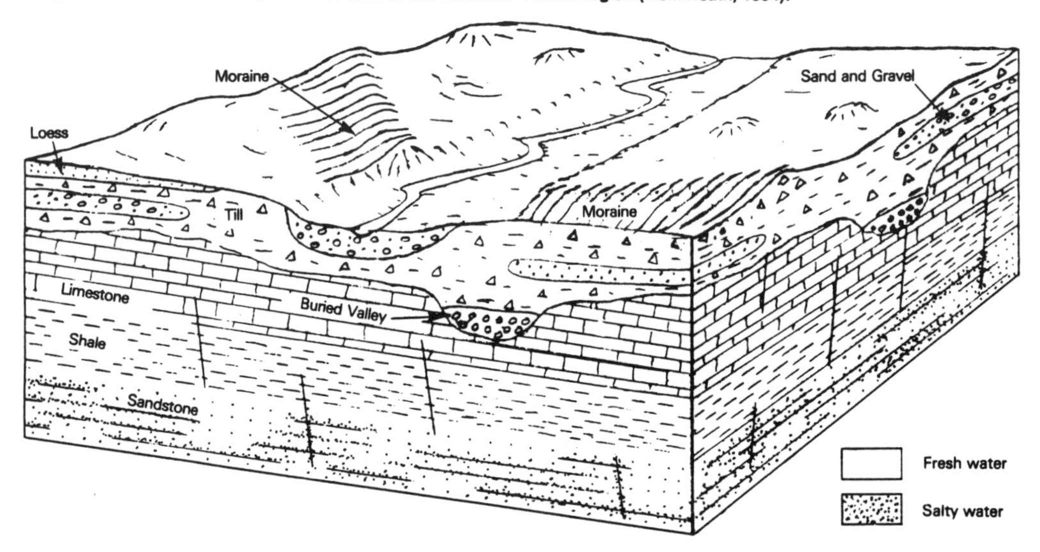

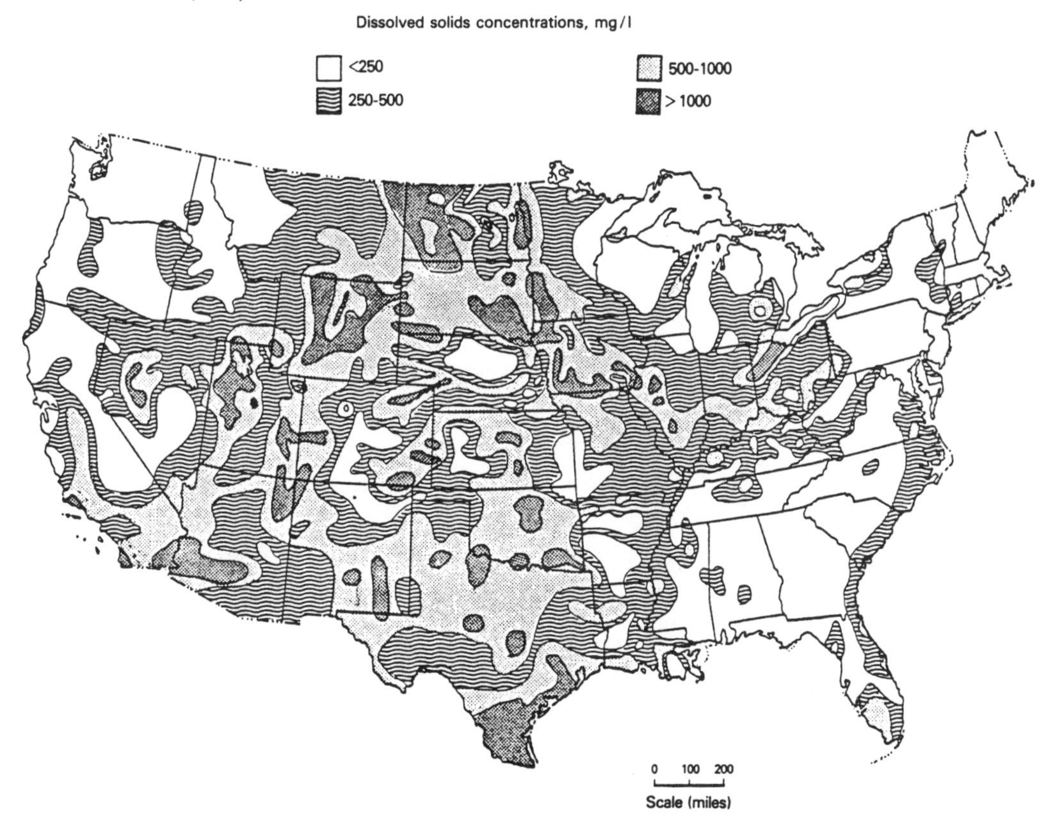

Figure 9-11 Dissolved solids concentrations in ground water used for drinking in the United States (from Pettyjohn and others, 1979).

harder and containing more dissolved minerals than does water along the coastal belt. Much of the difference in quality, however, is related to the abundance of carbonate aquifers which provide waters rich in calcium and magnesium.

Westward from the Appalachian Mountains to about the position of the 20 inch precipitation line (eastern North Dakota to Texas), dissolved solids in ground water progressively increase. They are generally less than 1000 mg/l and are most commonly in the 250 to 750 mg/l range. The water is moderately to very hard, and in some areas concentrations of sulfate and chloride are excessive.

From the 20 inch precipitation line westward to the northern Rocky Mountains, dissolved solids are in the 500 to 1500 mg/l range. Much of the water from glacial drift and bedrock formations is very hard and contains significant concentrations of calcium sulfate. Other bedrock formations may contain soft sodium bicarbonate, sodium sulfate, or sodium chloride water. Throughout much of the Rocky Mountains, ground-water quality is variable, although the dissolved solids concentrations commonly range between 250 and 750 mg/l. Stretching southward from Washington to southern California, Arizona, and New Mexico is a vast desert region. Here the difference in quality is wide and dissolved solids generally exceed 750 mg/l. In the central parts of some desert basins the ground water is highly mineralized, but along the mountain flanks the mineral content may be quite low.

Extremely hard water is found over much of the interior lowlands, Great Plains, Colorado Plateau, and Great Basin. Isolated areas of high hardness are present in northwestern New York, eastern North Carolina, the southern tip of Florida, northern Ohio, and parts of southern California. In general, the hardness is of the carbonate type.

On a regional level, chloride does not appear to be a significant problem, although it is troublesome locally due largely to industrial activities, the intrusion of seawater caused by overpumping coastal aquifers, or interaquifer leakage related to pressure declines brought about by withdrawals.

In many locations, sulfate levels exceed the Federal recommended limit of 250 mg/l; regionally, sulfate may be a problem only in the Great Plains, eastern Colorado Plateau, Ohio, and Indiana. Iron problems are ubiquitous; concentrations exceeding only 0.3 mg/l will cause staining of clothing and fixtures. Fluoride is abnormally high in several areas, particularly parts of western Texas, Iowa, Illinois, Indiana, Ohio, New Mexico, Wyoming, Utah, Nevada, Kansas, New Hampshire, Arizona, Colorado, North and South Dakota, and Louisiana.

A water-quality problem of growing concern, particularly in irrigated regions, is nitrate, which is derived from fertilizers, sewage, and through natural causes. When consumed by infants less than six months old for a period of time, high nitrate concentrations can cause a disease known as "blue babies." This occurs because the child's blood cannot carry sufficient oxygen; the disease is easily overcome by using low nitrate water for formula preparation. Despite the fact that nitrate concentrations in ground water appear to have been increasing in many areas during the last 30 years or so, the concern may be more imagined than real because there have been no reported incidences of "blue babies" for more than 20 years, at least in the states that comprise the Great Plains.

In summary, the study of geology is complex in detail, but the principles outlined above should be sufficient for a general understanding of the topic, particularly as it relates to ground water. If interested in a more definitive treatment, the reader should refer to the following sections and the references at the end of the chapter.

9.6 Minerals

The earth, some 7,926 miles in diameter at the equator, consists of a core, mantle, and crust, which have been defined by the analysis of seismic or earthquake waves. Only a thin layer of the crust has been examined by humans. It consists of a variety of rocks, each of which is made up of one or more minerals.

Most minerals contain two or more elements, but of all the elements known, eight account for nearly 98 percent of the rocks and minerals:

Oxygen 46%
Silicon 27.72%
Aluminum 8.13%
Iron 5%
Calcium 3.63%
Sodium 2.83%
Potassium 2.59%
Magnesium 2.09%

Without detailed study, it is usually difficult to distinguish one mineral from another, except for a few common varieties such as quartz, pyrite, mica, and some gemstones. On the other hand, it is important to have at least a general understanding of mineralogy because it is the mineral make-up of rocks that, to a large extent, controls the type of water that a rock will contain under natural conditions and the way it will react to contaminants or naturally occurring substances.

The most common rock-forming minerals are relatively few and deserve at least a mention. They can be divided into three broad groups: (1) the carbonates, sulfates, and oxides; (2) the rock-forming silicate minerals; and (3) the common ore minerals.

9.6.1 Carbonates, Sulfates, and Oxides

Calcite, a calcium carbonate ($CaCO_3$), is the major mineral in limestone. The most common mineral is quartz. It is silicon dioxide (SiO_2), hard, and resistant to both chemical and mechanical weathering. In sedimentary rocks it generally occurs as sand-size grains (sandstone) or even finer, such as silt or clay size, and it may also appear as a cement. Because of the low solubility of silicon, silica generally appears in concentrations less than 25 mg/l in water. Limonite is actually a group name for the hydrated ferric oxide minerals ($Fe_2O_3 \bullet H_2O$), which occur so commonly in many types of rocks. Limonite is generally rusty or blackish with a dull, earthy luster and a yellow-brown streak. It is a common weathering product of other iron minerals. Because limonite and other iron-bearing minerals are nearly universal, dissolved iron is a very common constituent in water, causing staining of clothing and plumbing fixtures. Gypsum, a hydrated calcium sulfate ($CaSO_4 \bullet 2H_2O$), occurs as a sedimentary evaporite deposit and as crystals in shale and some clay deposits. Quite soluble, it is the major source of sulfate in ground water.

9.6.2 Rock-Forming Silicates

The most common rock-forming silicate minerals include the feldspars, micas, pyroxenes, amphiboles, and olivine. Except in certain igneous and metamorphic rocks these minerals are quite small and commonly require a microscope for identification. The feldspars are alumino-silicates of potassium or sodium and calcium. Most of the minerals in this group are white, gray, or pink. Upon weathering they turn to clay and release the remaining chemical elements to water. The micas muscovite and biotite are platy alumino-silicate minerals that are common and easily recognized in igneous, metamorphic, and sedimentary rocks. The pyroxenes, a group of silicates of calcium, magnesium, and iron, as well as the amphiboles, which are complex hydrated silicates of calcium, magnesium, iron, and aluminum, are common in most igneous and metamorphic rocks. They appear as small, dark crystals of accessory minerals. Olivine, a magnesium-iron silicate, is generally green or yellow and is common in certain igneous and metamorphic rocks. None of the rock-forming silicate minerals have a major impact on water quality in most situations.

9.6.3 Ores

The three most common ore minerals are galena, sphalerite, and pyrite. Galena, a lead sulfide (PbS), is heavy, brittle, and breaks into cubes. Sphalerite is a zinc sulfide (ZnS) mineral that is brownish, yellowish, or black. It ordinarily occurs with galena and is a major ore of zinc. The iron sulfide pyrite (FeS), which is also called fools' gold, is common in all types of rocks. It is the weathering of this mineral that leads to acid-mine drainage which is nearly universal in coal fields and metal sulfide mining regions.

9.7 Rocks

Three types of rock make up the crust of the earth. Igneous rocks solidified from molten material either within the earth (intrusive) or on or near the surface (extrusive). Metamorphic rocks were originally igneous or sedimentary rocks that were modified by temperature, pressure, and chemically active fluids. Sedimentary rocks are the result of the weathering of preexisting rocks, erosion, and deposition. Geologists have developed elaborate systems of nomenclature and classification of rocks, but these are of little value in hydrogeologic studies and therefore only the most basic descriptions will be presented.

9.7.1 Igneous Rocks

Igneous rocks are classified on the basis of their composition and grain size. Most consist of feldspar and a variety of dark minerals; several others also contain quartz. If the parent molten material cools slowly deep below the surface, minerals will have an opportunity to grow and the rock will be coarse grained. Magma that cools rapidly, such as that derived from volcanic activity, is so fine grained that individual minerals generally cannot be seen even with a hand lens. In some cases the molten material began to cool slowly, allowing some minerals to grow, and then the rate changed dramatically so that the remainder formed a fine groundmass. This texture, consisting of large crystals in a fine grained matrix, is called porphyritic.

Intrusive igneous rocks can only be seen where they have been exposed by erosion. They are concordant if they more or less parallel the bedding of the enclosing rocks and discordant if they cut across the bedding. The largest discordant igneous masses are called batholiths and they occur in the eroded centers of many ancient mountains. Their dimensions are in the range of tens of miles. Batholiths usually consist largely of granite, which is surrounded by metamorphic rocks.

Discordant igneous rocks include dikes ranging in thickness from a few inches to thousands of feet. Many are several miles long. Sills are concordant bodies that have invaded sedimentary rocks along bedding planes. They are relatively thin. Both sills and dikes tend to cool quite rapidly and, resultingly, are fine grained.

Extrusive rocks include lava flows or other types associated with volcanic activity, such as the glassy rock pumice, and the consolidated ash called tuff. These are fine grained or even glassy.

With some exceptions, igneous rocks are dense and have very little porosity or permeability. Most, however, are fractured to some degree and can store and transmit a modest amount of water. Some lava flows are notable exceptions because they contain large diameter tubes or a permeable zone at the top of the flow where gas bubbles migrated to the surface before the rock solidified. These rocks are called scoria.

9.7.2 Metamorphic Rocks

Metamorphism is a process that changes preexisting rocks into new forms because of increases in temperature, pressure, and chemically active fluids. Metamorphism may affect igneous, sedimentary, or other metamorphic rocks. The changes brought about include the formation of new minerals, increase in grain size, and modification of rock structure or texture, all of which depend on the original rock's composition and the intensity of the metamorphism.

Some of the most obvious changes are in texture, which serves as a means of classifying metamorphic rocks into two broad groups, the foliated and nonfoliated rocks. Foliated metamorphic rocks typify regions that have undergone severe deformation, such as mountain ranges. Shale, which consists mainly of silt and clay, is transformed into slate by the change of clay to mica. Mica, being a platy mineral, grows with its long axis perpendicular to the principle direction of stress, forming a preferred orientation. This orientation, such as the development of cleavage in slate, may differ greatly from the original bedding.

With increasing degrees of metamorphism, the grains of mica grow larger so that the rock has a distinct foliation, which is characteristic of the metamorphic rock schist. At even higher grades of metamorphism, the mica may be transformed to a much coarser grained feldspar, producing the strongly banded texture of gneiss.

Nonfoliated rocks include the hornfels and another group formed from rocks that consist mainly of a single mineral. The hornfels occur around an intrusive body and were changed by "baking" during intrusion.

The second group includes marble and quartzite, as well as several other forms. Marble is metamorphosed limestone and quartzite is metamorphosed quartz sandstone.

There are many different types of metamorphic rocks, but from a hydrogeologic viewpoint they normally neither store nor transmit much water and are of only minor importance as aquifers. Their primary permeability is notably small, if it exists at all, and fluids are forced to migrate through secondary openings, such as faults, joints, or other types of fractures.

9.7.3 Sedimentary Rocks

Sedimentary rocks are deposited either in a body of water or on the land by running water, by wind, and by glaciers. Each depositional agent leaves a characteristic stamp on the material it deposits. The sediments carried by these agents were first derived by the weathering and erosion of preexisting rocks. The most common sedimentary rocks are shale, siltstone, sandstone, limestone, and glacial till. The change from a loose, unconsolidated sediment to a rock is the process of lithification. Although sedimentary rocks appear to be the dominant type, in reality they make up but a small percentage of the earth. They do, however, form a thin crust over much of the earth's surface, are the type most readily evident, and serve as the primary source of ground water.

The major characteristics of sedimentary rocks are sorting, rounding, and stratification. A sediment is well sorted if the grains are nearly all the same size. Wind is the most effective agent of sorting and this is followed by water. Glacial till is unsorted and consists of a wide mixture of material that ranges from large boulders to clay.

While being transported, sedimentary material loses its sharp, angular configuration as it develops some degree of rounding. The amount of rounding depends on the original shape, composition, transporting medium, and the distance traveled.

Sorting and rounding are important features of both consolidated and unconsolidated material because they have a major control on permeability and porosity. The greater the degree of sorting and rounding, the higher will be the water-transmitting and storage properties. This is why a deposit of sand, in contrast to glacial till, can be such a productive aquifer.

Most sedimentary rocks are deposited in a sequence of layers or strata. Each layer or stratum is separated by a bedding plane, which probably reflects variations in sediment supply or some type of short term erosion. Commonly bedding planes represent changes in grain size. Stratification provides many clues in our attempt to unravel geologic history. The correlation of strata between wells or outcrops is called stratigraphy.

Sedimentary rocks are classified on the basis of texture (grain size and shape) and composition. Clastic rocks consist of particles of broken or worn material and include such shale, siltstone, sandstone, and conglomerate. These rocks were lithified by compaction, in the case of shale, and by cementation. The most common cements are clay, calcite, quartz, and limonite. The last three, carried by ground water, precipitate in the unconsolidated material under specific geochemical conditions.

The organic or chemical sedimentary rocks consist of strata formed from or by organisms and by chemical precipitates from sea water or other solutions. Most have a crystalline texture. Some consist of well-preserved organic remains, such as reef deposits and coal seams. Chemical sediments include, in addition to some limestones, the evaporites, such as halite (sodium chloride), gypsum, and anhydrite. Anhydrite is an anhydrous calcium sulfate.

Geologists also have developed an elaborate classification of sedimentary rocks, which is of little importance to the purpose of this introduction. In fact, most sedimentary rocks are mixtures of clastic debris, organic material, and chemical precipitates. One should keep in mind not the various classifications, but rather the texture, composition, and other features that can be used to understand the origin and history of the rock.

9.8 Weathering

Generally speaking, a rock is stable only in the environment in which it was formed. Once removed from that environment, it begins to change, rapidly in some cases but more often slowly, by weathering. The two major processes of weathering are mechanical and chemical, but they usually proceed in concert.

9.8.1 Mechanical Weathering

Mechanical weathering is the physical breakdown of rocks and minerals. Some is the result of fracturing due to the volumetric increase when water in a crack turns to ice, some is the result of abrasion during transport by water, ice, or wind, and a large part is the result of gravity causing rocks to fall and shatter. Mechanical weathering alone only reduces the size of the rock; its chemical composition is not changed. The weathered material formed ranges in size from boulders to silt.

9.8.2 Chemical Weathering

Chemical weathering, on the other hand, is an actual change in composition as minerals are modified from one type to another. Many if not most of the changes are accompanied by a volumetric increase or decrease, which in itself further promotes additional chemical weathering. The rate depends on temperature, surface area, and available water.

The major reactions involved in chemical weathering are oxidation, hydrolysis, and carbonation. Oxidation is a reaction with oxygen to form an oxide, hydrolysis is reaction with water, and carbonation is a reaction with CO_2 to form a carbonate. In these reactions the total volume increases and, since chemical weathering is most effective on grain surfaces, disintegration occurs.

Quartz, whether vein deposits or individual grains, undergoes practically no chemical weathering; the end product is quartz sand. Some of the feldspars weather to clay and release calcium, sodium, silica, and many other elements that are transported in water. The iron-bearing minerals provide, in addition to iron and magnesium, weathering products that are similar to the feldspars.

9.9 Erosion and Deposition

Once a rock begins to weather, the by-products await erosion or transportation, which must be followed by deposition. The major agents involved in this part of the rock cycle are running water, wind, and glacial ice.

9.9.1 Waterborne Deposits

Mass wasting is the downslope movement of large amounts of detrital material by gravity. Through this process sediments are made available to streams that carry them away to a temporary or permanent site of deposition. During transportation some sorting occurs and the finer silt and clay are carried farther downstream. The streams, constantly filling, eroding, and widening their channels, leave materials in their valleys that indicate much of the history of the region. Stream valley deposits, called alluvium, are shown on geologic maps by the symbol Qal, meaning Quaternary age alluvium. Alluvial deposits are distinct but highly variable in grain size, composition, and thickness. Where they consist of glacially derived sand and gravel, called outwash, they form some of the most productive water-bearing units in the world. Sediments, either clastic or chemical/organic, transported to past and present seas and ocean basins spread out to form, after lithification, extensive units of sandstone, siltstone, shale, and limestone. In the geologic past, these marine deposits covered vast areas and when uplifted they formed the land surface, where they again began to weather in anticipation of the next trip to the ocean.

The major features of marine sedimentary rocks are their widespread occurrence and rather uniform thickness and composition, although extreme changes exist in many places. If not disturbed by some type of earth movement, they are stratified and horizontal. Furthermore, each lithologic type is unique relative to adjacent units. The bedding planes or contacts that divide them represent distinct differences in texture or composition. From a hydrologic perspective, differences in texture from one rock type to another produce boundaries that strongly influence ground-water flow. Consequently, ground water tends to flow parallel to these boundaries, that is, within a particular geologic formation rather than across them.

9.9.2 Windborne Deposits

Wind-laid or eolian deposits are relatively rare in the geologic record. The massively cross-bedded sandstone of the Navajo Sandstone in Utah's Zion National Park and surrounding areas is a classic example in the United States. Other deposits are more or less local and represent dunes formed along beaches of large water bodies or streams. Their major characteristic is the high degree of sorting. Dunes, being relatively free of silt and clay, are very permeable and porous, unless the openings have been filled by cement. They allow rapid infiltration of water and can form major water-bearing units, if the topographic and geologic conditions are such that the water does not rapidly drain.

Another wind-deposited sediment is loess, which consists largely of silt. It lacks bedding but is typified by vertical jointing. The silt is transported by wind from deserts, flood plains, and glacial deposits. Loess weathers to a fertile soil and is very porous. It is common along the major rivers in the glaciated parts of the United States and in China, parts of Europe, and adjacent to deserts and deposits of glacial outwash.

9.9.3 Glacial Deposits

Glaciers erode, transport, and deposit sediments that range from clay to huge boulders. They subdue the land surface over which they flow and bury former river systems. The areas covered by glaciers during the last Ice Age in the United States are shown in Figure 9-12, but the deposits extend far beyond the former margins of the ice. The two major types of glaciers include valley or mountain glaciers and the far more extensive continental glaciers. The deposits they leave are similar, differing for the most part only in scale.

As a glacier passes slowly over the land surface it incorporates material from the underlying rocks into the ice mass, only to deposit that material elsewhere when the ice melts. During this process it modifies the land surface, both through erosion and deposition. The debris associated with glacial activity is collectively termed glacial drift. Unstratified drift, usually deposited directly by the ice, is glacial till, a heterogeneous mixture of boulders, gravel, sand, silt, and clay. Glacial debris reworked by streams and in lakes is stratified drift. Although stratified drift may range widely in grain size, the sorting far surpasses

Figure 9-12 Areal extent of glacial deposits in the United States (from Heath, 1984).

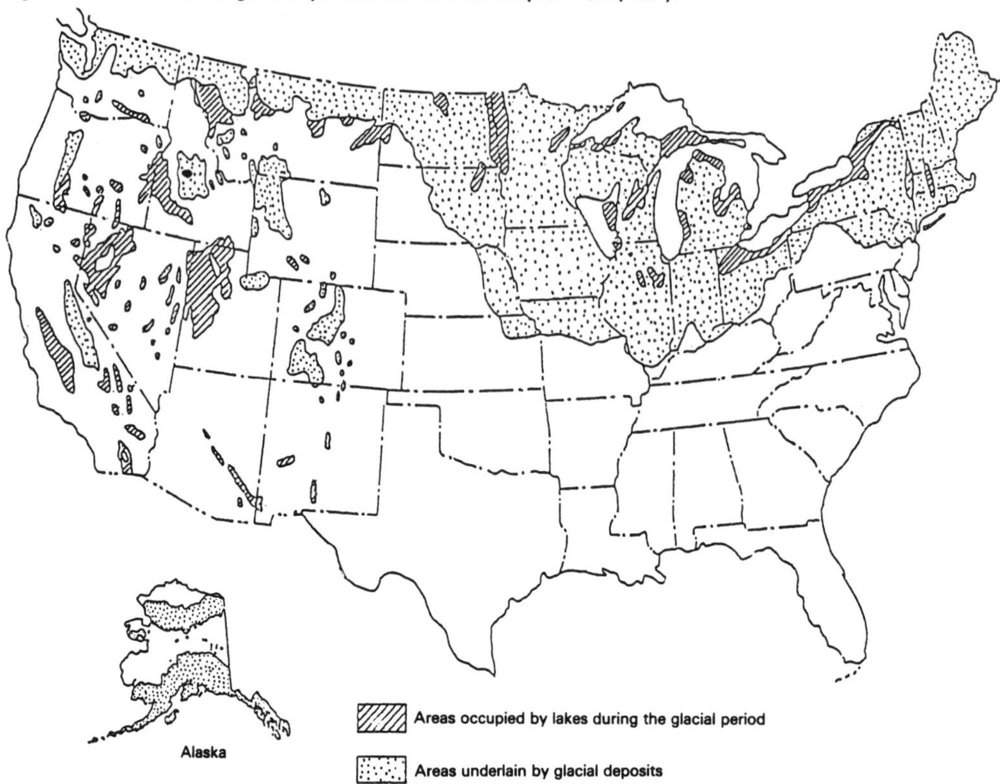

Figure 9-13 Dip and strike symbols commonly shown on geologic maps.

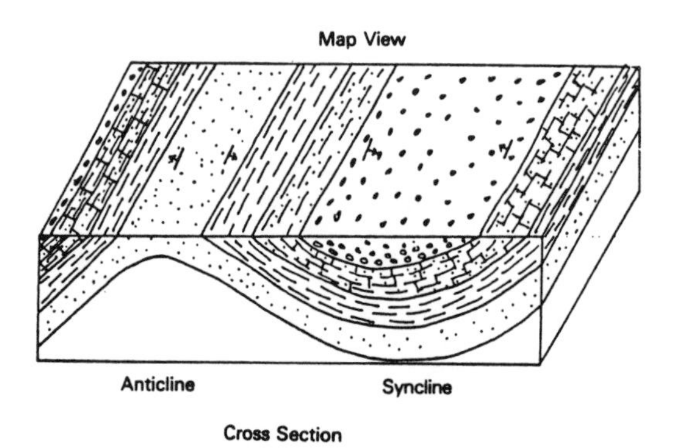

The arrow indicates the direction of dip. In an anticline, the rocks dip away from the crest and in a syncline they dip toward the center.

that of glacial till. Glacial lake clays are particularly well sorted.

Glacial geologists usually map not on the basis of texture but rather the type of landform that was developed, such as moraines, outwash, drumlins, and so on. The various kinds of moraines and associated landforms are composed largely of unstratified drift with incorporated layers of sand and gravel. Stratified drift is found along existing or former stream valleys or lakes that were either in the glacier or extended downgradient from it. Meltwater stream deposits are mixtures of sand and gravel. In places, some have coalesced to develop extensive outwash plains.

Glaciers advanced and retreated many times, reworking, overriding, and incorporating sediments from previous advances into the ice, subsequently redepositing them elsewhere. There was a constant inversion of topography as buried ice melted causing adjacent, waterlogged till to slump into the low areas. During advances, the ice might have overridden older outwash layers so that upon melting, these sand and gravel deposits were covered by a younger layer of till. Regardless of the cause, the final effect is one of complexity of origin, history, and stratigraphy. When working with glacial till deposits, it is nearly always impossible to predict the lateral extent or thickness of a particular lithology in the subsurface. Surficial stratified drift is more uniform than till in the thickness, extent, and texture.

9.10 Geologic Structure

A general law of geology is that in any sequence of sedimentary rocks that has not been disturbed by folding or faulting, the youngest unit is on the top. A second general law is that sedimentary rocks are deposited in a horizontal or nearly horizontal position. The fact that rocks are found overturned, displaced vertically or laterally, and squeezed into open or tight folds clearly indicates that the crust of the earth is a dynamic system. There is a constant battle between the forces of destruction (erosion) and construction (earth movements).

An unconformity is a break in the geologic record. It is caused by a cessation in deposition that is followed by erosion and subsequent deposition. The geologic record is lost by the period of erosion because the rocks that contained the record were removed.

If a sequence of strata is horizontal but the contact between two rock groups in the sequence represents an erosional surface, that surface is said to be a disconformity. Where a sequence of strata has been tilted and eroded and then younger, horizontal rocks are deposited over them, the contact is an angular unconformity. A nonconformity occurs where eroded igneous or metamorphic rocks are overlain by sedimentary rocks.

9.10.1 Folding

Rocks folded by compressional forces are common in and adjacent to former or existing mountain ranges (Figure 9-13). The folds range from a few inches to 50 miles or so across. Anticlines are rocks folded upward into an arch. Their counterpart, synclines, are folded downward like a valley. A monocline is a flecture in which the rocks are horizontal, or nearly so, on either side of the flecture.

Although many rocks have been folded into various structures, this does not mean that these same structures form similar topographic features. As the folding takes place over eons, the forces of erosion attempt to maintain a low profile. As uplift continues, erosion removes weathering products from the rising mass, carrying them to other places of deposition. The final topography is related to the erodibility of the rocks, with resistant strata such as sandstone forming ridges, and the less resistant material such as shale forming valleys. Consequently, the geologic structure of an area may bear little resemblance to its topography.

The structure of an area can be determined from field studies or a geologic map, if one exists. Various types of folds and their dimensions appear as unusual patterns on geologic maps. An anticline, for example, will be depicted as a series of rock units in which the oldest is in the middle, while a syncline is represented by the youngest rock in the center (Figure 9-13). More or less equidimensional anticlines and synclines are termed domes and basins, respectively.

The inclination of the top of a fold is the plunge. Folds may be symmetrical, asymmetrical, overturned, or recumbent. The inclination of the rocks is indicated by dip and strike symbols. The strike is perpendicular to the dip and the degree of dip is commonly shown by a number (Figure 9-13). The dip may range from less than a degree to vertical.

9.10.2 Fractures

Fractures in rocks are either joints or faults. A joint is a fracture along which no movement has taken place; a fault implies movement. Movement along faults is as little as a few inches to tens of miles. Probably all consolidated rocks and a good share of the unconsolidated deposits contain joints. Although not well recognized by most individuals involved in ground-water problems, joints exert a major control on water movement and chemical quantity. Characteristically joints are open and serve as major conduits or pipes. Water can move through them quickly, perhaps carrying contaminants, and, being open, the filtration effect is lost. It is a good possibility that the outbreak of many waterborne diseases that can be traced to ground-water supplies are the result of the transmission of infectious agents through fractures to wells and springs.

Faults are most common in the deformed rocks of mountain ranges, suggesting either lengthening or shortening of the crust. Movement along a fault may be horizontal, vertical, or a combination. The most common types of faults are called normal, reverse, and lateral (Figure 9-14). A normal fault, which indicates stretching of the crust, is one in which the upper or hanging wall has moved down relative to the lower or foot wall. The Red Sea, Dead Sea, and the large lake basins in the east African highlands, among many others, lie in a graben, which is a block bounded by normal faults. A reverse or thrust fault implies compression and shortening of the crust. It is distinguished by the fact that the hanging wall has moved up relative to the foot wall. A lateral fault is one in which the movement has been largely horizontal. The San Andreas Fault, extending some 600 miles from San Francisco Bay to the Gulf of California, is the most notable lateral fault in the United States. It was movement along this fault that produced the 1906 San Francisco earthquake.

Table 9-3 Geologic Time Scale

Era	Period	Epoch	Millions of Years Ago
Cenozoic	Quaternary	Recent	
		Pleistocene	0-1
	Tertiary	Pliocene	1-13
		Miocene	13-25
		Oligocene	25-36
		Eocene	36-58
		Paleocene	58-63
Mesozoic	Cretaceous		63-135
	Jurassic		135-181
	Triassic		181-230
Paleozoic	Permian		230-280
	Pennsylvanian		280-310
	Mississippian		310-345
	Devonian		345-405
	Silurian		405-425
	Ordovician		425-500
	Cambrian		500-600
Precambrian	Lasted at least 2.5 billion years		

9.11 Geologic Time

Geologic time deals with the relation between the emplacement or disturbance of rocks and time. The geologic time scale was developed in order to provide some standard classification (Table 9-3). It is based on a sequence of rocks that were deposited during a particular time interval. The divisions are commonly based on some type of unconformity. In considering geologic time, three types of units are defined. They are rock units, time and rock units, and time units.

9.11.1 Rock Units

A rock unit refers to some particular lithology. These may be further divided into geologic formations which are of sufficient size and uniformity to be mapped in the field. The Pierre Shale, for example, is a widespread and, in places, thick geologic formation that extends over much of the Northern Great Plains. Formations can also be divided into smaller units called members. Formations have a geographic name that may be coupled with a term that describes the major rock type. Two or more formations comprise a group.

9.11.2 Time and Time-Rock Units

Time-rock units refer to the rock that was deposited during a certain period of time. These units are divided into system, series, and stage. Time units refer to the time during which a sequence of rocks was deposited. The time-rock term "system" has the equivalent time term, "period." That is, during the Cretaceous Period, for example, rocks of the Cretaceous System were deposited, consisting of many groups and formations. Time units are named in such a way that the eras reflect the complexity of life forms that existed, such as the Mesozoic or "middle life." System or period nomenclature is largely based on the geographic location in which the rocks were first described, such as Jurassic, which relates to the Jura Mountains of Europe.

The terms used by geologists to describe rocks relative to geologic time are useful to the ground-water investigator in that they allow one to better perceive a regional geologic situation. The terms alone have no significance as far as water-bearing properties are concerned.

Figure 9-14 Cross sections of normal, reverse, and lateral faults.

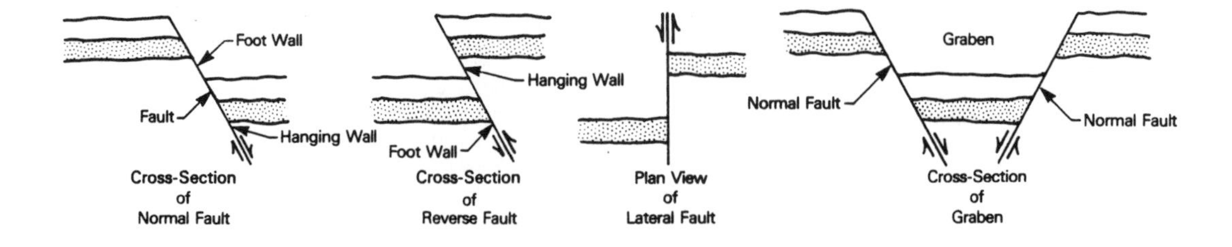

9.12 References

Blatt, H., G. Middleton, and R. Murray. 1980. Origin of Sedimentary Rocks. 2nd ed. Prentice-Hall Publishing Co., Inc., Englewood Cliffs, NJ.

Ernst, W.G. 1969. Earth Materials. Prentice-Hall Publishing Co., Inc., Englewood Cliffs, NJ.

Flint, R.F. 1971. Glacial and Quaternary Geology. John Wiley & Sons, New York, NY.

Foster, R.J. 1971. Geology. Charles E. Merrill Publishing Co., Columbus, OH.

Heath, R.C. 1984. Ground-Water Regions of the United States. U.S. Geological Survey Water-Supply Paper 2242, U.S. Government Printing Office, Washington, DC.

Pettyjohn, W.A., J.R.J. Studlick, and R.C. Bain. 1979. Quality of Drinking Water in Rural America. Water Technology 7/8.

Sawkins, F.J., C.G. Chase, D.G. Darby, and G. Rapp, Jr. 1978. The Evolving Earth, A Text in Physical Geology. Macmillan Publishing Co., Inc., New York, NY.

Spencer, E.W. 1977. Introduction to the Structure of the Earth. 2nd ed. McGraw-Hill Book Co., Inc., New York, NY.

Tarbuck, E.J., and F.K. Lutgens. 1984. The Earth, An Introduction to Physical Geology. Charles E. Merrill Publishing Co., Inc., Columbus, OH.

Tolman, C.F. 1937. Ground Water. McGraw-Hill Book Co., Inc., New York, NY.

Appendix: Sources of Information About Ground-Water Contamination

SOLID AND HAZARDOUS WASTE AGENCIES

ALABAMA
Daniel E. Cooper, Director
Land Division
Alabama Dept. of Environmental Management
1751 Federal Drive
Montgomery, AL 36130
Phone: (205) 271-7730

ALASKA
Stan Hungerford
Air and Solid Waste Management
Dept. of Environmental Conservation
Pouch O
Juneau, AK 99811
Phone: (907) 465-2635

AMERICAN SAMOA
Pati Faiai, Executive Secretary
Environmental Quality Commission
American Samoa Government
Pago Pago, American Samoa 96799
Phone: Overseas Operator 633-4116

Randy Morris, Deputy Director
Department of Public Works
Pago Pago, American Samoa 96799

ARIZONA
Ron Miller, Manager
Office of Water Quality Management
Arizona Dept. of Environmental Quality
2005 North Central Avenue
Phoenix, AZ 85004
Phone: (602) 257-2305

ARKANSAS
Vincent Blubaugh, Chief
Solid & Hazardous Waste Division
Dept. of Pollution Control and Ecology
P.O. Box 9583
8001 National Drive
Little Rock, AR 72219
Phone: (501) 562-7444

CALIFORNIA
Alex Cunningham, Chief Deputy Director
Toxic Substances Control Programs
Dept. of Health Services
714 P Street, Room 1253
Sacramento, CA 95814
Phone: (916) 322-7202

James Easton, Executive Director
State Water Resources Control Board
P.O. Box 100
Sacramento, CA 95801
Phone: (916) 445-1553

George Eowan
Chief Executive Officer
California Waste Management Board
1020 Ninth Street, Suite 300
Sacramento, CA 95814
Phone: (916) 322-3330

COLORADO
Joan Sowinski, Acting Director
Waste Management Division
Colorado Dept. of Health
4210 E. 11th Avenue
Denver, CO 80220
Phone: (303) 320-8333

COMMONWEALTH OF NORTHERN MARIANA ISLANDS
Russell Mechem, Director
Division of Environmental Quality
Dept. of Public Health and Environmental Services
Commonwealth of the Northern Mariana Islands
Saipan, CM 96950
Phone: Overseas Operator-6984

CONNECTICUT
Stephen Hitchock, Director
Hazardous Material Management Unit
Dept. of Environmental Protection
State Office Building
165 Capitol Avenue
Hartford, CT 06106
Phone: (203) 566-4924

Michael Cawley
Connecticut Resource Recovery Authority
179 Allyn Street, Suite 603
Professional Building
Hartford, CT 06103
Phone: (203) 549-6390

DELAWARE
Rick Folnsbee
Solid Waste Management Branch
Dept. of Natural Resources and Environmental Control
89 Kings Highway
P.O. Box 1401
Dover, DE 19901
Phone: (302) 736-4781

DISTRICT OF COLUMBIA
A. Padmanabha, Director
Division of Environmental Control
Room 112
5010 Overlook Avenue, S.W.
Washington, DC 20032
Phone: (202) 767-8422

FLORIDA
Robert W. McVety, Administrator
Solid & Hazardous Waste Section
Dept. of Environmental Regulation
Twin Towers Office Building
2600 Blair Stone Road
Tallahassee, FL 32301
Phone: (904) 488-0300

GEORGIA
John Taylor, Chief
Land Protection Branch
Environmental Protection Division
Dept. of Natural Resources
270 Washington Street, S.W., Room 723
Atlanta, GA 30334
Phone: (404) 656-2833

GUAM
Charles T. Crisostomo, Administrator
Guam Environmental Protection Agency
P.O. Box 2999
Agana, GU 96910
Phone: Overseas Operator 646-8863

HAWAII
Dr. Bruce Anderson, Deputy Director
Environmental Health Division
Dept. of Health
P.O. Box 3378
Honolulu, HI 96801
Phone: (808) 548-4139

IDAHO
Cheryl Koshuta, Manager
Hazardous Materials Bureau
Dept. of Health & Welfare
State House
Boise, ID 83720
Phone: (208) 334-2293

ILLINOIS
William Child, Deputy Manager
Division of Land Pollution Control
Environmental Protection Agency
2200 Churchill Road, Room A-104
Springfield, IL 62706
Phone: (217) 782-6760

INDIANA
David Lamm, Director
Land Pollution Control Division
Indiana Dept. of Environmental Management
105 South Meridian
Indianapolis, IN 46204
Phone: (317) 232-8603

IOWA
Ronald Kolpa
Hazardous Waste Program Coordinator
Dept. of Water, Air & Waste Management
Henry A. Wallace Building
900 East Grand
Des Moines, IA 50319
Phone: (515) 281-8925

KANSAS
Dennis Murphey, Manager
Bureau of Waste Management
Dept. of Health & Environment
Forbes Field, Building 321
Topeka, KS 66620
Phone: (913) 862-9360

KENTUCKY
J. Alex Barber, Director
Division of Waste Management
Dept. of Environmental Protection
Cabinet for Natural Resources and Environmental Protection
18 Reilly Road
Frankfort, KY 40601
Phone: (502) 564-6716

LOUISIANA
Paul Miller, Administrator
Solid Waste Management Division
Dept. of Environmental Quality
P.O. Box 44307
Baton Rouge, LA 70804
Phone: (504) 342-1216

Glenn Miller, Administrator
Hazardous Waste Management Division
Dept. of Environmental Quality
P.O. Box 44307
Baton Rouge, LA 70804
Phone: (504) 342-9072

MAINE
David Boulter, Director
Licensing and Enforcement Division
Bureau of Oil & Hazardous Materials
Dept. of Environmental Protection
State House -- Station 17
August, ME 04333
Phone: (207) 289-2651

MARYLAND
Bernard Bigham
Waste Management Administration
Dept. of Health & Mental Hygiene
201 W. Preston Street, Room 212
Baltimore, MD 21201
Phone: (301) 225-5649

Alvin Bowles, Chief
Hazardous Waste Division
Waste Management Administration
Dept. of Health & Mental Hygiene
201 W. Preston Street, Room 212
Baltimore, MD 21201
Phone: (301) 225-5709

Ronald Nelson, Director
Waste Management Administration
Office of Environmental Programs
Dept. of Health & Mental Hygiene
201 W. Preston Street, Room 212
Baltimore, MD 21201
Phone: (301) 225-5647

MASSACHUSETTS
William Cass, Director
Division of Solid & Hazardous Waste
Dept. of Environmental Quality
Engineering
One Winter Street
Boston, MA 02108
Phone: (617) 292-5589

MICHIGAN
Allan Howard, Chief
Waste Management Division
Dept. of Natural Resources
Box 30028
Lansing, MI 48909
Phone: (517) 373-2730

MINNESOTA
Richard Svanda, Director
Solid and Hazardous Waste Division
Pollution Control Agency
520 Lafayette Road
St. Paul, MN 55113
Phone: (612) 296-7282

MISSISSIPPI
Sam Mabry, Director
Division of Solid & Hazardous Waste Management
Bureau of Pollution Control
Dept. of Natural Resources
P.O. Box 10385
Jackson, MS 39209
Phone: (601) 961-5062

MISSOURI
Dr. David Bedan, Director
Waste Management Program
Dept. of Natural Resources
117 East Dunklin Street
P.O. Box 176
Jefferson City, MO 65102
Phone: (314) 751-3241

MONTANA
Duane L. Robertson, Chief
Solid Waste Management Bureau
Dept. of Health and Environmental Sciences
Cogswell Building
Helena, MT 59602
Phone: (406) 444-2821

NEBRASKA
Mike Steffensmeier
Section Supervisor
Hazardous Waste Management Section
Dept. of Environmental Control
State House Station
P.O. Box 94877
Lincoln, NE 68509
Phone: (402) 471-2186

NEVADA
Verne Rosse
Waste Management Program Director
Division of Environmental Protection
Dept. of Conservation and Natural Resources
Capitol Complex
201 South Fall Street
Carson City, NV 89710
Phone: (702) 885-4670

NEW HAMPSHIRE
John Minichello, Assistant Director
Dept. of Environmental Services
Waste Management Division
Health and Welfare Building
Hazen Drive
Concord, NH 03301
Phone: (603) 271-2905

NEW JERSEY
John Trela, Director
Division of Waste Management
Dept. of Environmental Protection
401 East State Street
Trenton, NJ 08625
Phone: (609) 292-1250

NEW MEXICO
Richard Mitzelfelt, Chief
Groundwater Bureau
Environmental Improvement Division
New Mexico Health & Environment Dept.
P.O. Box 968
Santa Fe, NM 87504-0968
Phone: (505) 827-0020

Jack Ellvinger, Program Manager
Hazardous Waste Section
Hazardous Waste Bureau
Environmental Improvement Division
New Mexico Health & Environment Dept.
P.O. Box 968
Santa Fe, NM 87504-0968
Phone: (505) 827-0020

NEW YORK
Norman H. Nosenchuck, Director
Division of Solid & Hazardous Waste
Dept. of Environmental Conservation
50 Wolf Road, Room 209
Albany, NY 12233
Phone: (518) 457-6603

NORTH CAROLINA
William L. Meyer, Head
Solid & Hazardous Waste Management Branch
Division of Health Services
Dept. of Human Resources
P.O. Box 2091
Raleigh, NC 27602
Phone: (919) 733-2178

NORTH DAKOTA
Martin Schock, Director
Division of Hazardous Waste
Management and Special Studies
Dept. of Health
1200 Missouri Avenue, 3rd Floor
Bismarck, ND 58501
Phone: (701) 224-2366

OHIO
Chuck Taylor, Chief
Division of Solid & Hazardous Waste Management
Ohio Environmental Protection Agency
PO Box 1049
1800 Water Mark Drive
Columbus, OH 43206-1049
Phone: (614) 481-7200

OKLAHOMA
Dwain Farley, Chief
Waste Management Service
Oklahoma State Dept. of Health
P.O. Box 53551
Oklahoma City, OK 73152
Phone: (405) 271-7047

OREGON
Mike Downs, Administrator
Hazardous & Solid Waste Division
Dept. of Environmental Quality
P.O. Box 1760
Portland, OR 97207
Phone: (503) 229-5356

PENNSYLVANIA
Donald A. Lazarchik, Director
Bureau of Solid Waste Management
Dept. of Environmental Resources
Bulton Building, 8th Floor
P.O. Box 2063
Harrisburg, PA 17120
Phone: (717) 787-9870

PUERTO RICO
Santos Rohena, Director
Solid, Toxics, & Hazardous Waste Program
Environmental Quality Board
P.O. Box 11488
Santurce, PR 00910-1488
Phone: (809) 725-0439

RHODE ISLAND
Thomas Goetz
Air & Hazardous Waste Management
Dept. of Environmental Management
204 Cannon Building
75 Davis Street
Providence, RI 02908
Phone: (401) 277-2797

SOUTH CAROLINA
Hartsill Truesdale, Chief
Bureau of Solid and Hazardous Waste Management
South Carolina Dept. of Health & Environmental Control
2600 Bull Street
Columbia, SC 29201
Phone: (803) 734-5200

SOUTH DAKOTA
Joel C. Smith, Administrator
Office of Air Quality & Solid Waste
Dept. of Water & Natural Resources
Joe Foss Building
Pierre, SD 57501
Phone: (605) 773-3329

TENNESSEE
Tom Tiesler, Director
Division of Solid Waste Management
Bureau of Environmental Services
Tennessee Dept. of Public Health
150 9th Avenue, North
Nashville, TN 37203
Phone: (615) 741-3424

TEXAS
Hector Mendieta, Director
Bureau of Solid Waste Management
Texas Dept. of Health
1100 West 49th Street, T-602
Austin, TX 78756-3199
Phone: (512) 458-7271

Sam Pole, Chief
Hazardous & Solid Waste Enforcement
Texas. Water Commission
1700 North Congress
P.O. Box 13087, Capitol Station
Austin, TX 78711
Phone: (512) 463-8177

UTAH
Brent Bradford, Director
Bureau of Solid & Hazardous Waste Management
Dept. of Health
P.O. Box 2500
150 West North Temple
Salt Lake City, UT 84110
Phone: (801) 533-4145

VERMONT
John Malter, Director
Solid & Hazardous Waste Management Programs
Agency of Environmental Conservation
State Office Building
103 South Main Street
Waterbury, VT 05676
Phone: (802) 244-8702

VIRGIN ISLANDS
Greg Rhymer, Director
Hazardous Waste Program
Division of Natural Resources
Planning and Natural Resources
179 Welgunst Altona
Charlotte Amalie, St. Thomas, VI 00802
Phone: (809) 774-3320

VIRGINIA
Cynthia Bailey
Dept. of Waste Management
Virginia Dept. of Health
Monroe Building, 11th Floor
101 North 14th Street
Richmond, VA 23219
Phone: (804) 225-2667

WASHINGTON
Earl Tower, Supervisor
Solid & Hazardous Waste Management Division
Dept. of Ecology
Olympia, WA 98504
Phone: (296) 459-6316

Phil Johnson and Mark Horton, Deputy Director
Office of Hazardous Substance & Air Quality Program
Dept. of Ecology
Olympia, WA 98504
Phone: (296) 459-6253

WEST VIRGINIA
Doug Steele, Chief
Division of Water Resources
Dept. of Natural Resources
1201 Greenbrier Street
Charleston, WV 25311
Phone: (304) 348-5935

WISCONSIN
Paul Didier, Director
Bureau of Solid Waste Management
Dept. of Natural Resources
P.O. Box 7921
Madison, WI 53707
Phone: (608) 266-1327

WYOMING
Dave Finley, Supervisor
Solid Waste Management Program
State of Wyoming
Dept. of Environmental Quality
Equality State Bank Building
401 West 19th Street
Cheyenne, WY 82002
Phone: (307) 777-7752

U.S. EPA OFFICE OF GROUND-WATER PROTECTION

Ms. Marian Mlay
Office of Ground-Water Protection (WH-550G)
U.S. EPA
401 M Street, SW
Washington, DC 20460
Phone: (202) 382-7077

Mr. Robert Mendosa
Office of Ground-Water
Water Management Division
U.S. EPA, Region I - Room 2113
JFK Federal Building
Boston, MA 02203
Phone: (617) 565-3600
FTS 835-3600

Mr. John Malleck
Office of Ground-Water
Water Management Division
U.S. EPA, Region II - Room 805
26 Federal Plaza
New York, NY 10278
Phone: (212) 264-5635
FTS 264-5635

Mr. Stuart Kerzner
Office of Ground-Water
Water Management Division
U.S. EPA, Region III
841 Chestnut Street
Philadelphia, PA 19106
Phone: (215) 597-2786
FTS 597-8826

Mr. James S. Kutzman
Office of Ground-Water
Water Management Division
U.S. EPA, Region IV
345 Courtland Street, N.E.
Atlanta, GA 30365
Phone: (404) 347-3866
FTS 257-3866

Ms. Jerri-Anne Garl
Office of Ground-Water
Water Management Division
U.S. EPA, Region V (MS-5WG-TUB9)
230 S. Dearborn Street
Chicago, IL 60604
Phone: (312) 353-1490
FTS 886-1490

Mr. Ken Kirkpatrick
Office of Ground-Water
Water Management Division
U.S. EPA, Region VI
1445 Ross Avenue
Dallas, TX 75202-2733
Phone: (214) 655-6446
FTS 255-6446

Mr. Timothy Amsden
Office of Ground-Water
Water Management Division
U.S. EPA, Region VII
726 Minnesota Avenue
Kansas City, KS 66101
Phone: (913) 236-2815
FTS 757-2815

Mr. Richard Long
Water Management Division
U.S. EPA, Region VIII (MC-8WMGW)
999 18th Street
Denver, CO 80202-2405
Phone: (303) 293-1543
FTS 564-1543

Ms. Pat Ekland
Office of Ground-Water
Water Management Division
U.S. EPA, Region IX (MC-W-1-G)
215 Fremont Street
San Francisco, CA 94105
Phone: (415) 974-0831
FTS 454-0831

Mr. William A. Mullen
Office of Ground-Water
Water Management Division
U.S. EPA, Region X (M/S WO-139)
1200 6th Avenue
Seattle, WA 98101
Phone: (206) 442-1216
FTS 399-1216

FEDERAL INTERAGENCY GROUND-WATER PROTECTION COMMITTEE

DEPARTMENT OF THE INTERIOR

Mr. James W. Zigklar
(Principal Agency contact)
Assistant Secretary for Water and Science
U.S. Dept. of the Interior
18th & C Street, NW
Washington, DC 20240
Phone: (202) 343-2186

Mr. Joe Findaro III
(EPA/Off. Gr. Wtr. Prot. Contact)
Deputy Assistant Secretary for Water and Science
U.S. Dept. of the Interior (Room 6652)
18th & C Street, NW
Washington, DC 20240
Attn: Nancy Lopez
Phone: (202) 343-2182

Mr. Roland Dolly
(Bureau of Reclamation representative)
Special Assistant to the Commissioner
Bureau of Reclamation (Attn: Code 104)
U.S. Dept. of the Interior (Room 7641)
18th & C Street, NW
Washington, DC 20240
Phone: (202) 343-4115

Mr. Al Perry
(Bureau of Mines representative)
Pittsburg Research Center
Bureau of Mines
U.S. Dept. of the Interior
Columbia Plaza
2401 E Street, N.W.
Washington, DC 20241
Phone: (202) 634-1245

Mr. Phillip Cohen
(U.S. Geological Survey representative)
Chief Hydrologist
U.S. Geological Survey
U.S. Dept. of the Interior
409 National Center
Reston, VA 22092

Mr. William Horn
(Principal Agency contact)
Assistant Secretary for Fish, Wildlife and Parks
U.S. Dept. of the Interior
18th & C Street, NW
Washington, DC 20240
Phone: (202) 343-4416

Mr. Donald S. Herring
(EPA/Off. Gr. Wtr. Prot. contact)
Engineering & Safety Service Division
National Park Service (610)
U.S. Dept. of the Interior
P.O. Box 37127
Washington, DC 20013-7127
Phone: (202) 343-7040

Mr. Hal O'Conner
(Fish & Wildlife Service representative)
Associate Director
Habitat Resources
Fish & Wildlife Service
U.S. Dept. of the Interior
18th & C Street, NW
Washington, DC 20240
Phone: (202) 343-4767

Mr. Steve Griles
(Principal Agency contact)
Acting Assistant Secretary for Lands & Mineral Management
U.S. Dept. of the Interior
18th & C Street, NW
Washington, DC 20240
Phone: (202) 343-2186

Mr. Dan Muller
(Bureau of Land Management representative)
Bureau of Land Management
U.S. Dept. of the Interior
Premier Building (WO222)
18th & C Street, NW
Washington, DC 20240
Phone: (202) 653-9210

Mr. Doug Growitz
(Office of Surface Mining representative)
Office of Surface Mining
U.S. Dept. of the Interior (Room 5101L)
1951 Constitution Avenue, NW
Washington, DC 20240
Phone: (202) 343-1507

DEPARTMENT OF AGRICULTURE

Mr. George Dunlop
(Principal Agency contact)
Assistant Secretary for Natural Resources and Environment
U.S. Dept. of Agriculture
Administration Building (217E)
Washington, DC 20250
Phone: (202) 447-7173

Mr. Dave Unger
(EPA/Off. Gr. Wtr. Prot. contact)
Forest Service
U.S. Dept. of Agriculture
Watershed and Air Management
Box 96090
Washington, DC 20013
Phone: (202) 235-8178

Mr. Louis Kirkaldie
(Soil Conservation Service representative)
Soil Conservation Service
U.S. Dept. of Agriculture
Room 6132
P.O. Box 2890
Washington, DC 20013
Phone: (202) 447-5858

Mr. Fred Swader
(Extension Service representative)
Extension Service
U.S. Dept. of Agriculture
Room 3340
South Agriculture Building
14th & Independence, SW
Washington, DC 20250
Phone: (202) 447-5369

DEPARTMENT OF JUSTICE

Mr. F. Henry Habicht II
(Principal Agency contact)
Assistant Attorney General
Land and Natural Resources
U.S. Dept. of Justice
10th Street & Constitution Avenue, NW
Washington, DC 20530
Phone: (202) 633-2701

Mr. Myles E. Flint
(EPA/Off. Gr. Wtr. Prot. contact)
Deputy Assistant Attorney General
Land and Natural Resources Division
U.S. Dept. of Justice
10th Street & Constitution Avenue, NW
Washington, DC 20530
Phone: (202) 633-2718

Mr. Chuck Sheenan
(representative)
Attorney
Policy, Legislation & Special Litigation Section
U.S. Dept. of Justice
Room 2615, Main Justice
10th Street & Constitution Avenue, NW
Washington, DC 20530
Phone: (202) 633-1442

DEPARTMENT OF THE ARMY

LTG E.R. Heiberg, III
(Principal Agency contact)
Commander U.S. Army
Corps of Engineers
U.S. Dept. of the Army
Pulaski Building
20 Massachusetts Avenue, NW
Washington, DC 20314-1000
Phone: (202) 272-0000

LTC Kit J. Valentine
(EPA/Off. Gr. Wtr. Prot. contact)
U.S. Army Corps of Engineers
U.S. Dept. of the Army (CECW-RE)
Pulaski Building
20 Massachusetts Avenue, NW
Washington, DC 20314-1000
Phone: (202) 272-0166

DEPARTMENT OF ENERGY

Ms. Mary Walker
(Principal Agency contact)
Assistant Secretary for Environment, Safety & Health
U.S. Dept. of Energy
Forrestal Building
1000 Independence Avenue, SW
Washington, DC 20585
Phone: (202) 252-4700

Mr. Ted Williams
(EPA/Off. Gr. Wtr. Prot. contact)
Director, Environmental Analysis
U.S. Dept. of Energy (Room 4G-036-EH-22)
1000 Independence Avenue, SW
Washington, DC 20585
Phone: (202) 252-2061

Mr. Raymond P. Berube
(Off. Env. Compliance representative)
Director
Office of Environmental Guidance and Compliance
U.S. Dept. of Energy (EG-23)
1000 Independence Avenue, SW
Washington, DC 20582
Attn: Tom Frangos
Phone: (202) 586-4600

DEPARTMENT OF TRANSPORTATION
Mr. Ray A. Barnhart
(Principal Agency contact)
Administrator, Federal Highway Administration
U.S. Dept. of Transportation
Nassif Building
400 7th Street, SW
Washington, DC 20590
Phone: (202) 426-0650

Mr. Charles R. DesJardins
(Off. Env. Compliance representative)
Ecologist
Office of Environmental Policy (HEV-20)
Federal Highway Administration
U.S. Dept. of Transportation
Nassif Building
400 7th Street, SW
Washington, DC 20590
Phone: (202) 366-20683

DEPARTMENT OF DEFENSE
Mr. Carl J. Schafer, Jr.
(Principal Agency contact)
Deputy Asst. Secretary of Defense (Environment)
DASD (A&L) (E)
U.S. Dept. of Defense
Room 3D 833
Pentagon
Washington, DC 20301-8000
Phone: (202) 695-7820

Mr. Peter Boice.
(Def. Environ. Leader. Proj. representative)
Environmental Engineering
U.S. Dept. of Defense
Suite 100
206 N. Washington Street
Alexandria, VA 22313
Phone: (202) 325-2215

TENNESSEE VALLEY AUTHORITY
Honorable Charles H. Dean, Jr.
(Principal Agency contact)
Chairman
Tennessee Valley Authority
TVA Building
400 West Summit Hill Drive
Knoxville, TN 37902
Phone: (615) 632-2101

Dr. Al Bruch
(EPA/Off. Gr. Wtr. Prot. contact)
Tennessee Valley Authority
215 Summer Place
Building 309
Walnut Street
Knoxville, TN 37902
Attn: Dr. John Crossman
Phone: (615) 632-2669

DEPARTMENT OF HEALTH & HUMAN SERVICES
Dr. James Mason
(Dept. Health & Human Serv. representative)
Acting Assistant Secretary, Public Health Service)
U.S. Dept. of Health & Human Services
Hubert H. Humphrey Building
200 Independence Avenue, SW
Washington, DC 20201
Phone: (202) 245-7694

Dr. Henry Falk
(Centers for Disease Control representative)
Centers for Disease Control
Center for Environmental Health
1600 Clifton Road
Atlanta, GA 30333
Phone: (404) 236-4095

NUCLEAR REGULATORY COMMISSION
Dr. Malcolm R. Knapp
(Principal Agency contact)
Director, Division of Low-Level
Waste Management and Decommissioning
Office of Nuclear Material Safety & Safeguards
U.S. Nuclear Regulatory Commission (M/S 623-SS)
Washington, DC 20555
Phone: (301) 427-4433

Mr. Michael Weber
(Nuclear Regulatory Commission representative)
U.S. Nuclear Regulatory Commission (M/S 623-SS)
Washington, DC 20555
Phone: (301) 427-4746

NATIONAL SCIENCE FOUNDATION
Mr. Nam P. Suh
(National Science Foundation representative)
Assistant Director for Engineering
National Science Foundation
1800 G Street, NW
Washington, DC 20550
Phone: (202) 357-7737

Mr. Edward H. Bryan
(National Science Foundation representative)
Program Director, Environmental Engineering
National Science Foundation
1800 G Street, NW
Washington, DC 20550
Phone: (202) 357-7737

Other Noyes Publications

PROTECTION OF PUBLIC WATER SUPPLIES FROM GROUND-WATER CONTAMINATION

Edited by

Wayne A. Pettyjohn
Oklahoma State University

Pollution Technology Review No. 141

This book provides an organized approach to acquiring necessary knowledge for effective and efficient management of ground-water supplies. The information provided is applicable to all regions of the United States, taking into account differences in geographic location, from the humid to the arid, and in geologic composition, from the porous to the impermeable.

The development of subsurface water supplies has little effect on land use and can usually be accomplished at relatively low cost compared to the development of surface supplies. However, the subsurface environment is a complex system subject to contamination from a host of sources. Restoration costs generally exceed the short-term value of the resource when compared to the costs of alternatives. For this reason, it is widely agreed that the most viable approach to ground-water quality protection is one of prevention rather than cure.

The most promising management option now available is to protect the groundwater resource from contamination. This will require many different "best management practices," including the development of protection plans at the local level to control activities that threaten the resource. The book covers the various options available to environmental planners in discussions of management alternatives; control of volatile organic compounds in drinking water; and in-ground treatment, restoration, and reclamation.

A condensed table of contents listing **chapter titles and selected subtitles** is given below.

ISBN 0-8155-1119-1 (1987) **177 pages**

Other Noyes Publications

GROUNDWATER CONTAMINATION AND EMERGENCY RESPONSE GUIDE

by

J.H. Guswa
W.J. Lyman
Arthur J. Little, Inc.

A.S. Donigian, Jr., T.Y.R. Lo,
E.W. Shanahan
Anderson-Nichols & Co., Inc.

Pollution Technology Review No. 111

An overview of groundwater hydrology; a technology review of equipment, methods, and field techniques; and a methodology for estimating groundwater contamination under emergency response conditions are provided in this book. It describes the state of the art of the various techniques used to identify, quantify, and respond to groundwater pollution incidents.

Interest in the causes and effects of groundwater contamination has increased significantly in the past decade as numerous incidents have brought the potential problems to public attention. Protection of our groundwater resources is of critical importance, thus making the book both timely and relevant.

Part I assesses methodology for investigating and evaluating known or suspected instances of contamination. Part II surveys groundwater fundamentals, state-of-the-art equipment, monitoring methods, and treatment and containment technologies. It will serve as a desk reference and guidance manual. Part III details possible emergency response actions at toxic spill and hazardous waste disposal sites.

A condensed table of contents listing **part and selected chapter titles** is given below.

ISBN 0-8155-0999-5 (1984) **490 pages**